Environmental and Criminal Geoforensics

The Geological Society of London
Books Editorial Committee

Chief Editor
RICK LAW (USA)

Society Books Editors
JIM GRIFFITHS (UK)
DAVE HODGSON (UK)
HOWARD JOHNSON (UK)
PHIL LEAT (UK)
DANIELA SCHMIDT (UK)
RANDELL STEPHENSON (UK)
ROB STRACHAN (UK)
MARK WHITEMAN (UK)

Society Books Advisors
GHULAM BHAT (India)
MARIE-FRANÇOISE BRUNET (France)
JAMES GOFF (Australia)
MARIO PARISE (Italy)
SATISH-KUMAR (Japan)
MARCO VECOLI (Saudi Arabia)
GONZALO VEIGA (Argentina)
MAARTEN DE WIT (South Africa)

Geological Society books refereeing procedures

The Society makes every effort to ensure that the scientific and production quality of its books matches that of its journals. Since 1997, all book proposals have been refereed by specialist reviewers as well as by the Society's Books Editorial Committee. If the referees identify weaknesses in the proposal, these must be addressed before the proposal is accepted.

Once the book is accepted, the Society Book Editors ensure that the volume editors follow strict guidelines on refereeing and quality control. We insist that individual papers can only be accepted after satisfactory review by two independent referees. The questions on the review forms are similar to those for *Journal of the Geological Society*. The referees' forms and comments must be available to the Society's Book Editors on request.

Although many of the books result from meetings, the editors are expected to commission papers that were not presented at the meeting to ensure that the book provides a balanced coverage of the subject. Being accepted for presentation at the meeting does not guarantee inclusion in the book.

More information about submitting a proposal and producing a book for the Society can be found on its website: www.geolsoc.org.uk.

It is recommended that reference to all or part of this book should be made in one of the following ways:

PIRRIE, D., RUFFELL, A. & DAWSON, L. A. (eds) 2013. *Environmental and Criminal Geoforensics*. Geological Society, London, Special Publications, **384**.

CARVALHO, Á., RIBEIRO, H., GUEDES, A., ABREU, I. & NORONHA, F. 2013. Geological and palynological characterization of a river beach in Portugal for forensic purposes. *In:* PIRRIE, D., RUFFELL, A. & DAWSON, L. A. (eds) *Environmental and Criminal Geoforensics*. Geological Society, London, Special Publications, **384**, 87–95. First published online May 16, 2013, http://dx.doi.org/10.1144/SP384.3

GEOLOGICAL SOCIETY SPECIAL PUBLICATION NO. 384

Environmental and Criminal Geoforensics

EDITED BY

D. PIRRIE
Helford Geoscience LLP, UK

A. RUFFELL
Queen's University Belfast, UK

and

L. A. DAWSON
James Hutton Institute, UK

2013
Published by
The Geological Society
London

THE GEOLOGICAL SOCIETY

The Geological Society of London (GSL) was founded in 1807. It is the oldest national geological society in the world and the largest in Europe. It was incorporated under Royal Charter in 1825 and is Registered Charity 210161.

The Society is the UK national learned and professional society for geology with a worldwide Fellowship (FGS) of over 10 000. The Society has the power to confer Chartered status on suitably qualified Fellows, and about 2000 of the Fellowship carry the title (CGeol). Chartered Geologists may also obtain the equivalent European title, European Geologist (EurGeol). One fifth of the Society's fellowship resides outside the UK. To find out more about the Society, log on to www.geolsoc.org.uk.

The Geological Society Publishing House (Bath, UK) produces the Society's international journals and books, and acts as European distributor for selected publications of the American Association of Petroleum Geologists (AAPG), the Indonesian Petroleum Association (IPA), the Geological Society of America (GSA), the Society for Sedimentary Geology (SEPM) and the Geologists' Association (GA). Joint marketing agreements ensure that GSL Fellows may purchase these societies' publications at a discount. The Society's online bookshop (accessible from www.geolsoc.org.uk) offers secure book purchasing with your credit or debit card.

To find out about joining the Society and benefiting from substantial discounts on publications of GSL and other societies worldwide, consult www.geolsoc.org.uk, or contact the Fellowship Department at: The Geological Society, Burlington House, Piccadilly, London W1J 0BG: Tel. +44 (0)20 7434 9944; Fax +44 (0)20 7439 8975; E-mail: enquiries@geolsoc.org.uk.

For information about the Society's meetings, consult *Events* on www.geolsoc.org.uk. To find out more about the Society's Corporate Affiliates Scheme, write to enquiries@geolsoc.org.uk.

Published by The Geological Society from:
The Geological Society Publishing House, Unit 7, Brassmill Enterprise Centre, Brassmill Lane, Bath BA1 3JN, UK

The Lyell Collection: www.lyellcollection.org
Online bookshop: www.geolsoc.org.uk/bookshop
Orders: Tel. +44 (0)1225 445046, Fax +44 (0)1225 442836

The publishers make no representation, express or implied, with regard to the accuracy of the information contained in this book and cannot accept any legal responsibility for any errors or omissions that may be made.

© The Geological Society of London 2013. No reproduction, copy or transmission of all or part of this publication may be made without the prior written permission of the publisher. In the UK, users may clear copying permissions and make payment to The Copyright Licensing Agency Ltd, Saffron House, 6–10 Kirby Street, London EC1N 8TS UK, and in the USA to the Copyright Clearance Center, 222 Rosewood Drive, Danvers, MA 01923, USA. Other countries may have a local reproduction rights agency for such payments. Full information on the Society's permissions policy can be found at: www.geolsoc.org.uk/permissions

British Library Cataloguing in Publication Data

A catalogue record for this book is available from the British Library.
ISBN 978-1-86239-366-0
ISSN 0305-8719

Distributors
For details of international agents and distributors see:
www.geolsoc.org.uk/agentsdistributors

Typeset by Techset Composition India (P) Ltd, Bangalore and Chennai, India
Printed by Berforts Information Press Ltd, Oxford, UK

Contents

Pirrie, D., Ruffell, A. & Dawson, L. A. Environmental and criminal geoforensics: an introduction — 1

Bowen, A. M. & Caven, E. A. Forensic provenance investigations of soil and sediment samples — 9

Bergslien, E. T. X-ray diffraction and field portable X-ray fluorescence analysis and screening of soils: project design — 27

Pirrie, D., Rollinson, G. K., Power, M. R. & Webb, J. Automated forensic soil mineral analysis; testing the potential of lithotyping — 47

Guedes, A., Murray, R. C., Ribeiro, H., Sant'ovaia, H., Valentim, B., Rodrigues, A., Leal, S. & Noronha, F. The potential application of magnetic susceptibility as a technique for soil forensic examinations — 65

Di Maggio, R. M. & Nuccetelli, L. Analysis of geological trace evidence in a case of criminal damage to graves — 75

Isphording, W. C. Using soil mineral signatures to confirm sources of industrial contaminant trespass — 81

Carvalho, Á., Ribeiro, H., Guedes, A., Abreu, I. & Noronha, F. Geological and palynological characterization of a river beach in Portugal for forensic purposes — 87

Guedes, A., Murray, R. C., Ribeiro, H., Rodrigues, A., Valentim, B., Sant'ovaia, H. & Noronha, F. Integration of different sediment characteristics to discriminate between sources of coastal sediments — 97

McKinley, J. How useful are databases in environmental and criminal forensics? — 109

Eby, G. N. Instrumental neutron activation analysis (INAA) and forensic applications — 121

Purvis, O. W., Williamson, B. J., Spiro, B., Udachin, V., Mikhailova, I. N. & Dolgopolova, A. Lichen monitoring as a potential tool in environmental forensics: case study of the Cu smelter and former mining town of Karabash, Russia — 133

Ruffell, A., Pirrie, D. & Power, M. R. Issues and opportunities in urban forensic geology — 147

Ruffell, A. Solid and drift geology in forensic investigations — 163

Donnelly, L. & Harrison, M. Geomorphological and geoforensic interpretation of maps, aerial imagery, conditions of diggability and the colour-coded RAG prioritization system in searches for criminal burials — 173

Donnelly, L. The design and implementation of a high-assurance forensic geology and police search following the discovery of the Staffordshire (Anglo Saxon) Gold Hoard — 195

Larizza, M. & Forbes, S. L. Detection of fatty acids in the lateral extent of the cadaver decomposition island — 209

Lowe, A. C., Beresford, D. V., Carter, D. O., Gaspari, F., O'Brien, R. C. & Forbes, S. L. Ground penetrating radar use in three contrasting soil textures in southern Ontario — 221

HANSEN, J. D. & PRINGLE, J. K. Comparison of magnetic, electrical and ground penetrating radar surveys to detect buried forensic objects in semi-urban and domestic patio environments   229

RICHARDSON, T. & CHEETHAM, P. N. The effectiveness of geophysical techniques in detecting a range of buried metallic weapons at various depths and orientations   253

Index   267

Environmental and criminal geoforensics: an introduction

DUNCAN PIRRIE[1]*, ALASTAIR RUFFELL[2] & LORNA A. DAWSON[3]

[1]*Helford Geoscience LLP, Menallack Farm, Treverva, Penryn, Cornwall TR10 9BP, UK*

[2]*School of Geography, Archaeology and Palaeoecology, Elmwood Avenue, Queens University, Belfast BT7 1NN, UK*

[3]*James Hutton Institute, Craigiebuckler, Aberdeen AB15 8Q, UK*

*Corresponding author (e-mail: dpirrie@helfordgeoscience.co.uk)

Many subdisciplines within the Earth Sciences make use of either geophysical instruments to investigate the subsurface environment or use analytical methods to determine the origin, or provenance, of geological materials. These same instruments and analytical methods can be used either directly, or adapted to suit, the acquisition of data that pertain to a wide range of forensic science investigations. Such approaches, as discussed below, are generally not new but, in recent years, there has been a significant resurgence globally in the application of geological and geophysical methods to aid forensic investigations. Traditionally, such methods were used in forensic investigations related to serious criminal cases such as terrorism, murder, abduction and serious sexual assaults, and to a lesser extent in the investigation of cases of fraud and theft. But with increasing concern into the environmental impact of human activity, with the release of contaminants into the atmosphere, hydrosphere and lithosphere, and their potential uptake into the biosphere, there has been an increased amount of environmental legislation. In turn, there has commonly been an increase in the costs associated with the legal discharge or disposal of wastes. Consequently, it is unsurprising that the illegal discharge and disposal of wastes has also increased. Identifying the distribution, impact and source of such waste materials can, in part, be addressed through the application of geological techniques, in much the same way as used traditionally, for example, in the investigation of murder cases. The diversification of the use of geological techniques into the investigation of environmental crime will, potentially, significantly increase the range of investigations in which geologists may be asked to assist but will also lead to a new array of research questions to be addressed, hence the need for this Special Publication.

In this short introduction, the aim is to place the current work into its broad historical context and summarize the key findings from the papers presented within this volume. The papers within this Special Publication were presented at two separate conferences in 2010: the Third International Soil Forensics Conference held at Long Beach California USA and organized by the SFI Group; and a conference on Environmental and Criminal Forensics, organized by the Forensic Geoscience Group of the Geological Society of London, and held at Burlington House, London, in December 2010.

Historical context

The first published account of the use of geology in investigating crime was around 150 years ago (Anon 1856), where barrels containing silver coins were loaded onto a train in Leipzig, bound for Berlin. On arrival, the barrels were found to be full of sand – a classic case of substitution, an activity that still occurs today. The famous microscopist, Hans Christian Ehrenberg, was asked to analyse the sand, which he compared to ballast and loose sand from railway stations along the route. Excluding all others, he regarded only one location as comparable. On questioning the station staff, a confession was obtained by the police and the silver coins recovered. This case was isolated in time, with the first renaissance in forensic geology occurring some 40 years later through the late 1900s and the early twentieth century. Accounts of the fictitious, theoretical and applied applications of forensic geology can be found in the available textbooks on forensic geology and geoscience (see the six books now published on the subject cited below), and need not be repeated here. In succession, Conan-Doyle's Sherlock Holmes stories, Hans Gross's *Handbook for Examining Magistrates*, Georg Popp's casework and publications, and Edmond Locard's casework and forensic principles all played a part in establishing the study of soil, sediment and rocks in criminal investigations. Gross also commented on the ways an investigator may conduct searches, being cognoscent of woods, ponds and other places objects (especially murder victims) were likely to be hidden. This is especially pertinent to this volume, in which the use of a geologist (and/or geophysicist) is considered in such searches, further demonstrating Gross's vision.

The lack of publications on the subject from the late 1920s through to the early 1960s could be interpreted as a hiatus in the use of soil and sediment as trace evidence (Fig. 1). However, in reality, government agencies such as the early FBI and even local police forces in England were either using such methods, or employing staff or outside experts such as the famous Berkeley scientist, Oscar Heinrich, to carry out such analyses: they simply didn't publish their work. Throughout the period 1962–2003, accounts of the use of soil and sediment analyses in comparing (or excluding) suspects and victims to scenes of crime, substitutions and engineering failures, for example, gradually increased in number, with the publication of Murray & Tedrow's *Forensic Geology* (Murray & Tedrow 1975) reflecting this steady interest; Ray Murray is widely regarded as the present-day father of forensic geology.

The mainstay of the subject continued to be soil and sediment analysis, generally by geologists and soil scientists, with other uses of the Earth Sciences being undertaken by the (then) relevant specialists. Most notable in this regard would be the means of searching for buried objects, commonly undertaken in the 1960s–1980s by archaeologists. The increased use of remote sensing and geophysics by geologists meant that, through the turn of the twentieth century, publications began appearing on the search methods used by geologists. The second renaissance in forensic geology was, thus, termed forensic geoscience (to include these diverse applications), and was generated by a 2 day meeting at the Geological Society (London) in 2003, where the established use of soil and sediment analysis was considerably strengthened, and a second branch of the uses of Earth Science techniques in forensics emerged – geophysics and the methods of searching the ground. The renaissance has continued following this 2003 meeting, with forensic geology and geoscience meetings held, at least annually, all around the world, and with six substantial books (Murray 2004; Pye & Croft 2004; Pye 2007; Ruffell & McKinley 2008; Ritz *et al.* 2009; Bergslien 2012) and numerous peer-reviewed journal articles now having been published.

As forensic geology expanded from the traditional area of soil and sediment analysis to search and remote sensing/geophysics through the early 2000s, so other applications of forensic geology have emerged, most notably their use in environmental forensics. The study of water, soil, aquifer and air pollution had (like traditional forensics) been the domain of chemists and biologists. However, the increasingly sophisticated means of analysing soil, sediment and waters, and of remotely sensing near-surface chemistry and deeper soils and geology (through geophysics), has resulted in geologists, remote sensors, geophysicists and

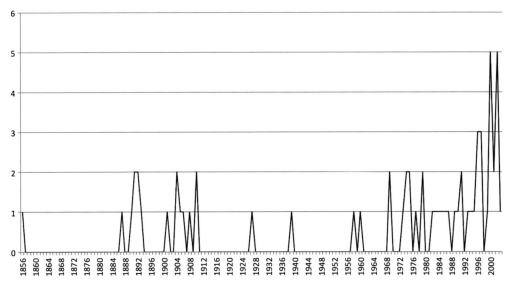

Fig. 1. The number of publications (peer-reviewed journal articles and textbooks) by year (1956–2003). Data provided by A. Ruffell and R. Murray.

geotechnical engineers becoming more and more involved in incidents and cases of environmental damage. In parallel, the now-established uses of soil and sediment analysis and search methods (geomorphology, geophysics, remote sensing) still continue to develop, with significant cross-over between the applications of the various Earth Sciences. We are at present in an interesting period in the application of the Earth Sciences to Forensic Science, with the diversification seen at the 2003 meeting leading to a coalescence where trace evidence, search and environmental applications are discussed together, as they so commonly use the same techniques for different applications, or a conjunction of techniques to provide as robust analysis of a problem as possible. However, these rapidly developing methods and new applications need to be underpinned, and supported by rigorous and sound science, not only in terms of the methods of detection or analysis, but also rigour in how the resultant data are presented within a Court of Law. In the UK, several cases where 'Expert Witness' evidence has been subsequently challenged and overturned as unsafe are well known and, sadly, this includes forensic geoscience evidence. Such work undermines the discipline as a whole, and it is hoped that volumes such as this can help redress the balance and highlight the value of Earth Science in forensic investigations.

Current research

The papers published within this volume fall within three broad groupings: (1) trace evidence or analytical approaches; (2) search techniques based on both geological and geophysical approaches; and (3) case studies that highlight the practical application of the approaches discussed.

The analysis of soil mineralogy still underpins many trace evidence investigations that may either aim to try to establish whether or not the available data can be used to exclude a potential association between an unknown sample and a known location or, alternatively, to try to describe an unknown location on the basis of the analysis of the trace evidence. To date, there is no internationally or, indeed, nationally adopted best practice for mineral analysis, and different approaches are advocated by different research groups. **Bowen & Caven (2013)** provide a rallying call to the advantages of detailed polarizing light microscopy, particularly related to the description and identification of unknown geological locations based on small amounts of trace evidence alone. Such an approach, although based on a very traditional methodology, can result in very-high-quality data sets and should clearly underpin many trace evidence investigations. **Bergslien (2013)** examines the use of hand-held X-ray fluorescence and detailed laboratory-based X-ray diffraction (XRD) approaches to identifying the source location of unknown soil samples when compared with regional soil geochemistry and mineralogy databases from the Buffalo–Niagara metropolitan area in New York. In particular, Bergslien highlights the importance of sample preparation and measurement in the use of XRD for the analysis of forensic soil samples. **Pirrie et al. (2013)** continue their work on automated mineral analysis, and demonstrate how it is possible to automatically characterize soil samples founded on the definition of 'lithotypes' based on an assemblage of minerals, and their grain size and texture. These 'lithotypes' are effectively the automated analysis of small rock fragments or other anthropogenic materials as shown in their experimental study. **Guedes et al. (2013a)** explore the possibility of using the magnetic susceptibility of soils and sediments as a means to discriminate between samples based on the analysis of beach and dune sediments, and soil samples, from Portugal. Preliminary results are promising in that the data are reproducible and the method appropriate for small samples, although additional research is needed before using the methodology in forensic casework. **Di Maggio & Nuccetelli (2013)** use a mineralogical approach in their case study investigating damage caused to Jewish graves in the Verano Monumental Cemetry in Rome. Trace evidence analysis showed that material recovered from a number of tools was consistent with the damaged areas on the graves and differed from suggested alibi locations where it had been claimed that the tools had been used. Similarly, the environmental forensic case studies presented by **Isphording (2013)** also utilize a mineralogical approach. First, the author considers the environmental impact of airborne particulates sourced from a steel mini-mill and demonstrates that the magnetite particles present at the affected addresses must have been derived from this industrial plant. Secondly, both significant health effects and rapid corrosion of metal surfaces were alleged as related to the emission from a paper mill of sulphur dioxide (SO_2), sulphur trioxide (SO_3) and hydrogen sulphide (H_2S), each of which when released to the atmosphere reacts with water vapour to form sulphuric acid (H_2SO_4). However, in addition to these gaseous compounds, was the release of calcite from the plant where it was used as a whitener. Calcite was present on the surfaces affected by corrosion but there was no local indigenous source for these particles.

Nearly all surface soils and sediments are complex mixtures of organic, inorganic (mineralogical) and anthropogenic components, and, as such, methods to analyse either the organic components or the overall characteristics of a sample can be as valid as examining the soil mineralogy. **Carvalho et al. (2013)** adopted an integrated approach, combining colour, grain size and mineralogy with palynology in an assessment of the characteristics of a river beach locality at Vila Nova de Gaia in northern Portugal. Eight samples were collected along a transect with a 15 m lateral spacing between the sampling locations. The colour, grain size and heavy mineral assemblage present was the same in all eight samples, and, although there was some variation in the relative abundance of the different pollen types present, all of the samples had the same assemblage of the dominant pollen grains present. Such studies are important when we come to assess how many samples are needed to characterize a crime scene. **Guedes et al. (2013b)** take this study a stage further, and evaluate whether colour, particle-size discrimination, chemical composition and magnetic susceptibility allow different coastal sediments collected from areas with different background geology to be differentiated. Colour analysis only allowed discrimination between different geographical areas. In contrast, particle-size analysis only allowed separation between dune and beach samples. Chemical composition allowed discrimination between both background geology and dune and beach settings. In combination, magnetic susceptibility, particle size and colour allowed the discrimination of both the geological setting and the local subenvironments. The work presented by both Guedes et al. (2013b) and Carvalho et al. (2013) is aimed at developing a database of Portugese coastal sediments that would be applicable in a forensic context. The value of such databases has been debated by numerous practitioners in forensic geoscience, and **McKinley (2013)** provides a thoughtful addition to this topic. It is concluded that, although such databases can be neither exhaustive nor precise enough to portray the spatial variability seen at a crime scene they do, however, when coupled with expert knowledge, play an invaluable role in providing background or reference material in a criminal investigation.

Eby (2013) provides a review of the applications of instrumental neutron activation analysis (INAA). This methodology can allow the detection of up to 40 chemical elements in the ppm–ppb detection range. To illustrate the potential of this method, particularly in environmental forensic investigations, Eby (2013) describes the analysis of: (i) small fragments of ceramics made using the same clay mixture but with different glazes; (ii) maple syrup; (iii) grass; and (iv) tree ring samples from trees growing downwind of the Palmerton, Pennsylvania, USA, zinc smelter. Distinctive geochemical differences allowed the various types of ceramic to be distinguished, the sources of maple syrup to be identified and the geographical/geological locations for the grass samples to be determined. Data from the tree rings can be used to map the history of environmental pollution associated with the zinc smelter. In a similar study into the environmental impact of Cu mining and smelting in the area around Karabash, in the Ural Mountains of Russia, **Purvis et al. (2013)** demonstrate the use of lichen in monitoring atmospheric contamination. Lichen collected from a reference site at a distance away from the smelter were transplanted onto 10 reference sites along a 65 km-long transect centred on Karabash. The transplants were then collected 2–3 months later, and geochemically analysed using inductively coupled plasma optical emission spectometry (ICP-OES) and inductively coupled plasma mass spectometry (ICP-MS), along with the analysis of the Pb isotopes. Particulates present on the surface of the lichens were characterized using scanning electron microscopy with energy-dispersive spectrometry (SEM-EDS) analysis. The lichen monitoring approach is discrete and comparatively low cost, enabling atmospheric deposition from various sources to be assessed over short duration periods.

Three papers consider the use of different geological approaches in both trace evidence investigations and also in the design and optimization of geological and geophysical searches. **Ruffell et al. (2013)** question why forensic geoscience is most commonly used within rural or semi-rural locations, and less commonly used within truly urban environments. The authors show how particulate trace evidence is common within the urban setting even though there is much less available data concerning its spatial distribution. Through a case study where an assailant contacted building materials, the authors demonstrate that current analytical techniques can provide evidence from very small sample sizes, a problem potentially common with the urban forensic investigation. In a separate paper, and illustrated through a series of case studies, **Ruffell (2013)** also considers the importance of traditional geological data, such as maps, cross-sections and measured sections, in both criminal and environmental forensic investigations. Using a traditional geological approach to read the landscape, valuable inferences can be drawn about the probability or ease with which items could be disposed of, or hidden, in the landscape. **Donnelly & Harrison (2013)** take this approach further, and discuss the value of aerial imagery, diggability surveys and the use of RAG (red–amber–green) maps as aids during ground searches for buried objects. Aerial imagery (when interpreted by those with a geological background) can assist in the identification of disturbed ground, where there is no likely geological or other documented (e.g. trial pits) explanation. Diggability surveys assess the ease with which different surface materials can be excavated in the light of both the time and implements available. Together these, and other data sets, can be integrated together as easy-to-read RAG maps, which can be used to prioritize search areas, so that available resources can be targeted effectively.

Donnelly (2013) documents a case study of a forensic search for remaining metal artefacts following the recovery of an Anglo-Saxon gold and silver hoard in Staffordshire, UK. Following the chance discovery of gold artefacts, an archaeological investigation resulted in the recovery of approximately 3940 metal items, which were sold in 2009 for £3.3 million. Before the location of the site became publicly known, a full forensic search was instigated to assess whether or not additional gold or silver items were present on the site. The author shows how the development of a conceptual geological model, aerial imagery and a diggability survey were combined to develop a RAG map of the site. Geophysical instruments were then selected on the basis of the size and composition of the likely targets to be acquired, depth of penetration required and the likely ground conditions. No further artefacts were located within the defined search area/depth; subsequent discoveries made in 2013 were from beyond the limits of the original search area.

The ability to identify biomarkers of decomposition may prove to be an important area of environmental and criminal forensics, and **Larizza & Forbes (2013)** present the results of an experimental study carried out in Canada. Pig carcasses, as analogues for human remains, were buried during two separate trials, and the fatty acids present in the soils containing decomposition fluids and the lateral extent of the fatty acids released in the cadaver decomposition island were characterized. Myristic, palmitic, palmitoleic, stearic and oleic fatty acids were detected, and these fatty acids extended up to 50 cm laterally away from the cadavers at higher levels of abundance than in the control samples. Work such as this has significant potential where cadavers are buried in one location but then subsequently re-excavated and the remains reburied elsewhere, as has been observed not only in single homicide cases but also in mass graves associated with war crimes.

The final set of three papers within the volume focus on shallow geophysical techniques used to identify areas of ground disturbance. **Lowe *et al.* (2013)** further evaluate the widely used method of ground penetrating radar (GPR) in the investigation of contrasting soil textures at three sites in southern Ontario, Canada. In each field trial, pig carcasses were buried, and GPR was used to detect and monitor the graves over a period of 14 months post-burial. In this study, whilst the GPR system used was able to locate the graves within silty clay loam and also a fine sandy loam, it was unable to detect the location of the graves in the site underlain by fine sandy soils.

Hansen & Pringle (2013) provide an experimental study comparing magnetic, electric and GPR surveys to detect a range of buried forensic objects including firearms, grenades, knives and a spade. A range of both metallic (forensic) and other common objects were buried within a semi-urban test site. Following burial, the site was geophysically surveyed with, subsequently, a 6 cm-thick layer of concrete slabs being laid over the area before being resurveyed to simulate a relatively common technique used to obscure illegal burial sites. The results showed that in the semi-urban environment the usable geophysical techniques gaining the highest buried forensic object target detection success rates were (in descending order): magnetic susceptibility; metal detection; 900 MHz GPR; electrical resistivity using 0.25 m fixed-offset probes; magnetic gradiometry; 450 MHz GPR; fluxgate gradiometry; and, finally, electrical resistivity with 0.5 m fixed offset probes. In contrast, once the concrete slabs were laid over the target items, the most successful detection techniques (in descending order) were: magnetic susceptibility; magnetic gradiometry; 900 MHz GPR; metal detection; 450 MHz GPR; and fluxgate gradiometry. Finally, **Richardson & Cheetham (2013)** continued the experimental geophysical approach, again using a range of geophysical tools to attempt to detect different types of buried weapons with different orientations. This study used magnetometry (fluxgate gradiometry), Slingram electromagnetic (both phases in horizontal magnetic dipole (HMD) and vertical magnetic dipole (VMD) orientations) and GPR (500 and 800 MHz). There were limitations for each instrument employed resulting from the orientation of the weapon in the ground, with none of the techniques in this work being able to detect all of the weapons in all orientations. A second limitation was the inability of all these instruments to detect smaller weapons. Clearly, further research is needed to help validate the most appropriate search tool, depending on the local ground conditions and the nature of the object(s) being looked for.

Concluding remarks

In forensic investigations of either the source and fate of environmental pollutants or the investigation of serious crime, such as murder and terrorism, investigating officers are willing and able to draw upon a very wide variety of scientific disciplines. Many offences will see offenders interacting in some way with Earth surface materials; hence, the potential application of forensic geoscience should be clear to see. However, where new equipment or analytical methodologies are being developed it is essential that their capability, accuracy and reproducibility are fully validated through rigorous scientific research before such methods can be safely deployed by practitioners working for, or alongside, the investigating team. Conferences such as those from which the papers presented in this Special Publication were sourced are an essential way to cross-pollinate and discuss best practice. The future looks vibrant for forensic geoscience globally, although the reality is that in any one geographical area there is likely to only be a small group of researchers and practitioners working in this complex area, which, due to its very nature, can be a challenging field personally. The establishment of the International Union of Geological Sciences Initiative on Forensic Geology is bringing researchers and practitioners from around the world together to share their common experiences; to develop best practice and standardized protocols appropriate for the different legislative frameworks; and to educate investigating officers around the world about the important role that Forensic Geoscience has to play.

The editors are very grateful to the authors who submitted their work to this volume, and to the referees who freely gave their time to provide insightful and constructive reviews. We owe a huge debt of gratitude to the team at the Geological Society, London, Publishing House, who helped us throughout the process, and in particular we acknowledge the help of Angharad Hills and Tamzin Anderson in the, at times, painful delivery of this volume, and Jo Armstrong during its production – thank you. Marianne Stam was the convenor of the California Soil Forensics Conference and aided the early stages of this volume. We are grateful to the Sponsors of the Forensic Geoscience Group Conference in London (Bartington Instruments, 3DRadar, Utsi Electronics Geoscan Research, Geomatrix Ltd, DW Consulting, Helford Geoscience LLP, Beta Analytic and FEI). Finally, we acknowledge our colleagues who globally make up the International Union of Geological Sciences, Initiative on Forensic Geology, who together aim to promote, educate and enhance the role of forensic geology around the world.

References

ANON 1856. Science and art: curious use of the microscope. *Scientific American*, **11**, 240.

BERGSLIEN, E. 2012. *An Introduction to Forensic Geoscience*. Wiley Blackwell, Chichester.

BERGSLIEN, E. T. 2013. X-ray diffraction and field portable X-ray fluorescence analysis and screening of soils: project design. *In*: PIRRIE, D., RUFFELL, A. & DAWSON, L. A. (eds) *Environmental and Criminal Geoforensics*. Geological Society, London, Special Publications, **384**. First published online July 16, 2013, http://dx.doi.org/10.1144/SP384.14

BOWEN, A. M. & CAVEN, E. A. 2013. Forensic provenance investigations of soil and sediment samples. *In*: PIRRIE, D., RUFFELL, A. & DAWSON, L. A. (eds) *Environmental and Criminal Geoforensics*. Geological Society, London, Special Publications, **384**. First published online May 16, 2013, http://dx.doi.org/10.1144/SP384.4

CARVALHO, Á., RIBEIRO, H., GUEDES, A., ABREU, I. & NORONHA, F. 2013. Geological and palynological characterization of a river beach in Portugal for forensic purposes. *In*: PIRRIE, D., RUFFELL, A. & DAWSON, L. A. (eds) *Environmental and Criminal Geoforensics*. Geological Society, London, Special Publications, **384**. First published online May 16, 2013, http://dx.doi.org/10.1144/SP384.3

DI MAGGIO, R. M. & NUCCETELLI, L. 2013. Analysis of geological trace evidence in a case of criminal damage to graves. *In*: PIRRIE, D., RUFFELL, A. & DAWSON, L. A. (eds) *Environmental and Criminal Geoforensics*. Geological Society, London, Special Publications, **384**. First published online May 16, 2013, http://dx.doi.org/10.1144/SP384.2

DONNELLY, L. 2013. The design and implementation of a high-assurance forensic geology and police search following the discovery of the Staffordshire (Anglo Saxon) gold Hoard. *In*: PIRRIE, D., RUFFELL, A. & DAWSON, L. A. (eds) *Environmental and Criminal Geoforensics*. Geological Society, London, Special Publications, **384**. First published online July 22, 2013, http://dx.doi.org/10.1144/SP384.8

DONNELLY, L. & HARRISON, M. 2013. Geomorphological and geoforensic interpretation of maps, aerial imagery, conditions of diggability and the colour-coded RAG prioritization system in searches for criminal burials. *In*: PIRRIE, D., RUFFELL, A. & DAWSON, L. A. (eds) *Environmental and Criminal Geoforensics*. Geological Society, London, Special Publications, **384**. First published online July 15, 2013, http://dx.doi.org/10.1144/SP384.10

EBY, G. N. 2013. Instrumental neutron activation analysis (INAA) and forensic applications. *In*: PIRRIE, D., RUFFELL, A. & DAWSON, L. A. (eds) *Environmental and Criminal Geoforensics*. Geological Society, London, Special Publications, **384**. First published online May 16, 2013, http://dx.doi.org/10.1144/SP384.5

GUEDES, A., MURRAY, R. C. ET AL. 2013a. The potential application of magnetic susceptibility as a technique for soil forensic examinations. *In*: PIRRIE, D., RUFFELL, A. & DAWSON, L. A. (eds) *Environmental and Criminal Geoforensics*. Geological Society, London, Special Publications, **384**. First published online July 8, 2013, http://dx.doi.org/10.1144/SP384.7

GUEDES, A., MURRAY, R. C., RIBEIRO, H., RODRIGUES, A., VALENTIM, B., SANT'OVAIA, H. & NORONHA, F. 2013b. Integration of different sediment characteristics to discriminate between sources of coastal sediments. *In*: PIRRIE, D., RUFFELL, A. & DAWSON, L. A. (eds) *Environmental and Criminal Geoforensics*. Geological Society, London, Special Publications, **384**. First published online May 16, 2013, http://dx.doi.org/10.1144/SP384.1

HANSEN, J. D. & PRINGLE, J. K. 2013. Comparison of magnetic, electrical and ground penetrating radar surveys to detect buried forensic objects in semi-urban and domestic patio environments. *In*: PIRRIE, D., RUFFELL, A. & DAWSON, L. A. (eds) *Environmental and Criminal Geoforensics*. Geological Society, London, Special Publications, **384**. First published online July 8, 2013, http://dx.doi.org/10.1144/SP384.13

ISPHORDING, W. C. 2013. Using soil mineral signatures to confirm sources of industrial contaminant trespass. *In*: PIRRIE, D., RUFFELL, A. & DAWSON, L. A. (eds) *Environmental and Criminal Geoforensics*. Geological Society, London, Special Publications, **384**. First published online August 1, 2013, http://dx.doi.org/10.1144/SP384.16

LARIZZA, M. & FORBES, S. L. 2013. Detection of fatty acids in the lateral extent of the cadaver decomposition island. *In*: PIRRIE, D., RUFFELL, A. & DAWSON, L. A. (eds) *Environmental and Criminal Geoforensics*. Geological Society, London, Special Publications, **384**. First published online July 22, 2013, http://dx.doi.org/10.1144/SP384.11

LOWE, A. C., BERESFORD, D. V., CARTER, D. O., GASPARI, F., O'BRIEN, R. C. & FORBES, S. L. 2013. Ground penetrating radar use in three contrasting soil textures in southern Ontario. *In*: PIRRIE, D., RUFFELL, A. & DAWSON, L. A. (eds) *Environmental and Criminal Geoforensics*. Geological Society, London, Special Publications, **384**. First published online July 11, 2013, http://dx.doi.org/10.1144/SP384.12

MCKINLEY, J. 2013. How useful are databases in environmental and criminal forensics? *In*: PIRRIE, D., RUFFELL, A. & DAWSON, L. A. (eds) *Environmental and Criminal Geoforensics*. Geological Society, London, Special Publications, **384**. First published online July 24, 2013, http://dx.doi.org/10.1144/SP384.9

MURRAY, R. 2004. *Evidence from the Earth*. Mountain Press, Missoula, MT.

MURRAY, R. & TEDROW, J. C. F. 1975. *Forensic Geology: Earth Sciences and Criminal Investigation*. Rutgers University Press, New York. (Republished in 1986.)

PIRRIE, D., ROLLINSON, G. K., POWER, M. R. & WEBB, J. 2013. Automated forensic soil mineral analysis; testing the potential of lithotyping. *In*: PIRRIE, D., RUFFELL, A. & DAWSON, L. A. (eds) *Environmental and Criminal Geoforensics*. Geological Society, London, Special Publications, **384**. First published online August 7, 2013, http://dx.doi.org/10.1144/SP384.17

PURVIS, O. W., WILLIAMSON, B. J., SPIRO, B., UDACHIN, V., MIKHAILOVA, I. N. & DOLGOPOLOVA, A. 2013. Lichen monitoring as a potential tool in environmental forensics: case study of the Cu smelter and former mining town of Karabash, Russia. *In*: PIRRIE, D., RUFFELL, A. & DAWSON, L. A. (eds) *Environmental and Criminal Geoforensics*. Geological Society, London, Special Publications, **384**. First published online June 26, 2013, http://dx.doi.org/10.1144/SP384.6

PYE, K. 2007. *Geological and Soil Evidence: Forensic Applications*. CRC Press, Boca Raton, FL.

PYE, K. & CROFT, D. (eds) 2004. *Forensic Geoscience: Principles, Techniques and Applications*. Geological Society, London, Special Publications, **232**.

RICHARDSON, T. & CHEETHAM, P. N. 2013. The effectiveness of geophysical techniques in detecting a range of buried metallic weapons at various depths and orientations. *In*: PIRRIE, D., RUFFELL, A. & DAWSON, L. A. (eds) *Environmental and Criminal Geoforensics*. Geological Society, London, Special Publications, **384**. First published online September 9, 2013, http://dx.doi.org/10.1144/SP384.18

RITZ, K., DAWSON, L. & MILLER, D. 2009. *Criminal and Environmental Soil Forensics*. Springer, Amsterdam.

RUFFELL, A. 2013. Solid and drift geology in forensic investigations. *In*: PIRRIE, D., RUFFELL, A. & DAWSON, L. A. (eds) *Environmental and Criminal Geoforensics*. Geological Society, London, Special Publications, **384**. First published online August 7, 2013, http://dx.doi.org/10.1144/SP384.15

RUFFELL, A. & MCKINLEY, J. 2008. *Geoforensics*. Wiley Blackwell, Chichester.

RUFFELL, A., PIRRIE, D. & POWER, M. R. 2013. Issues and opportunities in urban forensic geology. *In*: PIRRIE, D., RUFFELL, A. & DAWSON, L. A. (eds) *Environmental and Criminal Geoforensics*. Geological Society, London, Special Publications, **384**. First published online September 5, 2013, http://dx.doi.org/10.1144/SP384.19

Forensic provenance investigations of soil and sediment samples

A. M. BOWEN[1,2]* & E. A. CAVEN[1]

[1]*Stoney Forensic, Inc., 14101 Willard Road, Suite G, Chantilly, VA 20151, USA*

[2]*Present address: US Postal Inspection Service, National Forensic Laboratory, 22433 Randolph Drive, Dulles, VA 20104-1000, USA*

*Corresponding author (e-mail: AMBowen@uspis.gov)

Abstract: In recent decades, the polarizing light microscope has been slowly phased out of the geology curricula of many universities in the United States. This is unfortunate, as the polarizing light microscope is an instrument that is particularly well suited to overcoming problems typically encountered in forensic scenarios. Its versatility makes it uniquely powerful in provenance investigations in which sample size is limited and sample history is unknown. The polarizing light microscope can be used to determine the source rock type(s) of a sample of unconsolidated grains, either based on the mineral assemblage present in the sample or on properties observed for individual mineral grains or lithic fragments. Biological particles including pollen, spores, diatoms and botanical macerals can be identified and used to constrain the possible source of a sample. In addition, individual particles observed in the sample (geological, biological or anthropogenic) can be isolated and re-mounted for subsequent analyses that are useful for provenance determination. The inferences made can be used to produce maps illustrating potential source regions by means of geographical information systems (GIS). Several examples are provided to illustrate the various capabilities of the polarizing light microscope and its contributions to forensic soil analysis, specifically in forensic provenance investigations.

The authors have been involved in research and casework related to forensic provenance determination of dust, soil and sediment samples for over a decade. In this paper, we discuss the general approach taken when performing this type of analysis. The approach relies heavily on micro-analysis in general and on polarizing light microscopy in particular. That being said, the light microscope is often used to direct subsequent analyses by other instruments, and as such the analytical scheme is not limited to a single technique. The approach is largely multidisciplinary, and attempts to combine many different areas of expertise, namely geology, soil science, palynology, industrial particles, non-human DNA analysis and mapping techniques. Experts in other fields are often consulted when appropriate to the specific sample being analysed, and novel instrumental techniques are often considered when they might be applicable to the particle types present in a particular sample. Based on the data obtained from a sample, geographical areas that are unlikely sources for the material are excluded and remaining areas are displayed using geographical information systems (GIS) software. If inferences can be made from a variety of material types (geological, biological and anthropogenic), the result is often a very limited geographical region which is then investigated in greater detail by the examination of satellite images or by collection of comparative samples from the potential source region. The method is based on first principles and is not dependent on a comparative database in order to provide useful information regarding geographical origin.

The emphasis on the polarizing light microscope in this approach is due to several reasons, one of which is a bias resulting from the specific backgrounds of the scientists involved. Despite this admitted bias, there are a number of reasons why the polarizing light microscope is particularly well suited to examination of forensic soil samples. Many of the methods commonly employed in the Earth sciences are dependent on the availability of relatively large samples that are representative of their source. Owing to their typically unknown histories, forensic samples cannot be assumed to be representative of their source. Forensic samples are often too small to produce meaningful results using these methods. Any sample visible on evidence, however, is virtually guaranteed to contain more than enough material for microscopical analysis. Microscopical examination is, with few exceptions, non-destructive in nature and evidence is preserved for opposing experts to re-examine. The material is also available for analysis by other instruments. Identification based on optical crystallographical properties is highly appropriate for minerals. Mineral species are defined on the basis of both their crystal structure and their chemical composition. The optical properties of materials are related to both their structure and their composition, making these properties particularly well suited for mineral identification. Most other analytical techniques provide data related to only one of these properties. While appropriate for mineral identification, polarizing light microscopy is also applicable to a wide variety of additional materials including those of both biological and anthropogenic origin. Because microscopy involves direct observation of the individual particles, a number of particle properties other than identity can be observed and exploited for source determination (e.g. mineral varietal types, state of pollen preservation). Polarizing light microscopy can also direct subsequent analyses of specific particles towards the instruments most likely to provide data useful for source determination. Appropriate sample processing and sample preparation is a critical prerequisite for effective microscopical analysis of dust, soil and sediment samples. It is typically desirable to separate the sample into several fractions (sample size permitting), and to analyse each fraction in a manner appropriate to the materials concentrated in the specific fraction. The methods used are adapted from those described by Palenik (2007).

Geological materials

The geological materials in soil that have been found to be most useful for provenance determination are rock fragments, mineral grains and microfossils. These all contribute to a determination and description of the likely source rock(s) that contributed material to the sample. When present, rock fragments can be prepared into thin sections and identified by traditional petrographical analysis. This is the most efficient and effective means of identifying source rock type but it requires fairly coarse rock fragments, often unavailable in forensic scenarios (at least 0.5–1.0 mm and often larger depending on the rock type). The petrography can frequently be interpreted in combination with grain count analysis of the fine sand fraction to help constrain the rock type. A soil sample from a recent case example was dominated by volcanic rock fragments containing primarily mixed alkali feldspar microlites enclosed by larger quartz grains (Fig. 1). Accessory minerals, difficult to identify in thin section, were also observed in grain mounts, and were determined to be arfvedsonite (a sodic amphibole) and aegerine (a sodic pyroxene) by a combination of polarizing light microscopy, scanning electron microscopy (SEM) with energy-dispersive spectroscopy (EDS) and Raman microspectroscopy analysis. Together, these data suggested that the parent rock was comendite or a similar peralkaline rhyolitic rock (Winter 2009), significantly constraining the possible geographical source of the sample. There have been a number of published case examples where petrographical analysis of a rock assisted in the determination of its geographical source in a forensic investigation. In one case, a rock was left at the side of the road in southern New Jersey by a suspicious vehicle, presumably as a signal to co-conspirators. The rock was analysed by petrographical techniques and determined to be an unusual type of garnet schist. Similar metamorphic rocks were known to crop out in a small area in the highlands of western Connecticut. Investigators subsequently focused on that area and developed leads that ultimately led to the discovery of a nearby safe house containing explosives (Murray 2004).

The parent rock type can sometimes be inferred by grain count analysis of grain mounts of the sand-sized mineral grains present in a soil or sediment sample. While sediment samples are unlikely to ever be one-to-one images of their source rocks due to weathering, transport, mixing and diagenetic processes, among others, there is still quite a bit of useful provenance information to be gleaned from the mineral assemblage present (Weltje & von Eynatten 2004). For example, a sample submitted

Fig. 1. A thin section of volcanic lithic fragments, identified as comendite, shown (**a**) in plane polarized light and (**b**) in between crossed polars. The felsic phases are quartz and mixed alkali feldspar; the mafic phases are arfvedsonite and aegerine.

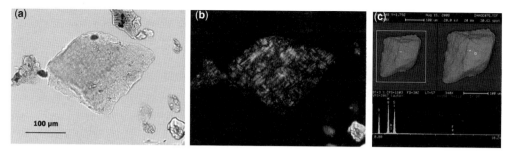

Fig. 2. A serpentine grain shown (**a**) in plane polarized light, (**b**) in between crossed polars and (**c**) by SEM with its EDS spectrum. The mounting medium is 1.540 for (a) and (b). The SEM image was taken by D. Exline.

for analysis contained significant amounts of serpentine (Fig. 2), clinopyroxene, olivine and chrome spinel in its fine sand fraction. This was a strong indication that at least one of the source rocks was a mafic or ultramafic rock. The presence of highly unstable mineral species provided evidence that the material was relatively close to its source, as extensive transport would probably have destroyed the olivine and other unstable minerals. In this case, geographical areas lacking outcrops of mafic or ultramafic rocks in their vicinity were excluded from being possible source areas for the material. In a well-known case example, analysis of the mineral assemblages present in ballast bags from balloon bombs deployed from Japan to the continental United States during World War II provided useful insights into their geographical source. The mineralogy of the sand was unusual, dominated by heavy minerals of volcanic origin. The mineralogy eliminated North America as a possible source of the sand and helped narrow down the proposed source area to two beaches on the east coast of Japan just north of Tokyo. One of these beaches was later confirmed to have been used to launch balloon bombs (Ross 1950; McPhee 1997). In some cases, the presence of a single mineral provides strong evidence of a specific source rock type or of a specific location, even when the mineral occurs in trace amounts and would be undetectable by powder X-ray diffraction. This is true for certain index minerals from metamorphic rocks or in cases of minerals with economic value and well-documented occurrences. Lazurite is one such mineral and, in an unpublished case example, the presence of lazurite on footwear submitted for analysis implied a very specific geographical source where this rare mineral has been mined for thousands of years (Fig. 3) (Stoney pers. comm. 2011). In a dramatic case example, the presence of a deep pink rhyolitic glass in soil recovered from the body of a murdered United States DEA agent helped narrow the search for the original burial site to a specific state park in Mexico. The original burial site was later located within this park (McPhee 1997). While not technically a mineral, the presence of this single particle type was extremely useful in constraining the source of this sample to a very specific geographical area.

In addition to the inferences possible from the identities and relative abundances of the minerals present, certain mineral properties can provide insights into the source rock. The properties

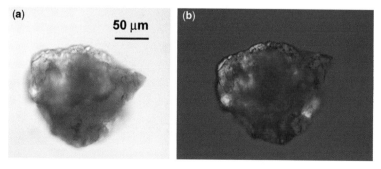

Fig. 3. A lazurite grain shown (**a**) in plane polarized light and (**b**) in between crossed polars with a 530 nm compensator. The mounting medium is 1.500.

observable with a polarizing light microscope that may be of value include mineral grain morphology, colour/pleochroic scheme, twinning, zoning, exsolution microstructures, inclusions, undulosity and polycrystallinity, among others. The particular properties that are useful in a given case depend on the specific minerals being considered and the available published research, typically in the sedimentary geology literature. For example, the proportions of polycrystalline quartz grains and monocrystalline quartz crystals in a sediment sample, along with the number of crystal units per polycrystalline grain, can be useful for distinguishing between plutonic, low-grade metamorphic and high-grade metamorphic source rocks (Basu et al. 1975). Quartz undulosity can be used to further distinguish between certain source rock types (Basu et al. 1975). The frequency and types of zoning observed in plagioclase grains in a sediment sample can provide clues to its source rock type (plutonic, volcanic or metamorphic). Zoning, in general, is rare in metamorphic rocks and therefore suggests an igneous source for the material, while oscillatory zoning is characteristic of volcanic rock sources (Pittman 1963). The relative abundance of different types of plagioclase twins can provide further insights into the likely source rock (Pittman 1970). Inclusions can prove useful for inferring source rocks, both by their identities and their textural relationship with their host. When present in soils, poikiloblasts can be particularly valuable for providing insights into the geological history of the source region, along with the grade of the metamorphic source rocks (Winter 2009). Sedimentary geologists have long recognized the value of mineral varietal studies for provenance determination (Brammall 1928), and single-grain studies (a modern analogy to mineral variety studies) are essential to modern sediment provenance studies (Mange & Maurer 1992; Weltje & von Eynatten 2004). Mineral varietal studies have also contributed to forensic provenance investigations. In one case, a large amount of silver being transported by rail was stolen from the steel cans containing the silver and replaced with sand. The theft was known to have taken place somewhere between the precious metals refinery near Denver, Colorado, and Rochester, New York, the location where the theft was first noticed. The variety of zircon present in the sand, along with its pollen content, led Skip Palenik to infer that the theft probably took place at the station near Denver where the silver was loaded onto the train. The zircon in the sample resembled varieties he had recently encountered in a reference sample from Pike's Peak in Colorado. Analysis of sand taken from the train station near Denver revealed that it was nearly identical to the sand from the robbery (Palenik pers. comm. 2004).

Microfossils are present in many sediments and soils, and if identified they can be extremely useful for constraining the geographical source of a sample. The age of the sediment and its palaeoenvironment can be determined in most cases in which the microfossils can be identified. The authors have encountered foraminifera, coccoliths, dinoflagellate cysts, ostracods, diatoms, pollen and other microfossils in samples submitted for analysis. Specimens of fossil foraminifera, coccoliths, dinoflagellate cysts, ostracods and pollen have been identified by consulting experts, and used to significantly constrain the source rock contributing those fossils to the sediment. In one case, the microfossil taxa present, including a specimen of the foraminifer *Pseudoguembelina hariaensis*, indicated a pelagic limestone source and constrained the age of the source rock to the Late Maastrichtian, between 66.8 and 65.5 Ma (Fig. 4). A large proportion of the geographical region of interest could be excluded based on these criteria. As evident in this example, the primary value of microfossils is in constraining the age of the source rock, along

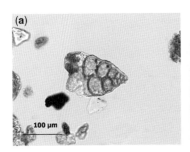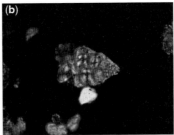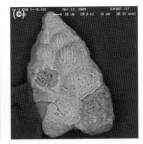

Fig. 4. A specimen of *Pseudoguembelina hariaensis*, a pelagic foraminifera taxon from the Late Maastrichtian, used to constrain the age of a sediment source to between 66.8 and 65.5 Ma. The fossil is shown (**a**) in plane polarized light, (**b**) in between crossed polars and (**c**) by SEM. The mounting medium is 1.540 for (a) and (b). The SEM image was taken by D. Exline.

with its depositional environment. Microfossils have contributed to several documented forensic investigations (Bowen 2010). Perhaps the most striking example, from Harrisburg, Pennsylvania, involved a missing police officer. After the officer failed to report to work, the officer's personal vehicle was found with blood in the trunk and conspicuous mud in one wheel well. The FBI analysed the mud and found, among other materials, microfossils so unusual that their limestone source rock cropped out in only two highly confined areas. One of the outcrops was in Appalachian Pennsylvania and intersected a country road. The FBI provided this information to local police who found the officer's body the following day at the location where the outcrop meets the road (McPhee 1997).

Sample properties that are related to features mapped by soil scientists can also be exploited, although generally to a lesser degree. Soils are classified on the basis of their colour, texture, mineralogy, the amount of organic matter present and a variety of other properties (Soil Survey Staff 2010). Many of the properties used to classify soils relate to subsurface horizons and cannot be inferred from typical forensic soil samples. However, it is often possible to exclude certain soil types on the basis of the soil colour, texture, the mineralogy of the finest sand fractions, the clay mineralogy and the relative abundance of organic matter. Inferences made about the parent rock type, as described above, can often also be useful for excluding certain soil types.

Biological materials

A tremendous variety of biological materials are regularly encountered in forensic soil samples. While each of these has potential to contribute to an investigation, it is efficient to focus on a small number of particle types that have proven their value in prior cases and only pursue the identification of other biological materials when dictated by the sample character. Among the various biological materials encountered in soils, palynomorphs, especially pollen, may well have the most to contribute on a consistent basis to forensic provenance investigations. Palynologists can identify a wide variety of palynomorphs, including but not limited to pollen and plant spores, and infer information about the climate, elevation, ecology and local vegetation in the source region of a sample. In addition, when pristine pollen grains are identified in a sample, it may be possible to determine the flowering period for the given plant and thus make inferences regarding the timeframe during which the sample could have been exposed to the source

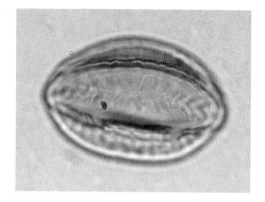

Fig. 5. A specimen of *Ephedra nevadensis* pollen shown in transmitted light. The grain is approximately 50 μm long. The mounting medium is glycerine. This image was taken by V. Bryant.

environment. In an unpublished provenance investigation, pollen from *Ephedra nevadensis* narrowed down the potential source region dramatically, and enabled all subsequent analyses to focus on a fairly limited geographical area (Fig. 5) (Bryant pers. comm. 2011). In another example, the presence of pristine pollen from *Dodonaea viscosa* indicated that this woody plant was growing within the local source environment. The potential ecological suitability for the growth of this taxon was modelled in GIS using growing season and climatic parameters within the geographical region of interest for the case, and was cross-referenced with data on the native distribution of the plant. The detection of this pollen type was useful in focusing the investigation on a relatively narrow region within areas not already excluded by other aspects of the analysis (Stoney *et al.* 2011). Pollen has been used to confirm the country of origin of commercial honey (Bryant *et al.* 1990). Pollen analysis has also contributed to source determination in a number of published forensic cases. One of the earliest cases involved pollen analysis of mud adhering to a murder suspect's boots. It occurred near Vienna, Austria, where the mud was analysed by Wilhem Klaus, a geologist with the Austrian Geological Survey. The boots were determined to contain both recent and Miocene fossil pollen taxa, and they were interpreted by Dr Klaus to exclude all but a very specific source location. When the suspect was confronted with the suspected location, he was shocked and confessed to the crime (Bryant & Mildenhall 1998).

Botanical taxa present in the source region can also be determined by microscopical examination of botanical macerals or by botanical DNA analysis. In one unpublished forensic provenance case that had a global scope, a small botanical maceral was observed on footwear and botanical DNA analysis identified the fragment as *Ceratocarpus arenarius*. The native distribution of this herbaceous weed is limited to areas of Asia and Europe, and it is not cultivated for economic uses within or outside of its range. Therefore, the presence of this botanical taxon in the sample significantly reduced the potential source region and allowed the global investigation to be focused on a much narrower geographical area within two continents (Stoney pers. comm. 2011). One case example published in the literature demonstrates how botanical macerals can be used for geographical source determination. In this case, a man was found dead in his vehicle in Butte County, California, killed by a self-inflicted gunshot wound. The man had a daughter with whom he had recently been seen, but the whereabouts of the child was unknown. Botanical macerals recovered from the automobile were sent to a botanist at a local university, who identified at least six different taxa (five to species level). Based on the botanical taxa present, the expert determined that the material had originated from a site with an elevation of between 980 and 1380 m on the western exposure of the Sierra Nevada under a fairly dense forest canopy. Police officials searched along roads for areas that matched the criteria described by the botanist and included the identified plant taxa. The police were able to locate the body of the missing child shortly thereafter with the help of this information (Ladd & Lee 2005).

Fungi are nearly ubiquitous in the environment, and are encountered in the vast majority of dust, soil and sediment samples. Mycologists can identify many of the fungal taxa present in a soil sample using light microscopy and infer information about the environment from which the sample originated that can prove useful in determining geographical origin. Some fungal taxa are plant pathogens that are specific to certain plant hosts and their presence can therefore be used to infer the presence of their host plant. Knowledge that the host plant is present at the sample origin can be used to constrain the geographical source in the same manner as for plants identified using other means. Some fungal taxa are found only in specific types of environments (near marine beaches, for example), and their presence could potentially prove valuable for constraining the source of the sample (Hawkesworth & Wiltshire 2011). In one unpublished case, DNA analysis of the sample detected the presence of *Wallemia sebi*, a fungus that has been associated with hay and farm buildings (Hanhela *et al.* 1995). This was one of several pieces of data suggesting that the source environment was associated with a region containing agricultural activities (Stoney pers. comm. 2011).

Other biological materials that have been exploited in forensic soil analysis include diatoms, plant opal phytoliths and insect fragments. In the aforementioned WWII Japanese balloon bombs case, there were more than 100 diatom species present in the sand from the ballast bags. The diatoms included both recent and fossil species, and the fossil diatoms were all Pliocene in age, significantly constraining the parent rock (McPhee 1997). Plant opal phytoliths have long been used for palaeoecology reconstructions as a complementary technique to palynology, and there seems to be no reason why the ecology of the geographical source region of a recent sediment or soil could not be determined in an analogous manner and used to constrain the origin of the sample (Rovner 1971). The potential offered by plant opal phytoliths for forensic soil comparison has been investigated, and the results suggested that inferences regarding the local land use may be possible on the basis of the phytolith types present in a sample (Marumo & Yanai 1986). In a case discussed in a television documentary, an entomologist was able to constrain the geographical region in which a vehicle was driven based on the insect fragments found on the radiator of the vehicle. Several of the insect taxa identified are found only in the western United States, and their presence was inconsistent with the alibi of the suspect in the case who claimed that the vehicle had been only driven in Columbus, Ohio (Garvey 2007). Animal hair is another type of material with the potential to contribute to a provenance investigation; however, no published case examples could be found in which animal hair contributed to source determination. All of these biological materials can be identified with a polarizing light microscope and have tremendous potential to contribute to forensic provenance investigations of future samples.

Non-human DNA analysis has the potential to identify fragments of insects and other animals in addition to the plants and fungi discussed earlier. These methods can be applied to individual insect parts or animal hairs that are manually isolated, or applied to a fraction of the soil, sediment or dust sample. One provenance investigation case involved both insect and mammal DNA in a complementary manner. Vacuumed dust from a rug originating in the rural near-east was examined with a number of DNA primers designed for detection of both plant and animal DNA. Plant taxa detected in the sample included grasses, weeds, and pines along with chick pea (*Cicer arietum*) and alfalfa (*Medicago sativa*). Animal taxa included the water buffalo (*Bubalis bubalis*) and a cattle-biting louse (*Bovicola*). These findings contributed to inferences of exposure to an environment where, among other

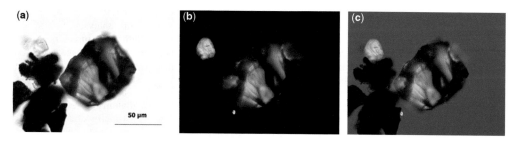

Fig. 6. A carborundum particle shown (**a**) in plane polarized light, (**b**) in between crossed polars and (**c**) in crossed polars with a 530 nm compensator. The mounting medium is 1.660.

features, water buffalo are kept and fed with alfalfa (Stoney pers. comm. 2011).

Anthropogenic materials

Polarizing light microscopy has the capability of identifying a tremendous variety of anthropogenic materials (McCrone & Delly 1973). Anthropogenic substances can be used to make inferences regarding land use, local human activities, population density and other mappable features of a source location. Materials of interest include building materials, combustion products, fertilizers, pesticides, foodstuffs, cosmetics and drugs. The presence of anthropogenic materials can provide insights into the industrial activities in the vicinity of the sample source. For example, the presence of carborundum grains (Fig. 6), welding spheres and metal turnings found together in one sample were interpreted as being from nearby metal-working activities (Stoney pers. comm. 2011). Synthetic fibres are typical of indoor environments, as are individual starch grains (Fig. 7). The type of starch present can be indicative of a particular plant or foodstuff. Fibreglass and other building materials can provide insights into the human activities in the source area of a sample. Power plant flyash (Fig. 8) and other combustion materials are indicators of specific types of human activity, such as high-efficiency coal burning. Abraded rubber particles are particularly common near roads, coming from tyre wear. When found in combination with asphalt particles, often along with glass spheres and white and/or yellow paint fragments, the presence of a paved road can be inferred. Certain botanical matter, such as large amounts of pollen from cultivated plants, implies agricultural activities, especially if evidence of fertilizers or pesticides can be detected in the sample. One case example from Russia has been published in which anthropogenic materials were used to help constrain the source of soil found on the body of a murder victim. The victim's body appeared to have been moved from the original crime scene based on the dissimilarity between soil found on the victim and soil from the site where the body was discovered. The presence of power plant flyash spheres on the body was used to infer that the soil had originated from near a large power plant in the south or SE of Moscow. Flyash had been reported from soils within 10–15 km of these power plants, and areas further away from these power plants were excluded as possible sources for the soil material (Gradusova & Nesterina 2009).

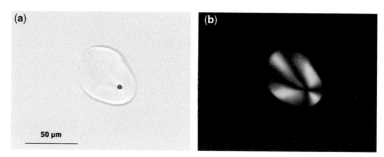

Fig. 7. A starch grain shown (**a**) in plane polarized light and (**b**) in between crossed polars. The mounting medium is 1.540.

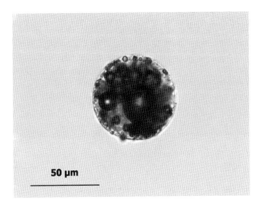

Fig. 8. A flyash sphere shown in transmitted light. The mounting medium is 1.540.

Other analytical techniques

One of the primary advantages of the polarizing light microscope is that it provides the analyst with helpful information for determining what subsequent tests (if any) would be appropriate for gaining additional data useful for constraining provenance. A wide variety of analytical instruments and techniques that are regularly used by geologists and soil scientists can help provide information pertinent to provenance determination. These include but are not limited to SEM, EDS, cathodoluminescence imaging, isotopic dating methods, electron microprobe analysis, X-ray diffraction, Raman microspectroscopy, stable isotope analysis and Fourier transform infrared microspectroscopy.

Scanning electron microscopy provides high-resolution, high depth-of-field images and topographical information that cannot be obtained with a polarizing light microscope. These types of images are critical for studying the surface morphology of certain minerals (especially quartz) (Bull & Morgan 2006). Quartz grain surface morphology can be useful for determination of the source rock type and transport history of the quartz, as well as for describing the depositional environment of the sediment. Figure 9 shows a desert sand grain, a river sand grain, a beach sand grain and a colluvial grain. Coatings of clay or iron oxides and adhering particles, such as diatoms or precipitated salt crystals, can be observed and identified on the surfaces of mineral grains. These can be useful for inferring

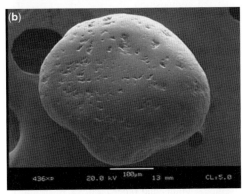

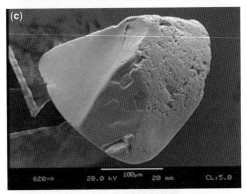

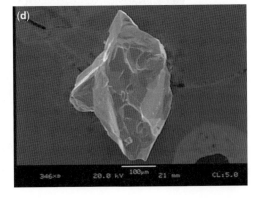

Fig. 9. Four quartz grains imaged by SEM. They come from a variety of depositional environments, including (**a**) a desert, (**b**) a river, (**c**) a marine beach and (**d**) a deposit of colluvium. These images were taken by S. Sparenga.

the depositional environment of the grains (e.g. marine beach or freshwater lake beach). Scanning electron microscopy is also critical for the identification of a variety of microfossils, many of which are classified based on fine surface features that are difficult to observe by light microscopy (Fig. 4).

Energy dispersive spectroscopy is a technique used to obtain semi-quantitative data on the chemical composition of a sample, typically as an accessory to an SEM. This type of data is useful in combination with optical properties determined using a polarizing light microscope for the identification of a wide variety of minerals and anthropogenic materials, and for obtaining information about the composition of a particular mineral in order to better describe its source rock. Garnet, for example, has several compositional end members that occur in different geological settings, and EDS analysis can provide insights into the garnet composition and therefore its source rock (Morton 1985). Opaque grains are particularly challenging to identify with a polarizing light microscope. Their morphology, colour in reflected light and magnetic susceptibility can provide some insights into their identity but they generally require additional analyses to be definitively identified. Analysis by EDS is a rapid method for distinguishing between many different opaque minerals. Correct identification of mineral grains and other particles is critical for subsequent data interpretation.

There are a number of dating methods available for rock fragments and mineral grains, most of which are based on the decay of unstable parent nuclei to stable daughter isotopes. The half-life of the reactions can be used to calculate the time required to produce the observed number of daughter isotopes from the parent nuclei present in a sample. One common example is the U–Pb dating method, most often applied to the mineral zircon (although several other minerals are also appropriate for the method). Age distributions for detrital zircon populations can be produced and used to gain insights into the age of a sediment sample's source rock(s). This is a method commonly employed by sedimentary geologists, especially those engaged in sediment provenance determination. In some cases, insights into the geological history of a single grain can be gained by dating different spots on a single grain; locations are typically selected based on textures observed by cathodoluminescence imaging (Fig. 10). This method proved particularly useful in a provenance investigation where a customer was trying to determine the more likely of two source areas. One region was largely underlain with Proterozoic (1.8 Ga) rocks that cropped out in numerous locations in the area, while the other region was essentially devoid of crystalline rocks of this age. The U–Pb age distribution of igneous zircon grains from the sample, as determined by multicollector laser ablation inductively coupled plasma mass spectrometry (MCLA-ICP-MS), exhibited a large contribution of zircons roughly 1.8 Ga, excluding one of the two possible source regions.

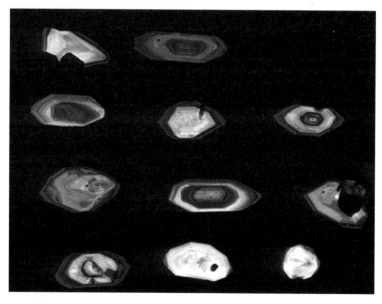

Fig. 10. A cathodoluminescence image of several polished detrital zircon grains of igneous origin. The SEM image was taken by Washington State University.

While the approach being advocated here is based on first principles and is not dependent on any single comparative database, it is advisable to take advantage of such databases whenever possible. For example, Washington State University maintains a database of electron microprobe analysis data from North American tephra samples. Using their database, it is often possible to determine the volcano from which a particular tephra sample was erupted and, in certain instances, it may be possible to determine the precise eruptive event (Foit pers. comm. 2010). For some of these volcanoes/eruptions the airfall distribution of their ash has been documented. A small portion of a soil sample containing tephra (Fig. 11) was submitted to Washington State University for analysis. The elemental composition data were compared to their database using a similarity coefficient to determine whether any reference samples were extremely similar to the submitted tephra sample (Borchardt et al. 1972). The analysis determined that the submitted sample probably originated from one of several eruptions of the Mono Craters in east central California near the Nevada border. Exact airfall distributions of the Mono Crater tephras are not published but they have been reported from the area of California around Mono Craters, southern Nevada and SW Utah, and are also likely to occur in NW Arizona. These data were useful in significantly constraining the potential source region for that particular sample.

Raman microspectroscopy has proven to be particularly useful as an aid in mineral identification. In combination with optical properties and EDS spectra, Raman is an extremely powerful identification tool. There are large, publicly accessible on-line databases of mineral Raman spectra, along with free software that enables the user to compare their own Raman spectra with those in the database (Downs 2006). Mineral grains can be analysed by Raman microspectroscopy directly in grain mounts without removing the coverslip (provided the mounting medium does not fluoresce). Raman is also extremely useful for identifying a wide range of anthropogenic materials.

X-ray diffraction is regularly used to identify minerals in the clay-sized fraction. These data are often very useful for excluding possible soil types when using soil maps for interpretation purposes. It can also be helpful for identifying minerals that cannot be identified with other available instrumentation.

A number of other analytical techniques are available and used by sedimentary geologists involved in provenance studies, as well as by scientists in other fields. These can be extremely useful for certain samples, and groups involved in provenance determination should be willing to explore new techniques when the character of the sample is such that these techniques may prove useful. The selection of analytical instrumentation used for any given sample is largely driven by the specific particles present in the sample and may differ significantly from sample to sample.

Spatial analysis and mapping

A wide range of spatial analysis and data resources are available for producing and utilizing geographically referenced data in support of forensic provenance research and investigations. The Esri ArcGIS® software provides powerful tools for visualizing and analysing spatial data and related attributes. Data types of varying subject matters and resolutions are represented as separate layers in a digital map. Map layers containing individual data types, such as soil units, rock types and elevation, can be overlain to display a wealth of information about where features of interest are located within a geographical region, along with attribute data regarding the types of features located at a specific site of interest. This provides significant analytical capabilities for focusing and visualizing forensic provenance investigations. By analysing relevant sample data with GIS, the investigation can be concentrated in geographical areas with features that are consistent with inferences made about a sample's exposure site and, similarly, by excluding regions having features inconsistent with various aspects of the inferred exposure site.

The geospatial capabilities and resources available are diverse and robust enough to handle significant data-processing requirements, address unique analysis methods, deal with varying subject matter and cover areas of the globe at varying

Fig. 11. An example of tephra recovered from a soil sample shown in transmitted light. The mounting medium is 1.540.

spatial resolutions depending on project needs. In support of this type of casework, the authors have accumulated a very large database containing over 4300 maps and GIS datasets, along with over 6700 books, theses and articles in searchable formats. These resources are often used in the production of maps based on the data obtained from submitted samples. A large number of GIS datasets are available for public use from government, non-profit and commercial sources, and many of these are free to download from various geospatial data portals. In addition, hard-copy maps can be converted to digital format and georectified for accurate display within a GIS. These maps can also be digitized to extract features of interest, which enables a vast quantity of map products to be displayed and utilized in GIS. This capability has proven invaluable when working with obscure resources relevant to a given provenance investigation, such as an out of print geological map for a region of interest or a plant species' distribution map published in a regional flora guide.

A variety of different particles types and inferences that can be made from them were discussed in the previous sections. The particles of primary interest in this type of spatial analysis are those that can be translated into a mapped property. For geological materials, the particles of interest and analytical tests of choice are those that can provide insights into the source rock type and age. The legends of geological maps can be ranked in a manner that communicates the likelihood that a particular legend unit (rock type and/or age) is the source for the material observed in the sample.

Legend units that cannot be excluded as being likely sources can be illustrated in maps using GIS. For example, in one United States-based case the sample contained an abundance of felsic volcanic rock fragments, mineral grains of volcanic origin and tephra fragments. The distribution of volcanic rocks was mapped within the continental United States to display areas of correspondence with the inferred source environment (Fig. 12). This was useful in focusing the provenance investigation to specific areas of the country where volcanic rocks occur. In some cases, age can be inferred from rock type if the possible source region being considered is limited in size. For example, if the source rock is clearly volcanic, and there are no volcanic rocks older than 50 Ma in a suitably detailed geological map for the potential source region, all rocks older than 50 Ma can be excluded from being possible source rocks for the sample.

It is often possible to exclude some soil taxonomic types as being potential sources for a sample submitted for analysis, as well as to identify some soil types that are consistent with mineral traces observed in a sample. Soil maps and GIS datasets can be used in a manner analogous to the geological maps, and often serve to further constrain the possible source region for the sample. For example, soil types as described for a 1:5 000 000 scale digital soil map were assessed for correspondence with the mineralogy and several other properties of a soil sample originating from an unknown location within the United States (UN-FAO 2003). The sample was a well-drained soil with relatively low soil organic matter content. It was a sandy soil

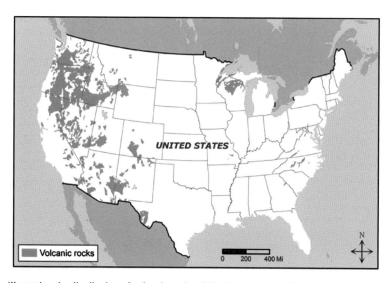

Fig. 12. Map illustrating the distribution of volcanic rocks within the continental United States, using data obtained from the 'Generalized Geologic Map of the Conterminous United States' (Reed & Bush 2004).

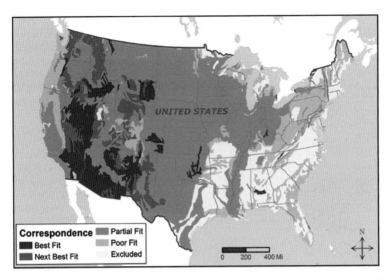

Fig. 13. Map illustrating a soil sample's mineral correspondence within the continental United States with soil types as described for the 'Digital Soil Map of the World' (UN-FAO 2003).

with little silt and clay; the clay present was primarily illite with some smectite and a small amount of kaolinite. There were abundant weatherable minerals present and little evidence of weathering in general. There were only trace amounts of carbonates and tephra. The resulting soil correspondence map was useful for focusing further investigative efforts primarily on the western United States, which had the greatest representation of soil types ranked as 'Best Fit' and 'Next Best Fit' correspondence with the sample data (Fig. 13). In another case focused on the African continent, soil properties were assessed for preliminary correspondence with dominant soil types as described in the aforementioned digital soil map (UN-FAO 2003). The presence of weatherable minerals with little evidence of alteration on their surfaces indicated that the soil was young and relatively unweathered. There was no calcite, gypsum or any soluble salt present in the soil, suggesting sufficient precipitation for leaching, and the clay fraction was dominated by illite. While these data were not particularly powerful in terms of describing the soil taxonomy, 20 of the 87 dominant soil types occurring in Africa could still be excluded as possible sources, and seven were classified as being a poor fit with the inferred source location, making it possible to considerably constrain the potential source regions within the continent (Stoney et al. 2011). Even properties that exclude only small regions should still be mapped, as overlaying a large number of these has the potential to considerably constrain the source.

A variety of inferences useful to provenance investigations may be made when botanical materials are identified from a given sample. Greater specificity in the level of taxonomic classification (i.e. family, genus, species or subspecies) identified for a given botanical taxon will yield more accurate and precise representation of the plant's actual or potential distribution within a geographical region of interest, which can be mapped using GIS. For example, if a taxon is identified to the species or subspecies level, flora treatises and other relevant botanical reference materials can be consulted to determine the plant's native distribution and habitat requirements, along with information on whether the plant is cultivated or occurs outside of its native range. In addition, it may be possible to obtain data on growth thresholds, such as altitudinal range, precipitation requirements and minimum–maximum temperature requirements during the growth season. In the absence of well-documented geographical distribution data for a botanical taxon, a taxonomic viability estimate can be created and used to constrain the source of the sample. In one provenance investigation focused on the African continent, the presence of pollen from *Diospyros mespiliformis* was used to constrain the potential source region by creating a taxonomic viability estimate for the species using GIS (Stoney et al. 2011). This climate suitability model was based on the DIVA-GIS EcoCrop modelling approach, and utilized long-term monthly precipitation and temperature data from the World-Clim database to model the potential growth of the

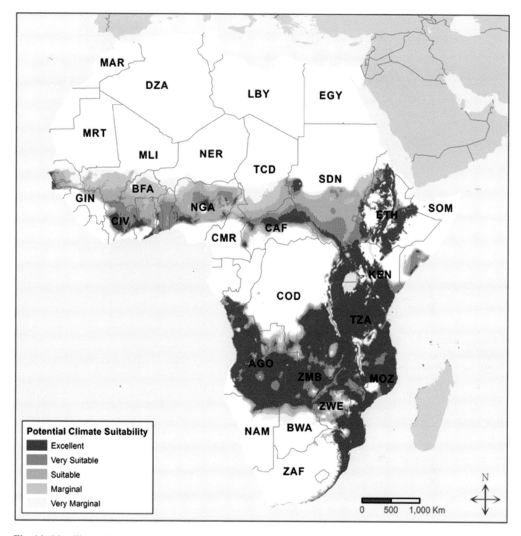

Fig. 14. Map illustrating the potential climate suitability of *Diospyros mespiliformis* on the African continent, as modelled using the DIVA-GIS EcoCrop approach and associated data (Hijmans *et al.* 2005a, b; UN-FAO 2007).

species over the geographical region of interest based upon growing season, precipitation and temperature thresholds (Hijmans *et al.* 2005a, b; UN-FAO 2007). The model generated six categories for the potential climate suitability of the species: Excellent; Very Suitable; Suitable; Marginal; Very Marginal; and Unsuitable (Fig. 14). This modelling approach can be useful to a forensic provenance investigation in predicting where a given plant species may be able to grow within a particular region of interest.

When provenance investigations include inferences regarding the vegetative elements expected within the source environment, it may also be of use to consult biogeographical- and ecosystem-related resources. These types of regional treatments often include descriptions of dominant and/or characteristic vegetative elements within a defined region. For example, maps of vegetation zones or ecosystems can be classified based upon their degree of correspondence with the inferred source environment, and these data can be mapped in a manner comparable to that described for geological and soil maps. In one unpublished case example, the inferred source environment was expected to be located in China, and contain *Quercus* spp. (oak), *Pinus* spp. (pine), *Betula* spp. (birch) and *Ulmus* spp./*Zelkova* spp. (elm) trees within the region of

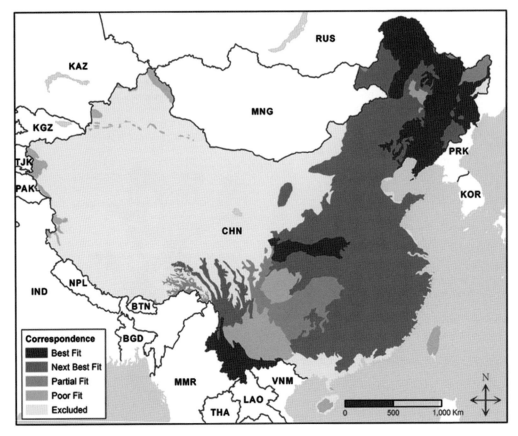

Fig. 15. Map illustrating the correspondence of a sample's major botanical elements with ecoregions as described for the 'Terrestrial Ecoregions of the World' database (Olson *et al.* 2001, WWF 2001*a*, *b*).

interest. Ecosystems as described for a suitably detailed terrestrial ecoregion database and digital map were assessed for the recorded presence of these tree genera and classified as follows: Best Fit (all four genera recorded); Next Best Fit (three of four genera recorded); Partial Fit (two of four genera recorded); Poor Fit (one of four genera recorded); or Excluded (no genera of interest recorded) (Olson *et al.* 2001; WWF 2001*a*, *b*). The resulting map was useful in excluding a significant portion of the country and focusing investigative efforts on several disjunct regions of interest (Fig. 15).

Data from other biological materials, such as animals, fungi and bacteria, can be assimilated into the spatial analysis of a sample in a manner similar to that described for botanical elements. In one provenance investigation, several types of freshwater green algae and aquatic protozoa were identified via DNA analysis. The collective presence of these organisms supported inferences that the sampled material was exposed to a source of freshwater containing protozoa and algae, rather than filtered or treated municipal water or saline water. As the case was focused on a narrow geographical region of interest, the locations of freshwater hydrological features could be mapped and assimilated into the provenance investigation (Stoney pers. comm. 2011).

When an organism is identified in a given sample, it is often possible to make inferences regarding the source environment based upon available data regarding the taxon's distribution, habitat, life cycle and so on. If biological material is identified from a bird, mammal or insect species, for example, data can be utilized to map or model the organism's native and potential distributions. In addition, information regarding potential economic uses or impact of the organism on society (e.g. livestock, agricultural pests or endangered species) can be incorporated into the analysis as relevant. For example, agricultural land use can often be inferred based upon the presence of certain taxa, either as botanical or animal species that are cultivated or tended for economic uses or as associated pest organisms such as weeds, fungi and insects. In addition, the detection in a sample of organisms that are either endangered or on watch lists can have a significant impact on a provenance investigation given that the geographical distribution and ecological requirements of these taxa are often fairly limited and/or well documented.

Sample data associated with anthropogenic materials can also be visualized and analysed using GIS, and this frequently results in valuable contributions to provenance investigations. The presence of particles derived from synthetic products and human activities in a given sample can lead to various inferences about activities and anthropogenic features within the source environment that can be mapped. For example, when botanical elements are detected in a sample and identified as various types of ornamental or cultivated plants, inferences can be made regarding land cover and land use within the source environment. Agricultural growing regions, gardens and urbanized environments with various ornamental plants may be implicated depending on the assemblage of botanical taxa in a sample. These types of features are often well documented in land-cover and land-use maps, readily detectable in satellite images, described in a variety of publications or can be inferred based upon proximity to human settlements, making this human activity a particularly valuable one for constraining the source of a sample.

The detection of combustion products and other particles associated with industrial facilities can contribute to the spatial analysis of a sample's provenance, as facility locations are often well-documented and it may be possible to generate particle dispersion models to estimate the facility's proximity to the source location. In addition, the proximity of paved roads has been used to significantly constrain the potential source area in several provenance cases, and these features are readily mappable and/or detectable in satellite imagery. In one unpublished case example, the source location was expected to contain non-native deciduous trees and various ornamental plants as evidenced by pollen and botanical DNA data, and the sample also contained the presence of carborundum grains, welding spheres, and metal turnings, which indicated metal-working activities at the source location. Within a narrowly defined region of interest, these human activity inferences were extremely useful in contributing to a review of satellite images for one or more buildings that could potentially house metal-working activities that had proximity to deciduous trees and a garden feature (Stoney pers. comm. 2011).

In general, the presence of large amounts of anthropogenic particles indicates proximity to densely populated areas, while a complete absence of anthropogenic particles suggests the opposite. This can be integrated into the spatial analysis of a provenance investigation in various ways. Population density data and maps can be reviewed for relative correspondence with the inferred environment, such as by excluding all sparsely populated regions as potential sources of a sample containing a significant quantity and diversity of anthropogenic particles. Similarly, land-cover and land-use data and maps can be reviewed for classification types that are similar (or dissimilar) to the inferred environment, such as by the inclusion of cultivated areas (e.g. non-native woody orchards or herbaceous croplands) or developed lands (e.g. high-intensity development such as urban areas or low-intensity development such as rural areas) when relevant particles are detected in a sample.

Geographical information systems and associated mapping technologies are invaluable tools for forensic provenance investigations. All available data should be incorporated into the mapping process when possible. Even properties that exclude only very small possible source areas should be mapped, as the combination of a large number of independent mapped properties, each of which does little to constrain the source, may significantly limit the potential source region when overlapped. Depending on the specific case scenario, it is often useful to begin a round of mapping with small-scale, lower-resolution maps to focus on broader regions where the sample properties being mapped occur most frequently. Subsequent rounds of mapping can be performed as needed using more large-scale, higher-resolution maps focused on narrower regions of interest. This is often performed in an iterative fashion, occasionally requiring different types of analytical data to use the maps at different scales. The highest resolution geological maps often come with explanatory texts, and these may offer greater insights into the character of the rocks than the maps themselves. These should be consulted during the final iteration of mapping once the source has been constrained as much as possible. It might then be possible to eliminate additional source rocks based on their descriptions in the text. The incorporation of many layers of mapped sample data in the spatial analysis efforts of a provenance investigation can often result in a very limited potential source region.

Discussion and conclusions

In forensic provenance investigations it is generally not possible to provide answers to which one can attach absolute confidence. However, in these cases the results are not meant to stand up to scrutiny as evidence in a court of law but simply to increase the efficiency of searches conducted by law enforcement officials by narrowing the search region as much as possible. For intelligence purposes, the goal is a similar one, narrowing down the source region as much as possible to increase the efficiency of a search or to corroborate statements provided. While absolute answers are generally unobtainable, if reasonable assumptions are made it is almost always possible to provide very useful investigative leads in these types of cases. The method has proven to be surprisingly effective in a small number of field tests where the true source location was known to the customer but unknown to the examiners.

The approach described in this paper relies heavily on microscopical analysis, with subsequent analytical techniques selected based on the specific sample character. The approach is multidisciplinary, with a wide range of materials regularly exploited, including those of geological, biological and anthropogenic origin. Experts in a variety of fields are consulted on a case-by-case basis depending on the nature of the sample. One of the strengths of the polarizing light microscope is that it can be used to efficiently determine the presence or absence of a large variety of substances in a sample and to direct the next steps to be taken. For example, if a sample is determined to contain large amounts of diatoms, then an expert in this field could be contacted for assistance. If a sample contains a large amount of industrial dusts, an expert on environmental pollution studies might be able to contribute to the interpretation. Many other methods employed for forensic provenance investigations (e.g. X-ray diffraction, bulk chemical analysis or particle size analysis) would not detect the presence of these particles and they could not be exploited. The microscopical approach enables an analyst with relatively limited training to examine a sample non-destructively and to assess its overall character. The amount and variety of rock fragments, mineral grains, microfossils, pollen and building materials, etc., can be observed, and the particle types most useful for source determination can be isolated for subsequent analyses.

Once the data have been collected, the interpretations in terms of geographical source are based on first principles and do not require comparison with reference samples in a database, although they do require access to a large number and variety of GIS datasets and maps. All of the inferences relevant to a mapped property (e.g. rock type, rock age, depositional environment, soil type, climate, ecosystem, topography, elevation and local human activities) can be overlain using GIS software to constrain the source region as much as possible. This method has contributed to a wide range of casework samples, sometimes in dramatic fashion. This type of approach has a tremendous amount to offer in forensic provenance investigations involving soil, sediment and dust samples.

The authors would like to thank D. Stoney, P. Stoney, M. Cimino and V. Bryant for their material contributions to many of the provenance investigations discussed in this text. They would also like to thank S. Palenik, B. Huber, P. McDaniel and Washington State University for contributing their expertise to one or more of the case examples described above.

References

BASU, A., YOUNG, S. W., SUTTNER, L. J., JAMES, W. C. & MACK, G. H. 1975. Re-evaluation of use of undulatory extinction and polycrystallinity in detrital quartz for provenance interpretation. *Journal of Sedimentary Petrology*, **45**, 873–882.

BORCHARDT, G. A., ARUSCAVAGE, P. J. & MILLARD, H. T. 1972. Correlation of the Bishop Ash, a Pleistocene marker bed, using instrumental neutron activation analysis. *Journal of Sedimentary Research*, **42**, 301–306.

BOWEN, A. M. 2010. Forensic applications of foraminifera. *The Microscope*, **58**, 3–18.

BRAMMALL, A. 1928. Dartmoor detritals: a study in provenance. *Proceedings of the Geological Association of London*, **39**, 27–48.

BRYANT, V. M. & MILDENHALL, D. C. 1998. Forensic palynology: a new way to catch crooks. *In*: BRYANT, V. M. & WRENN, J. W. (eds) *New Developments in Palynomorph Sampling, Extraction, and Analysis*. American Association of Stratigraphic Palynologists, Contributions Series, **33**, 145–155.

BRYANT, V. M., JONES, J. G. & MILDENHALL, D. C. 1990. Forensic palynology in the United States of America. *Palynology*, **14**, 193–208.

BULL, P. A. & MORGAN, R. M. 2006. Sediment fingerprints: a forensic technique using quartz sand grains. *Science & Justice*, **46**, 107–124.

DOWNS, R. T. 2006. The RRUFF Project: an integrated study of the chemistry, crystallography, Raman and infrared spectroscopy of minerals. *In: Program and Abstracts of the 19th General Meeting of the International Mineralogical Association in Kobe*, Japan, Abstract O03–13. http://rruff.info/.

GARVEY, K. K. 2007. *Landmark Criminal Case: UC Davis Entomologist Links Insects in Car Radiator with Defendant's Whereabouts*. UC Mosquito Research Program, Department of Entomology, University of California, Davis. http://entomology.ucdavis.edu/news/kimseyinsecttestimony.html.

GRADUSOVA, O. & NESTERINA, K. 2009. The current status of forensic soil examination in the Russian Federation. *In*: RITZ, K., DAWSON, L. & MILLER, D. (eds) *Criminal and Environmental Soil Forensics*. Springer, Dordrecht, 61–73.

HANHELA, R., LOUHELAINEN, K. & PASANEN, A. L. 1995. Prevalence of microfungi in Finnish cow barns and some aspects of the occurrence of *Wallemia sebi* and *Fusaria*. *Scandinavian Journal of Work, Environment & Health*, **21**, 223–228.

HAWKESWORTH, D. L. & WILTSHIRE, P. E. J. 2011. Forensic mycology: the use of fungi in criminal investigations. *Forensic Science International*, **206**, 1–11.

HIJMANS, R. J., GUARINO, L., BUSSINK, C., MATHUR, P., CRUZ, M., BARRANTES, I. & ROJAS, E. 2005a. *DIVA-GIS, Version 5.2. A Geographic Information System for the Analysis of Biodiversity Data. Manual.* http://www.diva-gis.org.

HIJMANS, R. J., CAMERON, S. E., PARRA, J. L., JONES, P. G. & JARVIS, A. 2005b. Very high resolution interpolated climate surfaces for global land areas. *International Journal of Climatology*, **25**, 1965–1978.

LADD, C. & LEE, H. C. 2005. The use of biological and botanical evidence in criminal investigations. In: COYLE, H. M. (ed.) *Forensic Botany: Principles and Applications to Criminal Casework.* CRC Press, New York, 97–115.

MANGE, M. M. & MAURER, H. F. W. 1992. *Heavy Minerals in Colour.* Chapman & Hall, London.

MARUMO, Y. & YANAI, H. 1986. Morphological analysis of opal phytoliths for soil discrimination in forensic science investigation. *Journal of Forensic Sciences*, **31**, 1039–1049.

MCCRONE, W. C. & DELLY, J. G. 1973. *The Particle Atlas, Edition Two.* Ann Arbor Science, Ann Arbor, MI.

MCPHEE, J. 1997. *Irons in the Fire.* Farrar, Straus and Giroux, New York.

MORTON, A. C. 1985. A new approach to provenance studies – electron-microprobe analysis of detrital garnets from Middle Jurassic sandstones of the northern North Sea. *Sedimentology*, **32**, 553–566.

MURRAY, R. C. 2004. *Evidence from the Earth: Forensic Geology and Criminal Investigation.* Mountain Press Publishing Company, Missoula, MT.

OLSON, D. M., DINERSTEIN, E. ET AL. 2001. Terrestrial ecoregions of the world: a new map of life on Earth. *BioScience*, **51**, 933–938. http://dx.doi.org/10.1641/0006-3568(2001)051[0933:TEOTWA]2.0.CO;2.

PALENIK, S. 2007. Heavy minerals in forensic science. In: MANGE, M. A. & WRIGHT, D. T. (eds) *Heavy Minerals in Use.* Elsevier, New York, 937–961.

PITTMAN, E. D. 1963. Use of zoned plagioclase as an indicator of provenance. *Journal of Sedimentary Petrology*, **33**, 380–386.

PITTMAN, E. D. 1970. Plagioclase feldspar as an indicator of provenance in sedimentary rocks. *Journal of Sedimentary Petrology*, **40**, 591–598.

REED, J. C. & BUSH, C. A. 2004. *U.S. Geological Survey Generalized Geologic Map of the Conterminous United States, Version 1.1. National Atlas of the United States.* http://nationalatlas.gov.

ROSS, C. S. 1950. The dark-field stereoscopic microscope for mineralogic studies. *American Mineralogist*, **35**, 906–910.

ROVNER, I. 1971. Potential of opal phytoliths for use in paleoecological reconstruction. *Quaternary Research*, **1**, 343–359.

SOIL SURVEY STAFF. 2010. *Keys to Soil Taxonomy*, 11th edn. USDA-Natural Resources Conservation Service, Washington, DC. http://soils.usda.gov/technical/classification/tax_keys/.

STONEY, D. A., BOWEN, A. M., BRYANT, V. M., CAVEN, E. A., CIMINO, M. T. & STONEY, P. L. 2011. Particle combination analysis for predictive source attribution: tracing a shipment of contraband ivory. *Journal of the American Society of Trace Evidence Examiners*, **2**, 13–72.

UN-FAO. 2003. *Food and Agriculture Organization of the United Nations Digital Soil Map of the World and Derived Soil Properties, Rev. 1* [CD-ROM]. Food and Agriculture Organization of the United Nations, Rome.

UN-FAO. 2007. *Food and Agriculture Organization of the United Nations Data Sheet:* Diospyros mespiliformis. http://ecocrop.fao.org/ecocrop/srv/en/dataSheet?id=5466.

WELTJE, G. J. & VON EYNATTEN, H. 2004. Quantitative provenance analysis of sediments: review and outlook. *Sedimentary Geology*, **171**, 1–11. http://dx.doi.org/10.1016/j.sedgeo.2004.05.007.

WINTER, J. D. 2009. *Principles of Igneous and Metamorphic Petrology.* 2nd edn. Prentice-Hall, New York.

WWF 2001a. *World Wildlife Fund Indo-Malayan Ecoregion Profiles.* http://www.worldwildlife.org/wildworld/profiles/terrestrial_im.html.

WWF 2001b. *World Wildlife Fund Palearctic Ecoregion Profiles.* http://www.worldwildlife.org/wildworld/profiles/terrestrial_pa.html.

X-ray diffraction and field portable X-ray fluorescence analysis and screening of soils: project design

ELISA T. BERGSLIEN

SUNY Buffalo State, Earth Sciences, 1300 Elmwood Ave, 271 Science Building, Buffalo, NY 14222 USA (e-mail: bergslet@buffalostate.edu)

Abstract: This paper presents the preliminary results of a study investigating the application of laboratory X-ray diffraction (XRD) and field portable X-ray fluorescence (XRF) analysis of soils as screening methods for forensic comparison and generalized provenancing. The study area is the Buffalo–Niagara metropolitan region in New York, a glacially draped area of the northeastern USA. For the initial stages of this study, soils are being collected from publicly accessible areas (parks, playgrounds, etc.) for mineralogical and elemental analysis. Initially, minimal sample preparation is being applied to create specimens. An investigation of the published literature reveals that there are a number of different suggested approaches for forensic application of XRD data, not all of which appear to be appropriate to the task at hand. The data being generated in this study are being used to build a reference set for comparison studies. In addition, two simulated forensic samples were collected to test the usefulness of XRD and XRF screening for determining possible regional source areas. For one sample this method was reasonably successful in identifying the general source area, while the results for the second sample were somewhat less satisfactory. In future additional sample handing and analysis protocols will be added.

Soil is a common form of trace evidence found adhering to almost everything, including people, vehicles and physical evidence. The ability to compare soil samples from a crime scene with a suspect's effects in order to determine if there is a relationship between them can prove valuable. Even with decades' worth of work (Murray & Tedrow 1975; Junger 1996; Ruffell & Wiltshire 2004; Pye *et al.* 2006), there are still significant unanswered questions about how soil varies geographically within an area and over different provinces and how the properties of a soil change when material is transferred from the ground to a carrier such as automobile tyres, shoe treads or clothing.

The most robust approach to comparative soil analysis employs a highly trained analyst utilizing a detailed, stepwise microscopic examination of delineated size fractions (Ruffell & McKinley 2008; Murray 2011). There is currently no substitute for such an approach; however, it is usually quite time and labour intensive. This study proposes utilizing X-ray diffraction (XRD) and X-ray fluorescence (XRF) analysis as screening methods to narrow down the range of samples to be examined and to potentially aid in determining provenance. XRD is a powerful tool for the identification of crystalline materials, including differentiating components of mixtures. It is used to determine the bulk mineralogy of soils and to evaluate changes to the mineral profile when soil is transferred. XRF is used to determine trace elemental composition ($>$K). Both techniques are non-destructive and can be employed with little or no sample preparation. The goal of this ongoing project is to systematically employ XRD and XRF analysis under a variety of conditions, starting with conditions of minimal sample preparation, to assess their robustness for screening of transferred soil materials and usefulness in determining their relationships to possible source zones. This project is being carried out in conjunction with an environmental assessment delineating regional anthropogenic heavy metal contaminant zones.

Materials and methods

Study area

The Buffalo–Niagara Falls Metropolitan Area, with a population of c. 1 135 509, is the second-largest metropolitan area in the state of New York (US Census Bureau 2010). The population is considered to be 89% urban and 11% rural. The area has four principal cities: Buffalo (23% of population), Cheektowaga (6.6%), Tonawanda (5.1%) and Niagara Falls (4.4%) (US Census Bureau 2012*a*, *b*, *c*). Owing to its location at the western terminus of the Erie Canal, the region has a long history as an industrial corridor, home to steel mills, tanneries, colourant manufacturers and chemical production facilities. The region lies in the Erie–Ontario Lowlands physiographic province, which is characterized by nearly horizontal beds of shale, limestone and dolostone. The topography has been modified substantially by Pleistocene glacial activity and deposition of glacial drift transported from Canada; thus in many areas the overlying soil is compositionally different from the underlying bedrock. Regionally overburden materials are glaciolacustrine clays, glacial outwash and till, typically ranging in thickness from 3 to 30 m (Isachsen et al. 2000).

Sample collection

A range of publicly accessible areas, such as parks, playgrounds and beaches, were sampled as part of an ongoing project to delineate the trace element signature of localized anthropogenic inputs (steel plants, factories, colourworks, etc.). Using a 10 × 10 cm frame, soil samples were collected directly adjacent to existing play structures to a depth of c. 2.5 cm with stainless steel collection spoons or palette knives. The steel tools were thoroughly cleaned between each use with deionized water and laboratory tissues, which has been shown to minimize the risk of cross-contamination (Pye 2007, p. 206). The samples were packed into new zip-top collection bags and stored in an ice chest for transport to the laboratory. Upon arrival each sample was debrided and oven-dried at 65 °C (150 °F) for 24 h. Moisture content affects the accuracy of X-ray analysis and should be <20% to ensure that this source of error is minimized; thus, the use of drying or toaster ovens is strongly recommended (US EPA 1998). The US Environmental Protection Agency (EPA) recommends against microwave drying of samples as their field studies have shown that it increases variability in elemental analysis as compared with laboratory results (US EPA 1998).

Collection of simulated forensic samples

These preliminary results concern analysis of two simulated forensic samples (SBCB and XSBC) generated by having a volunteer run through mud puddles in one of the publicly accessible parks in the study area, although not in one of the areas used to generate samples for the reference collection. Initially the volunteer was the only person with knowledge of the 'crime scene' so that the comparison of the simulated forensic samples with the reference collection was blind. The volunteer was 'apprehended' once they reached the college and their clothing collected. The clothing was air-dried in the laboratory.

Sample preparation

The dry, bagged, bulk samples were homogenized, which typically involved rolling the back of a stainless steel spoon over any clumps, while they were still protected inside the bag, to disaggregate them and then shaking the sample bag. For soils with a high clay content, more aggressive methods of disaggregation were necessary, such as gentle crushing by impact with a mortar and pestle. Each sample was then poured through a nested stack of 7.62 cm (3 inch) stainless steel sieves consisting of the following: no. 35 (500 μm), no. 60 (250 μm), no. 120 (120 μm), no. 200 (75 μm) and a catch pan. Between each use the sieves were cleaned with sieve cleaning brushes (for sieves with mesh >100 μm) and low-pressure compressed air, and dusted clean with laboratory tissues. The sieving procedure had been tested using well-characterized standard soil samples and no evidence of heavy metal contamination from the sieves had been found. Ideally disposable nylon mesh sieves would be used for each sample, but given the volume of samples being collected and monetary constraints, this is currently not possible for this study. The author notes that in the future additional sample preparation procedures (milling, etc.) will be incorporated but the intent with the current approach was to elucidate the level of information that can be repeatedly extracted with minimal sample handling.

Randomly oriented XRD specimens were created from subsamples of each of the no. 200 sieve (<120–75 μm) and catch pan (>75 μm) fractions using a modified back-loading technique and aluminium sample holders (Koninklijke 2001). The traditional powder loading techniques (front, back and side) all create some level of orientation to the samples (Kleeberg et al. 2008), with side loading typically seen as the preferred method while back loading, which is more susceptible to minerals that have strong tendencies to orient, is seen as a less desirable but still acceptable approach (Bish & Reynolds 1989; Moore & Reynolds 1997). Top loading is clearly the least consistent mounting technique as the effects of preferential orientation can render the reproducibility of the pattern questionable; thus most practitioners recommend its avoidance (Kleeberg et al. 2008).

Random power mounts are useful for preliminary screening of samples and identification of the non-clay fraction (Moore & Reynolds 1997). As the specimens described for this study were being used for preliminary assessment of the bulk mineralogy of the study area, minimal sample handing was being used. It has been shown that qualitative and semi-quantitative identification of bulk mineralogy (quartz, calcite, feldspars) with no sample pre-treatment is generally robust and comparable results can be achieved under a variety of conditions (Ottner et al. 2000). However, the application of different pre-treatment methods (use of chemicals, ultrasonic treatments, use of dispersants, etc.), including those used to isolate or generate the

<20 μm fraction (sedimentation, centrifugation, milling, etc.), actually increases variation and in some cases may have inadvertently introduced errors in clay mineral identification (Ottner et al. 2000).

To create XRF specimens, subsamples of each sieve and pan fraction were poured to a depth of at least 1 cm into plastic XRF sample cups with 2.5 μm thin-film Mylar® supports. The sample cups were gently tapped against the table surface for a minute or two to improve packing and additional sample added if the minimum depth criteria were not met. These field samples served as the core reference materials for comparison of our simulated forensic materials. Ottner et al. (2000) found that for XRF elemental analysis the results of the bulk sample (<2 mm) showed a small range of variation while the data collected for the clay fraction varied to a much greater extent. Pye & Blott (2004) reported that 'for a majority of circumstances, analysis of the <150 μm fraction allows adequate discrimination between samples and provides a relatively consistent indicative measure of the composition of a sample' (p. 185).

Preparation of simulated forensic samples

Four methods of sample preparation were used to create specimens for XRD analysis. First, the clothing was positioned on a frame inside the XRD chamber for *in situ* analyses of mud spatter. Next tape pulls were taken from the clothing and mounted on glass slides. Finally scrapings were taken from different locations on the clothing using new, alcohol-cleaned, straight-edge razor blades. The scrapings were used either to fill 2 ± 0.1 mm poly(ethylene terephthalate) capillary tubes (MicroRT brand by MiTeGen) or to create typical back-loaded random power specimens mounted in aluminium sample holders, depending on the amount of material available. For analysis, the capillary tubes were sealed and under-mounted in sample holders such that the top of the tube was level with the top of the sample holder and then analysed while spinning.

In addition, for each article of clothing, the mud spatter was analysed *in situ* (on the clothing) using the field-portable XRF system, and a subsample of the scrapings taken was prepared in standard XRF sample holders, using the same technique as applied to the bulk reference samples, although the depth requirement was occasionally violated owing to recovered sample size.

Sample analysis background

X-ray diffraction, although less common in a forensic setting, is a standard analytical technique used in geology and soil science laboratories for the analysis of fine-grained materials; thus there is a large body of literature to draw from when developing an applied analysis approach. Kugler (2003) and Abraham et al. (2007) provide an introduction to forensic application of the method and additional detailed background on the technique can be found in Moore & Reynolds (1997), Jenkins & Snyder (1996) and Jenkins (1989). XRD is attractive for forensic analysis because, depending on the sample preparation method(s) used, it is non-destructive, repeatable, reasonably fast and can be used for small samples (Dawson & Hillier 2010). However there are so many variables involved that there is no standard methodology for analysis of materials by XRD and a wide range of methods have been developed over time to address specific types of problems (Moore & Reynolds 1997; Kahle et al. 2002).

An examination of the forensic literature shows that there are multiple approaches to acquisition and application of XRD data. Some authors suggest minimal sample handling and employing diffractograms in a manner conceptually similar to fingerprints, relating the similarity of the diffraction pattern to the similarity of the specimen being examined (Kugler 2003; Isphording 2004; Fitzpatrick et al. 2009). Other authors propose detailed characterization of the clay mineral fraction of a sample through the use of multi-stage sample preparation procedures such as chemical decomposition of organic matter, treatment with a deflocculant, magnesium saturation, potassium saturation, calcium saturation, extraction of iron oxides, solvation with glycerol or ethylene glycol, and/or heating the sample to specific set points (Marumo et al. 1986; Rawlins et al. 2006; Melo et al. 2008). Still others advocate the use of quantitative analysis procedures such as the addition of an internal standard and use of mineral intensity factors to facilitate quantitative analysis of a set of common minerals based on the intensity of the 060 peaks relative to the standard (Ruffell & Wiltshire 2004).

Simple mineral identification is for the most part relatively straightforward, but accurate and comparable quantitative analysis can be a significant challenge (Środoń et al. 2001). One common misconception is that the intensity of the diffraction peaks relates to the relative concentration of the materials in the sample, as with X-ray fluorescence. Actually, some crystalline structures diffract X-rays much more efficiently than others and variations in chemical composition will alter peak heights even within the same mineral group. A sample with equal parts of smectite, quartz and muscovite will not produce a diffractogram with equal 001 peak areas. Variations in crystal orientation also affect peak intensity, thus the importance of employing a uniform sample preparation technique for comparisons. Replicate analyses should have a standard deviation of no more than ±5%, so precision of this method is quite high, but since accuracy is directly related to such factors as the sample preparation method used, data collection time and the protocols employed for quantification, errors of ±10% for major constituents and ±20% for minerals at concentrations <20% are not uncommon (Moore & Reynolds 1997).

A diffraction pattern records X-ray intensity, in counts, as a function of degrees 2θ. The precision of counting depends simply on the number of counts and the standard deviation of the count for a peak is equal to ± the square root of the total count. Thus the longer the count time, the greater the signal-to-noise ratio and the lower the detection limit (Moore & Reynolds 1997). Few papers have substantive discussions of the detection limits for powder X-ray diffraction. For minerals like quartz, which have no cleavage, no substitution and a tight crystalline structure it is theoretically possible to have detection limits in the 10 parts per million range with a long enough collection time (>999 s) (Moore & Reynolds 1997). According to Bish & Chipera (1991, p. 439),

> a peak is considered observed if it is at least three times the standard deviation (σ_b) of the background intensity at a given 2θ position ... where the standard deviation of background intensity is the square root of the background counts accumulated in t seconds (C_b). The lower limit of detection ... in intensity is then $3\sigma_b$ or $3\sqrt{C_b}$.

At common collection rates, where matrix interference occurs, the lower limit of detection usually lies in the 1–2% range for non-clay minerals with high intrinsic diffraction intensities and for poorly crystalline materials with broad diffraction peak, like clay minerals, in the 3–5% range, although this can be substantially improved upon with the use of higher count times and specialized preparation methods (Bish & Chipera 1991). These rules of thumb also only apply to samples of standard size. Smaller samples require longer acquisition times, up to 2000–2500 s per step, for total acquisition times of 10 h or more, in order to produce usable diffraction patterns (Švarcová et al. 2010). They also have their own host of special considerations.

The other important running condition is the step size in degrees, 2θ, which describes the area over which the signal is collected (step mode) or integrated (scanning mode). Smaller step sizes allow for greater pattern resolution. For qualitative analysis Moore & Reynolds (1997) suggest a 0.1° step size for clay minerals, which have broad peaks, and a 0.05° step size for reconnaissance of bulk samples; for non-clay minerals a step size of 0.02° is considered adequate, although a step size of 0.01° should be used for development of lattice parameters. Unfortunately few of the papers published in the forensic literature describe the running conditions for data collection and most lack significant procedural details.

A recent paper by Eckardt et al. (2012) describes using the EN ISO/IEC 17025 method for validation of powder X-ray diffraction employing a variety of sample materials such as zinc oxide, calcite, soil standard PS-3 COD no. 312a-98 from MBH Analytical Ltd and 2686 Portland Cement Clinker from the National Institute of Standards and Technology. Data for the project were collected using a D8 Advance with a high speed Våntec detector (Bruker AXS GmbH, Karlsruhe, Germany) and their reported measurement parameters for data collection in scanning mode were 0.1 s per 0.0086° step (0.086° s^{-1}) with an overall measurement time of 14.55 min utilizing a 6 mm-diameter aperture, although for some samples count time was increased.

Powders were 'prepared between two Kapton® foils (8 μm®) on a specimen holder', but no other information is given. The authors also mention testing a 'grinding phase' for some samples, with no mention of how the grinding was accomplished. Such details are important because the authors report that, even by increasing measurement time to 4 h, they were unable to discern the peaks for chlorite in a cement standard though the sample contained 9 wt% of the mineral. This is suggestive of problems with sample preparation with regard to platy minerals, which produce intrinsically broad peaks and possible issues with the data collection protocol (Moore & Reynolds 1997).

Based on their investigation, Eckardt et al. (2012) include a detailed discussion of detection limits in which they suggest that the $3\sigma_b$ criterion is inadequate owing to a lack of peak resolution. However, the information provided suggests a somewhat backward approach, with arbitrary selection of sample preparation methods and running conditions, and more emphasis on post-processing of data rather than optimizing the running conditions to produce crisp diffraction patterns. Even so, they report detection limits on the order of 2 wt% for calcite in a quartz matrix and 3 wt% for quartz in a calcite matrix. One key point about X-ray diffraction analysis that the authors reiterate is that 'every substance or mixture is individual, so there can be no general statement as [a given system] "has a detection limit of x weight percent"' (p. 735).

As a complementary analytical technique to XRD, XRF also has a long history of use in the geological sciences for determination of elemental composition. An introduction to the method can be found in Jenkins (1988), while discussion of forensic application can be found in Pye & Croft (2007). Field portable X-ray fluorescence (FPXRF) spectrometry, a somewhat more recent technological development, has become a common analytical technique for acquiring rapid, high-quality results in the field (Bernick et al. 1995; Clark et al. 1999; Kalnicky & Singhvi 2001). Typical modern FPXRF systems use X-ray tubes as excitation sources to irradiate samples and have energy dispersive analysers that convert the X-rays incident on the solid-state detector into an electronic signal based on photon energy.

The major limitations of this technique are lower detector resolution than wavelength dispersive systems and more significant spectral overlap issues that also effect resolution, especially on light elements. Laboratory-based EDX systems can analyse for elements from sodium to uranium, but FPXRF systems, because they are typically working in air, are unable to detect elements lighter than phosphorus. This latter limitation has significance from a geological perspective, as the most common elemental constituents of minerals, silicon, oxygen, aluminium and magnesium, are not detectable. However, heavier elements, such as toxic metals, can be readily detected, some with great sensitivity, making the method useful for environmental and forensic screening.

For FPXRF the relative standard deviation (RSD) of the sample mean is used to access precision, where RSD is equal to the standard deviation of the specimen's analyte concentration divided by the mean concentration for the analyte multiplied by 100. According to the US EPA, for environmental applications utilizing an FPXRF, the RSD should not be greater than 20% for all elements except chromium. For chromium, owing to spectral peak overlaps, RSD should not be greater than 30%. The method limit of detection (MLD) is defined as three times the standard deviation. Measurements that are above the manufacturer's reported detection limit but below the MLD should not be considered quantitative measurements, while measurements that are below the manufacturer's reported detection limit but are above the MLD can be reported as estimated valves (US EPA 1998).

Accuracy of the method is assessed using reference standards with known compositions and characteristics (particle size distribution, texture, etc.) similar to the specimens to be analysed. A minimum of seven non-consecutive replicate measurements are taken of each standard and the percentage difference (%D) calculated for each target analyte. Percentage difference is calculated as follows:

$$\%D = [(C_s - C_k)/C_k] \times 100$$

where C_s is the measured concentration and C_k is the certified concentration of the standard. The %D should be within $\pm 20\%$ of the certified value for each analyte.

Sample analysis running conditions

For this study, each prepared sample was analysed in the laboratory using a θ–θ PANalytical X'pert MPD X-ray diffractometer using CuKα radiation produced by a Cu-tube and a nickel filter. The bulk samples spin at a rate of 4 revolutions per second. The system uses an X'Celerator high speed detector and has a diffracted beam monochromator. Each sample is analysed at 45 kV and 40 mA from 3 to 80° 2θ in scanning mode with a collection time of 150 s over a 0.0167° scanning arc, or roughly $0.0135° \text{ s}^{-1}$, which is on a par with the running conditions of 2 s per 0.02° step, or $0.01° \text{ s}^{-1}$ used by Środoń et al. (2001) for analysis of non-clay minerals. For analysis of clay minerals, Środoń et al. (2001) increased the collection time to 5 s per 0.02° step, or $0.004° \text{ s}^{-1}$. Mineral identification and semi-quantification was performed using the full pattern matching software package X'pert Highscore linked to the ICDD Powder Diffraction database (ICDD 2007).

The prepared XRF specimens were analysed in the laboratory using an enclosed bench-top sample holder to ensure consistent placement. All of the samples were analysed using a Niton XLt field portable X-ray fluorescence unit with a low power (1.0 W) Ag anode X-ray tube and a Si PiN detector. The unit was set in bulk soil mode and sample data collected for 120 s. Measurement times of 30 s could be used for initial screening of samples and identification of hot spots, while longer collection times, up to 300 s, were used to improve precision and accuracy of the data (US EPA 1998). At around 300 s, or less, the procedure reached a point of diminishing returns and there was nothing to be gained by an increased sampling time. The measurement time selected is reflective of the condition and texture of the samples being analysed. For the soils commonly found in the study area, replicate analyses at 30, 60, 90, 120, 150 and 180 s demonstrated that there was little improvement to be gained with a sampling time greater than 120 s. A minimum of five non-consecutive replicate measurements were performed for each prepared specimen. Soil mode automatically analyses for Ag, As, Cd, Co, Cr, Cu, Fe, Hg, Mn, Ni, Pb, Rb, Sb, Se, Sn, Sr and Zn. Additional information can be acquired by downloading the raw spectral data and using software for analysis; however, this paper focuses on the use of FPXRF as a screening method so only the elemental subset listed above was used in analysis.

In general, based on measurements of soils standards with characteristics similar to the soils in the study area, the lower detection limits for this approach were 10–50 ppm for titanium to plutonium, 250 ppm for potassium to scandium, and between 1 and 5% for phosphorous to argon. A series of standards tests were periodically preformed with the Niton to assess its precision and accuracy. Typical results are presented in Table 1.

Data

Regional XRD

Regional mineralogy of the study area is fairly consistent as determined per XRD analysis. The soils examined from the majority of locations in the study area so far are roughly 30–60% quartz, 10–40% muscovite, 3–20% plagioclase feldspars, <20% potassium feldspars, 0–25% carbonates and some traces of a potentially useful clay mineral component based on the HighScore SemiQuantification 100% method (Table 2). However, some places in the study area show a significant deviation. Areas further outside the Buffalo–Niagara urban corridor have more varied mineralogy while, in general, many of the urban areas cannot be easily differentiated using XRD diffraction pattern comparison. Large sections (>90%) of Buffalo are classified by the Natural Resource Conservation Service as some form of urban land, composed of fill materials and urban debris. Certain areas, especially in the northern towns, have a distinctive carbonate component (e.g. see Newstead, Table 2), but for much of the urban corridor, the intra-area variation is of the same scale as the inter-area variation. Areas outside the urban corridor show more variation in mineralogy, allowing for some degree of discrimination (Fig. 1).

Regional XRF

Broadly speaking there are two major sources of trace elements in the environment. The first source is the local bedrock and surficial geological material. The second source is anthropogenic releases into the environment via such activities as manufacturing, mining and power generation. At this point in the project, over 140 samples, collected from 43 different sites, have been analysed (Table 3). The samples come from areas that overlie six different stratigraphic units, all of which were draped by glacial materials. Geologically speaking, all of the samples collected thus far are from approximately the same basic underlying parent materials: dolomite and/or limestone interbedded with shale.

In soils mode the Niton FPXRF automatically produces quantified results for a suite of 17 elements (Ag, As, Cd, Co, Cr, Cu, Fe, Hg, Mn, Ni, Pb, Rb, Sb, Se, Sn, Sr and Zn). Only two samples showed traces of mercury, both from inside the city of Buffalo, while approximately three dozen showed traces of selenium at approximately the detection limit. All of the samples contained detectable amounts of iron, manganese, lead, rubidium, strontium and zinc. For the remainder of the elements, a non-parametric Spearman's rank correlation shows a strong correlation between iron and

Table 1. *Precision and Accuracy of Niton FPXPF measuring standards for 120 s (n = 15)*

Standard element	CCRMP Till-4			NSC 73308			GBW 7411			RCRA	
	Pb	As	Cu	Pb	As	Cu	Pb	As	Cu	Pb	As
Average FPXRF measurement (ppm)	56	97	230	29	21	<LOD	2700	180	69	520	450
Standard deviation	4.2	5.2	13	3.8	0.8		70	25	11	15	9.8
Relative standard deviation (%)	7.5	5.4	5.8	13	3.8		2.6	14	17	2.9	2.2
Method limit of detection (ppm)	13	16	40	11	2.4		210	74	34	45	30
Certified standard value (ppm)	50	111	237	27	25	22.6	2700	205	65.4	500	450
Difference	+6	−14	+7	+2	−4		0	−25	+4	+20	0
Percentage difference	12	13	3	7	16		–	12	6	4	–

All values in parts per million.
<LOD, Less than the limit of detection.

Table 2. Summary of XRD bulk soil analysis for selected locations in the Buffalo–Niagara Falls Metropolitan Area

	Albite	Anorthite	Calcite	Dolomite	Kaolinite	Microline	Muscovite	Orthoclase	Quartz	Vermiculite	Chlorite	Lead Oxide	Magnetite
Alden	3	8		8		6	31	6	35				2
Amherst	5	6			16	6	28	5	35				
Aurora	3		3	1			40		53				
Boston	8				5		30		58				
Buffalo BIP	5	6	4	4		4	43	4	30	Trace			
Cheektowaga	5	4				5	35	5	45				
Clarence	12	7	3	10		7	13	6	42			Trace	
Colden	3			1		5	34	5	51				
Collins	4	6				5	27	4	54	Trace			
Concord	8				5	6	24	5	51	Trace			
Eden	6	7			11	4	28	4	40	Trace			
Eden 2	6	4			6	6	26	5	47	Trace			
Elma	4	4		1	3	5	39	5	39				
Evans	7	7				4	32	6	44	Trace			
Grand Island	3	5				5	32	4	51	Trace			
Holland	3	3		4		5	40	9	35	Trace	Trace		1
Lackawana	3	4				5	34	6	47				
Lancaster	4	4	1	2		5	36	5	37				
Love Canal	6	7		1	6	8	26	11	42		Trace		
Marilla	3					4	38	4	52	Trace			
Newstead	3		11	12		7	11	7	49				
North Collins	5	3		1			37	6	47	Trace			
Orchard Park	3						41		56	Trace			
Sardinia	8	4				6	21	11	49	Trace	Trace		1
SBCS-unk1	2	4				7	32	14	37				
Tonawanda	10	3				3	34	3	46				
Wales	4					4	30	11	51	Trace	Trace		1
West Seneca	7	10		1		5	27	4	46	Trace			
X-SBD-unk2	7	6		1	8	5	25	5	42				2

All values percentages based on HighScore Semi-Quantification 100% method. Trace indicates the presence of diagnostic peaks but the software was unable to calculate a percentage value.

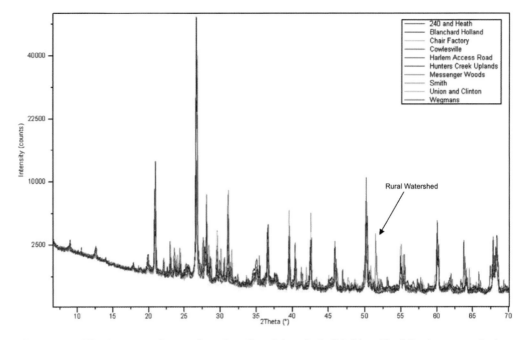

Fig. 1. X-Ray diffraction patterns for a set of samples collected along the Buffalo River. The diffraction patterns for the urban areas are generally quite similar, while the rural patterns show some significant points of distinction.

rubidium, manganese, zinc and lead, as well as correlations between arsenic and iron and rubidium, between lead, zinc, manganese and iron, and one between rubidium and cobalt (Fig. 2).

Iron concentration in general appears to be geologically controlled, with shale areas toward the south of the study area tending to have higher concentrations and limestone/dolostone, in the north of the study area, having lower concentrations (Fig. 3). The remainder of the elements analysed for do not appear to correlate with underlying geology. Elemental distribution was also compared with the relative urbanization of an area. In general, the urban samples show significantly higher levels of lead, arsenic and zinc than the suburban samples, which could potentially become a useful discriminator. Lead especially varies significantly, from <13 to >700 ppm. Also of interest, arsenic and chromium show localized anthropogenic highs associated with proximity to sources such as chromated copper arsenate-treated wooden structures. There were a few other points in the study area that had highly distinctive elemental signatures. For example, an area that is now a nature preserve, but was formally the site of a municipal dump, has areas of soil that are 12% Fe and also show elevated levels of lead, cobalt and chromium. These are presumably the remains of slag piles left over from the steel plants that formerly occupied the area. The other elements in the soil set have too few data points for a general characterization.

Simulated forensic samples

In situ analysis of soil on objects or clothing, or transferred to substrates like tape samples, can be fraught with difficulty, as the incident radiation will almost inevitably be out of plane, the sample surface will not be flat and the sample may not be 'infinitely' thick. The effective penetration depth of X-rays depends on the angle of incidence, mass attenuation, sample homogeneity and the mean particle size of the sample. For the Cu Kα radiation commonly used for XRD, sample depth of penetration ranges from a few micrometres to nearly 200 μm. Thus the diffraction signature of the host substrate may show up in the analysis (Fig. 4, note circled peaks). Samples that are out of plane and lack flat incident surfaces will show effects like peak shift (specimen displacement error), broadening of the peak shape and/or changes in peak intensity, all of which can complicate sample comparisons (Jenkins & Snyder 1996). It is also worth noting that clay minerals, which generally make up a sizeable fraction of a soil sample, will intrinsically have weak, broad peaks in contrast to the sharp peaks generated by tightly crystalline materials (Fig. 5; Moore & Reynolds 1997).

Table 3. *Soil data collected using Niton FPXPF measuring for 120 s averaged over five runs*

Sample ID	Ag	As	Cd	Co	Cr	Cu	Fe	Hg	Mn	Ni	Pb	Rb	Sb	Se	Sn	Sr	Zn
T207	–	4	–	–	–	–	3486	–	120	–	12	8	–	–	–	22	31
T208	–	–	–	–	–	–	16616	–	251	–	54	49	–	–	–	126	159
T209	44	17	–	–	–	–	29001	–	711	60	41	91	88	–	91	136	127
T210	–	16	–	–	–	–	17870	–	210	–	56	48	–	–	–	101	84
T213	–	–	–	–	–	–	14829	–	337	–	166	37	–	–	–	143	183
BS215	–	155	–	–	68	436	6983	–	438	–	26	17	–	–	–	42	248
BS216	–	–	–	–	–	–	13097	–	198	–	34	44	–	–	–	141	73
BS217	–	58	–	–	200	–	3692	–	117	–	7	7	–	–	–	15	43
BS218	–	130	–	–	371	–	5264	–	330	–	14	14	–	–	–	39	59
BS219	–	–	–	–	–	–	12729	–	187	–	36	45	–	–	–	122	79
BS221	–	–	–	–	–	29	15825	–	240	39	68	52	–	–	–	127	92
BR222	–	9	–	–	–	–	22352	–	612	–	25	45	–	–	–	80	45
BR223	–	8	–	–	–	–	13335	–	244	–	–	26	–	–	–	126	37
BR224	–	9	–	–	–	–	20395	–	352	–	12	40	–	–	–	92	56
BR225	–	9	–	–	–	103	31586	–	549	–	34	100	–	–	–	118	133
BR226	28	–	–	–	99	–	18026	–	367	–	15	59	–	–	–	137	47
BR227	39	–	–	–	–	–	11969	–	253	–	15	52	110	–	85	162	43
BR228	–	–	–	–	–	42	17224	–	511	–	20	49	–	–	–	134	58
BR229	–	–	–	–	–	67	27058	–	803	58	61	68	–	–	–	134	152
BR230	37	–	–	–	–	–	16377	–	221	51	24	52	81	–	60	153	61
BR231	52	–	–	–	–	–	10682	–	242	43	10	43	136	–	96	167	41
T240	–	–	–	–	–	–	15175	–	298	–	31	38	–	–	–	119	81
T241	39	–	–	–	–	37	16143	–	249	–	35	49	–	4	–	157	73
T242	–	9	–	–	–	33	19211	–	362	–	56	60	–	–	–	93	148
T243	–	–	–	–	–	–	18942	–	323	56	38	62	–	–	–	86	85
T244	–	13	–	–	–	–	24407	–	492	55	38	75	–	–	–	117	105
T245	–	10	–	–	–	30	17713	–	601	–	35	44	–	–	–	90	62
T246	–	–	–	–	–	–	19753	–	192	–	26	59	–	–	–	88	57
T247	38	–	–	–	–	–	23312	–	133	48	33	72	65	–	65	99	67
T248	–	–	–	–	–	–	20694	–	308	62	29	62	87	–	–	131	101
T249	–	11	–	–	–	–	20440	–	255	–	26	19	–	–	–	103	77
T254	–	8	–	–	–	–	13986	–	500	–	28	39	–	–	–	75	58
T255	–	10	–	–	–	21	11057	–	377	–	42	26	–	–	–	65	100
T256	–	–	–	–	–	38	17770	–	385	–	201	41	–	–	–	163	184
T257	35	–	–	–	97	–	17858	–	293	47	48	52	99	–	–	143	161
T258	–	10	–	–	–	–	29696	–	696	47	42	91	76	–	–	132	139
T259	–	–	–	–	–	37	22148	–	229	–	74	64	–	–	–	132	114
T260	–	23	–	–	–	–	25724	–	308	–	25	17	–	–	–	91	94
T261	–	7	–	–	–	–	16194	–	268	–	21	52	–	–	–	103	98
BR262	48	–	–	–	–	32	14263	–	327	47	20	50	131	–	85	165	40
T263	–	11	–	–	–	–	20001	–	169	48	22	59	–	–	–	99	61
T264	–	–	–	–	–	–	18395	–	393	–	25	58	–	–	–	133	102
T265	42	–	–	–	–	–	15605	–	303	–	32	45	71	–	89	169	78
T266	–	18	–	–	–	45	17090	–	332	–	179	38	–	–	–	163	190
UG851	–	–	–	–	–	–	14443	–	154	–	29	53	–	–	–	140	56
UG852	–	–	–	149	–	–	19812	–	385	–	259	57	–	–	–	167	213
UG853	–	–	–	267	–	–	21400	–	357	55	197	70	–	–	–	134	155
UG856	–	27	–	–	–	50	23423	–	537	42	388	55	–	5	–	135	309
UG857	–	–	–	–	–	30	16764	–	329	–	114	43	–	–	–	171	140
UG858	–	–	–	–	112	36	23389	–	394	45	272	62	–	–	–	149	173
UG859	–	–	–	–	–	–	17964	–	524	–	31	55	–	5	–	146	76
UG860	–	13	–	–	–	–	18721	–	305	48	79	64	–	5	–	127	100
UG861	–	15	–	–	–	–	15774	–	184	–	125	25	–	–	–	111	150
UG862	–	12	–	197	–	24	16757	–	145	–	120	30	–	–	–	119	141
UG863	–	8	–	151	–	–	15452	–	322	–	29	59	–	–	–	152	84
UG864	–	–	–	–	90	–	19778	–	315	–	460	54	–	–	–	166	271
UG865	–	–	–	–	–	–	10821	–	123	–	30	35	–	–	–	111	45
UG866	–	–	–	–	–	29	19946	–	312	–	557	60	–	–	–	133	226

(Continued)

Table 3. *Continued*

Sample ID	Ag	As	Cd	Co	Cr	Cu	Fe	Hg	Mn	Ni	Pb	Rb	Sb	Se	Sn	Sr	Zn
UG867	–	–	–	–	–	–	17826	–	280	46	129	51	–	–	–	128	93
UG868	–	–	–	–	–	–	17241	–	287	36	117	57	–	–	–	108	109
UG869	–	–	–	–	–	–	17941	–	317	–	105	66	–	–	–	107	121
UG871	–	–	–	138	–	–	16722	–	345	–	93	45	–	–	–	125	113
UG872	–	–	–	157	–	–	17114	–	528	–	44	52	–	–	–	155	81
UG873	–	–	–	–	–	–	16838	–	452	44	37	55	–	–	–	149	95
UG874	–	–	–	–	–	–	14282	–	361	–	51	46	–	–	–	116	87
UG875	–	–	–	–	–	–	16391	–	247	–	193	52	–	4	–	115	136
UG877	–	21	–	–	–	–	18870	–	381	55	255	55	–	–	–	190	177
UG878	–	–	–	–	90	–	15385	–	425	–	204	47	–	4	–	164	172
UG879	–	–	–	283	124	131	26910	–	415	–	322	21	–	5	–	130	527
UG880	–	–	–	–	96	–	21396	–	405	–	93	61	–	–	–	130	122
UG881	–	–	–	–	–	36	16058	–	598	42	73	48	–	–	–	116	138
UG882	–	13	–	147	–	38	15641	–	260	–	170	31	–	–	–	131	175
DP884	–	35	–	–	–	–	21622	–	350	53	131	71	–	4	–	128	98
DP887	–	21	–	232	–	–	19492	–	603	–	171	49	–	–	–	132	240
DP890	–	–	–	–	–	–	19684	–	420	–	46	48	–	–	–	136	84
DP893	–	22	–	146	–	31	18482	–	466	–	100	45	–	5	–	144	105
DP896	–	16	–	–	–	34	19885	–	460	41	188	56	–	–	–	119	164
DP897	–	28	–	–	–	–	20807	–	532	–	195	61	–	4	–	136	182
DP900	–	–	–	–	–	45	27895	–	576	–	131	84	–	5	–	138	176
DP903	–	–	–	–	–	79	27330	–	447	51	779	60	–	–	–	138	662
DP906	–	40	–	–	–	–	27866	7	513	66	132	86	–	–	–	133	115
DP909	–	15	–	–	–	–	18612	–	246	–	85	53	–	–	–	153	90
DP912	–	–	–	–	–	33	17381	–	305	66	158	60	–	4	–	175	210
DP916	–	22	–	–	–	–	22107	–	607	64	152	75	–	–	–	131	109
DP921	–	14	–	–	–	–	17398	–	344	52	48	55	–	5	–	126	91
DP925	73	–	76	–	–	–	25308	–	354	–	136	71	315	–	233	1930	184
DP928	–	–	–	–	–	36	26040	–	439	44	114	74	76	4	–	1553	129
DP931	–	–	–	–	–	–	24163	–	494	–	138	82	59	–	69	134	254
DP934	–	–	–	136	–	–	16371	–	339	–	46	45	–	4	–	132	101
DP937	29	35	–	–	–	–	22651	–	316	56	296	68	80	–	64	141	243
DP940	–	18	–	–	82	–	16915	–	435	59	72	52	–	4	–	141	101
DP944	–	–	–	–	–	–	31044	–	404	64	53	98	–	–	–	138	86
DP947	–	25	–	–	94	31	19096	–	454	–	138	54	–	–	–	139	138
DP951	–	24	–	–	137	–	21744	–	531	89	109	57	–	–	–	158	99
DP954	–	–	–	–	–	–	19740	–	581	–	53	55	–	4	–	128	85
DP957	–	17	–	192	–	–	27924	–	325	–	108	89	–	–	–	115	97
DP660	–	9	–	168	–	–	21095	–	381	–	38	58	–	–	–	126	96
DP963	–	–	–	–	–	–	24397	–	331	–	54	74	–	–	–	115	88
T975	–	12	–	164	–	–	22548	–	309	–	63	68	–	6	–	112	136
T976	–	–	–	–	–	–	13793	–	167	–	46	50	–	4	–	128	89
T977	–	–	–	–	–	–	27862	–	405	–	102	76	–	–	–	98	183
T978	–	16	–	–	–	–	23691	–	648	–	71	81	–	4	–	108	161
T979	–	–	–	169	–	–	26263	–	811	43	41	90	–	4	–	97	122
T980	–	11	–	238	–	–	25983	–	510	–	48	102	–	7	–	93	141
T981	–	–	–	–	–	–	15912	–	483	54	113	40	–	–	–	165	159
T982	–	–	–	265	109	–	21931	–	443	–	69	73	–	5	–	101	158
T983	–	21	–	–	–	–	27022	–	329	–	41	95	–	6	–	108	155
T984	–	14	–	–	–	40	23736	–	483	–	75	70	–	5	–	105	200
T985	–	8	–	–	–	–	23867	–	565	–	25	66	–	–	–	171	56
T986	–	12	–	164	–	–	22930	–	214	–	19	80	–	–	–	129	41
T987	–	–	–	198	–	–	20666	–	354	–	72	48	–	–	–	137	172
T988	–	–	–	–	–	–	15417	–	245	–	99	37	–	–	–	146	139
T989	–	10	–	–	–	–	21073	–	467	55	39	54	–	–	–	120	122
T991	–	–	–	–	–	–	24943	–	810	–	38	59	–	–	–	107	111
T992	–	–	–	–	–	–	18130	–	208	–	49	65	–	–	–	136	81
T993	–	9	–	205	–	–	18659	–	194	–	49	68	–	–	–	137	73

(*Continued*)

Table 3. *Continued*

Sample ID	Ag	As	Cd	Co	Cr	Cu	Fe	Hg	Mn	Ni	Pb	Rb	Sb	Se	Sn	Sr	Zn
T994	–	–	–	–	–	–	20246	–	288	*45*	42	66	–	5	–	124	101
T995	–	11	–	–	–	–	24068	–	471	*49*	30	74	–	–	–	124	89
T996	–	9	–	*176*	99	–	21354	–	220	–	20	60	–	–	–	133	52
T997	–	22	–	–	–	–	21011	–	433	–	172	45	–	–	–	166	326
T998	23	–	–	–	–	–	11608	–	113	*43*	20	41	117	–	71	154	52
T999	–	11	–	–	–	–	21889	–	235	–	23	70	–	–	–	110	69
T1000	–	32	–	–	–	–	18340	–	428	51	31	57	–	–	–	138	113
T1001	–	9	–	–	–	–	22711	–	849	–	38	97	–	–	–	83	118
T1002	–	–	–	–	–	51	17649	–	319	–	147	46	–	–	–	146	217
jenas blank	–	–	–	–	–	–	58	–	–	–	–	–	–	–	–	6	–
jeans unk 1	–	–	–	–	–	–	9825	–	112	–	18	33	–	–	–	107	42
jeans unk 2	–	–	–	–	–	–	4182	–	68	–	*4*	2	–	–	–	13	21

All values in parts per million. Values in italics are near the limit of detection.

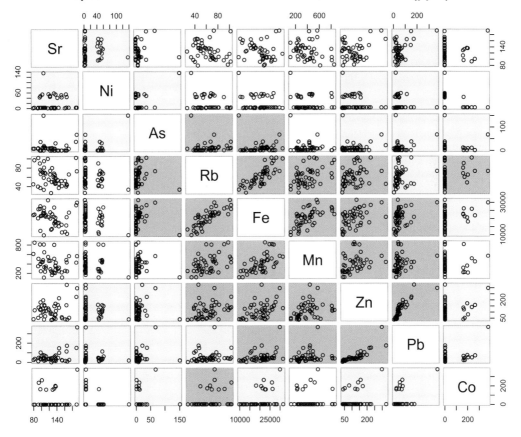

Fig. 2. Scatterplots of XRF of paired elemental data arranged by Spearman's rank correlation strength. Red indicates a strong correlation (>0.4), green a weak correlation (<0.4 to >0.15) and yellow indicates the data are uncorrelated (<0.15). Plot created using R and the gclus library package (Wickham 2009; R Core Team 2012).

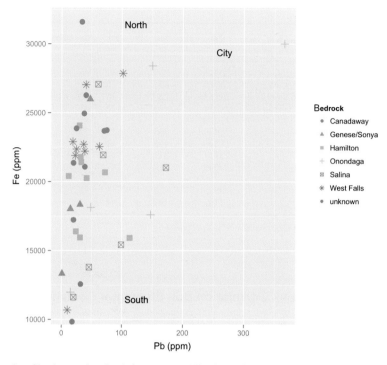

Fig. 3. Scatterplot of lead verses iron levels in parts per million in relation to underlying bedrock geology. In general iron concentrations increase from north to south in the study area. Lead levels increase significantly in the urban corridor. Plot created using R and the ggplot2 library package (Hurley 2012; R Core Team 2012).

In situ XRD analysis using standard powder diffraction equipment was fairly unsatisfactory. Such an analysis is generally only possible on relatively thick spatter marks, which may not be very common in a true forensic setting. Mounting the clothing inside the XRD chamber was awkward and time-consuming, and depending on the location of the spatter, it could prove quite difficult, or impossible, to create an appropriately level surface for analysis. Creation of reusable mounting frames could help address this concern. However, the appearance of extraneous peaks, either from the fabric or from the underlying mounting frame, which could completely mask the soil peaks of interest, was more concerning (Fig. 5). Tape samples gave reasonable results, as long as it was ascertained that the tape used for the creation of the mounts did not itself generate peaks (Figs 5 & 6). For both of these approaches there is also a concern with preferential orientation of minerals, which can intensify some peaks and make others disappear, making comparisons with bulk powder samples more complicated. Analysis of samples using poly-capillary tubes was quite satisfactory, showing little peak-shift, no appearance of extraneous peaks, and better random orientation of the specimen. Small amounts of sample were needed, although longer analysis times were required (300–600 s over a 0.0167° scanning arc or roughly 0.027–0.054° s^{-1}). One downside of this approach is that the signal from weaker peaks can be masked (Fig. 7). Of the approaches tried, the capillary tube specimens had characteristics most similar to diffraction patterns generated using traditional back mounted powder samples.

For XRF analysis of bulk soils, depth of X-ray penetration into a soil matrix is on the order of 2–3 mm, so a thickness of >5 mm is considered to be infinitely thick (i.e. the depth at which 99% of the analyte X-rays have been generated). *In situ* soil measurements taken directly on the ground are always considered infinitely thick, but in soil cups it is considered best practice to fill the cup at least three-quarters full. Low-density soils with highly porous textures present additional difficulties. The simulated forensic samples were analysed in place with the clothing positioned on a thick sheet of poly(methyl methacrylate) to absorb X-rays that penetrate through the sample, and the scrapings used for XRD analysis were placed inside standard sampling cups and analysed inside the bench-top sample holder. In this case, *in situ*

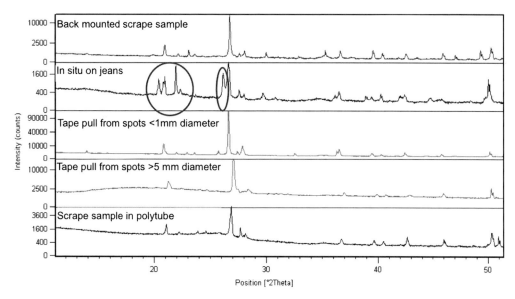

Fig. 4. Diffraction patterns for Unknown 1 (SBCB) on different substrates (from top to bottom): back mounted power sample in an aluminium tray, *in situ* on jeans, a tape pull from small (<1 mm diameter) spatter marks, a tape pull from large spatter marks (>5 mm diameters) and a sample packed into a poly tube.

analysis was more satisfactory than it proved for XRD. The signal strength from *in situ* samples was low, but the elemental ratios generated were similar to the signatures from bulk samples (Fig. 8).

Discussion

The goal of this project is to evaluate the utility of XRD and FPXRF as screening methods for analysis of transferred soil materials and their usefulness in

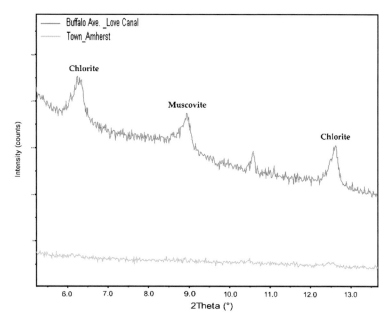

Fig. 5. XRD diffraction plot showing the broad diffraction peaks characteristic of the clay mineral component.

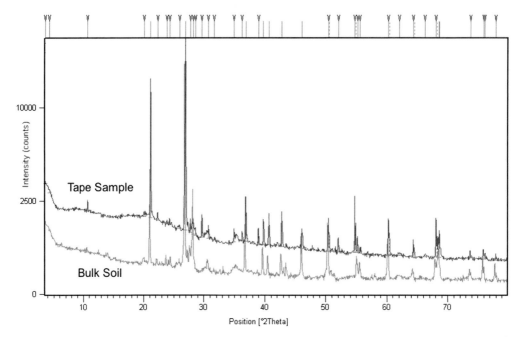

Fig. 6. Comparison of the diffraction patterns created using a tape sample supported on glass slide and a standard back mounted powder sample in an aluminium sample tray.

locating possible source zones as part of a larger environmental assessment delineating regional anthropogenic heavy metal contaminant zones. Before large-scale analysis of hundreds of samples, it was deemed advantageous to determine what analysis protocols would provide robust results with a minimum of time and expense, thus the project started under conditions of minimal sample handling. Therefore, two unknowns were collected as simulated forensic samples from places within the study area that met the same criteria as the locations used for the collection of database samples but were not part of the initial dataset.

Based on the XRD analysis of simulated forensic samples 'unknown 1' and 'unknown 2', it was possible to eliminate several portions of the study area from consideration as possible source zones. Forensic sample SBCB, designated unknown 1, is composed of roughly 37% quartz, 32% muscovite, 21% potassium feldspar and 6% plagioclase feldspar, with indications of a clay mineral component, while unknown 2 (X-SBC) is composed of roughly 42% quartz, 25% muscovite, 13% plagioclase feldspar, 10% potassium feldspar, 8% kaolinite and 2% magnetite (Table 2; Fig. 9).

The soils in the Lackawanna region of the county show significant iron oxide content, as this area was historically the home of the Bethlehem Steel Plant, once the fourth largest such plant in the world. The plant was in operation from 1902 (under the name Lackawanna Steel) until 1983, when the steel-making facility closed, although the coke ovens, galvanizing department and a few other associated divisions operated for a while longer. Soils from this region tend to have a detectable high-temperature magnetite signature (in the 2–5% range). Soils from Tonawanda and the outer harbour area also typically have a detectable iron oxide component, again owing to a long history of manufacturing and industry. The presence of magnetite in unknown 2 potentially places its source as somewhere in the urban region of the northern portion of the study area.

In general, soils in the northern portion of the study area, which is underlain predominantly by limestone and dolostone, typically have a significant carbonate signature, while those in the south have little or no carbonate content. This also suggests that unknown 2 is from the northern portion of the study area. In contrast, much of the southern portion of the study area shows a detectable kaolinite component, which is lacking in most of the north, with the notable exception of Amherst. Based on the combination of factors from above and the significant kaolinite component, this suggests that unknown 2 is from somewhere in the NW portion of the study area, potentially on the Tonawanda or Buffalo side of Amherst.

Unknown 1 lacks carbonates, magnetite and kaolinite, which tends to eliminate the northern- and

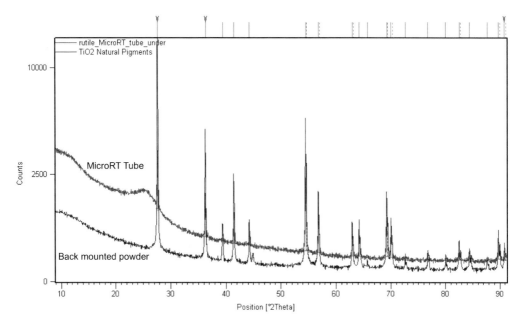

Fig. 7. Comparison of the diffraction patterns created (using a rutile standard) using a MicroRT poly-capillary tube and a standard back mounted powder sample in an aluminium sample tray.

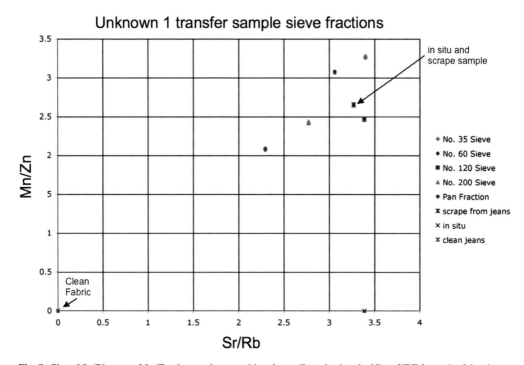

Fig. 8. Plot of Sr/Rb verses Mn/Zn elemental composition data collected using the Niton XRF for each of the sieve fractions of a soil sample collected from the source area of simulated forensic sample Unknown 1 (SBCB) and the results from the Unknown 1 spatter sample.

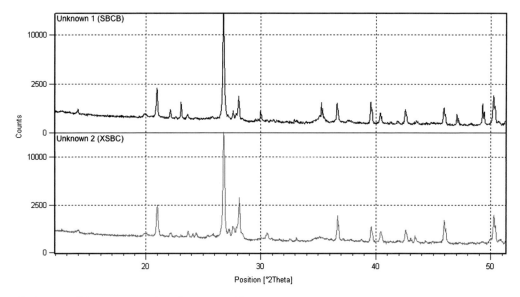

Fig. 9. X-ray diffraction patterns for unknowns 1 and 2.

southern-most sections of the study area, as well as the urban industrial corridor. The only distinctive point about the specimen is that it has a higher potassium feldspar content than most areas in the region. There are variations in the relative proportions of feldspars across the study area, which could potentially be useful for comparison, but intra-town variations are sometimes of a similar scale to inter-town variations, making further discrimination difficult. A Ward hierarchical clustering dendrogram of the mineralogy places unknown 1 closest to Alden and somewhat similar in nature to one of the Buffalo parks samples (Fig. 10). Further discrimination using the existing XRD database was not possible; at this point it would be necessary to gather additional reference samples to further isolate possible source zones or to invoke additional sample preparation procedures such as isolating the clay mineral fraction of the sample. This leaves most of the north-central portion of the study area as possible source areas for unknown 1.

Using the XRF dataset, a Ward hierarchical clustering analysis of the simulated forensic samples was performed utilizing As, Co, Fe, Mn, Ni, Pb, Sr, Rb and Zn levels. The mobility of elements in the surface environment, or roughly speaking the likelihood of an element transferring from the soil to another reservoir, such as water or the biosphere, is governed by factors such as pH, eH, the kind and amount of organic matter, and the sorption capacity of soil minerals. Under the oxidizing and neutral to alkaline conditions common in local soils, arsenic, strontium and zinc are considered moderately mobile in the surface environment, while cobalt, manganese, nickel and lead are considered slightly mobile (Kabata-Pendias & Mukherjee 2007). Iron is generally considered to be immobile in the surface environment except under acidic or reducing conditions. The iron may change oxidation state but will form insoluble compounds that remain in the soil, tending to serve as adsorption sites for other heavy metals, such as As, Cu, Pb, Co, Zn, Ni and Mn (Kabata-Pendias & Mukherjee 2007). The cluster analysis puts both unknowns in the north-central region of the study area, closer to the city (Fig. 11). Thus the XRD and XRF results are in general agreement. For unknown 1, the areas in the cluster suggest that the sample came from somewhere in the northeastern part of Buffalo or the nearby northern suburbs. Unknown 2 clusters most closely with Tonawanda, which corresponds well to the results of the XRD analysis, and in the same general region as unknown 1.

Both unknowns were collected from distinct areas that were within a few miles of one or more reference collection points with the goal of determining whether it was possible to identify their points of origin using the XRD/XRF dataset. Unknown 1 actually did come from the northeastern outer circle area of Buffalo, so in general the results of the combined XRD and XRF analysis were reasonably accurate, potentially narrowing down a search area. Unknown 2, however, came from a park in the northern town of Clarence, an area that

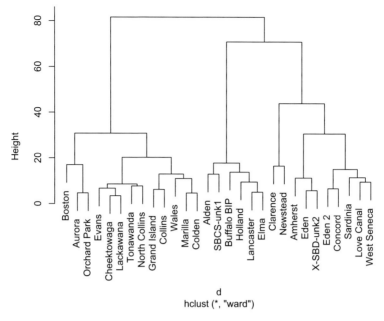

Fig. 10. Ward Hierarchical Clustering Dendrogram of XRD bulk mineral composition created using R and the cluster library package (Maechler *et al.* 2012; R Core Team 2012).

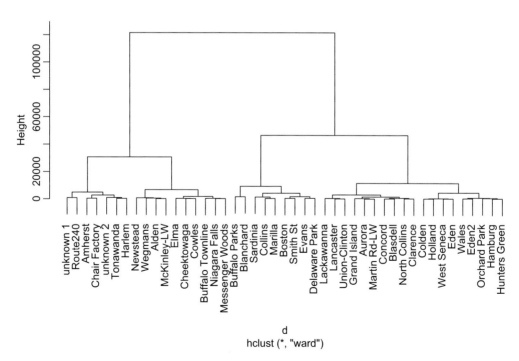

Fig. 11. Ward Hierarchical Clustering Dendrogram of XRF elemental data created using R and the cluster library package (Maechler *et al.* 2012; R Core Team 2012).

was not successfully identified by either approach. Clarence lies on the eastern side of Amherst rather than the western (or city) side.

The reference samples collected in Clarence all showed a significantly higher carbonate component (16% on average) than the transferred mud spatter sample. There was also a clear lead oxide signature to those samples and no detected magnetite, as well as significant differences in trace elemental signature. Examining the XRF reference dataset, it appears that the samples collected from Clarence have unusually high levels of lead, zinc and manganese, which are more typical of the urban sections of the study area and the industrial corridor in particular. Additional samples need to be collected from Clarence to see if the reference samples were anomalous. It is also worth noting that the trace elemental values for the mud spatter collected from the jeans were more similar to those of the no. 60 and no. 120 sieve fractions than the finer no. 200 sieve or pan fractions.

Based on these results, overall the proposed method shows promise. Ideally, additional sample preparation procedures can now be added in order to improve our results. The traditional next steps for XRD analysis would be preparation of oriented specimens for positive identification of the clay mineral fraction. Creation of these specimens generally involves removal of non-platy minerals and application of chemical and/or heat treatments. As such treatments can be time- and labour-intensive, and often will irreversibly alter a sample, the necessity of such treatment needs to be carefully weighed on a case-by-case basis against the value of the data gained. Another concern is that, as with sample pre-treatment, new sources of variance and potential error are introduced with each new procedure.

An additional consideration is that many, if not most, forensic soil samples are quite small, so some traditional sample preparation techniques cannot be employed however '*there must be a single method of preparation for a given set of samples if X-ray diffraction data are to be compared within the set,* particularly for quantitative analysis' (Moore & Reynolds 1997, p. 205, emphasis in the original). As the intent underlying forensic use of diffraction data is quite different than when it is employed by a soil scientist or clay mineralogist, some of the procedures developed for elucidating information from a sample may not be appropriate for a forensic examination and it is worth exploring where the point of diminishing returns might lie. For example, the efficacy of using quantitative techniques such as mineral intensity factors or 100% quantification approaches is unclear as the recovered forensic samples are often so small that they cannot be prepared in such a fashion that they will comply with the specimen length and thickness parameters necessary to minimize errors in peak height and area as required for reasonable quantification (Moore & Reynolds 1997; Kahle *et al.* 2002).

These concerns are especially true for quantitative clay mineral analysis. Ottner *et al.* (2000) eliminated variations owing to sample preparation, mineral identification and peak-area fitting by utilizing the clay sample and diffraction pattern and then utilizing three different quantification protocols (Weir *et al.* 1975; Reynolds 1989; Tributh 1991). The resulting weight percentages for smectite, illite, kaolinite and chlorite varied significantly (10–16%), demonstrating 'that the choice of quantification method has a strong influence on the results' (Ottner *et al.* 2000, p. 240).

Conclusions

Based on these preliminary results, applying a combination of XRD and XRF approaches does allow for some discrimination of potential source areas, although clearly reference sets should be used with caution. It is anticipated that the accuracy of this approach will improve as the reference dataset grows. In addition, several distinctive components of the urban environment were revealed, such as iron oxide components, high zinc levels in an area where tyres were vulcanized and industrially related titanium dioxide signatures, all of which suggest that soil forensics can prove to be useful even in areas that are composed of seemingly unpromising 'urban fill'.

Based on the results of the XRD analysis using bulk powders, additional sample preparation procedures specifically to elucidate the clay mineral content are clearly needed, but the development of appropriate protocols will require great care. The addition would increase the discriminating power significantly. To this end, a critical evaluation of a variety of sample preparation and data collection methods will be performed.

The author would like to thank former students of the GES 460/529 course for their assistance in sample collection over the years, as well as the reviewers for their comments and questions, which inspired significant improvements in this article. She thanks Duncan Pirrie for his forbearance as well. XRD purchase partially funded by a National Science Foundation Course, Curriculum & Laboratory Improvement (CCLI) Program grant DUE 0410466.

References

ABRAHAM, J. T., SHUKLA, S. K. & SINGH, A. K. 2007. Application of x-ray diffraction techniques in forensic science. *Forensic Science Communications*, **9**, April, http://www.fbi.gov/about-us/lab/forensic-science-communications/fsc/april2007/index.htm/research/2007_04_research03.htm

BERNICK, M. B., KALNICKY, D. J. & PRINCE, G. 1995. Results of field-portable X-ray fluorescence analysis of metal contaminants in soil and sediment. *Journal of Hazardous Materials*, **43**, 101–110.

BISH, D. L. & CHIPERA, S. J. 1991. Detection of trace amounts of erionite using x-ray powder diffraction: erionite in tuffs of Yucca Mountain, Nevada and central Turkey. *Clays and Clay Minerals*, **39**, 437–445.

BISH, D. L. & REYNOLDS, R. C. JR. 1989. Sample preparation for x-ray diffraction. *In*: BISH, D. L. & POST, J. E. (eds) *Modern Powder Diffraction*. Reviews in Mineralogy. Mineralogical Society of America, Washington, DC, **20**, 73–99.

CLARK, S., MENRATH, W., CHEN, M., RODA, S. & SUCCOP, P. 1999. Use of a field portable X-Ray fluorescence analyzer to determine the concentration of lead and other metals in soil samples. *Annals of Agricultural and Environmental Medicine*, **6**, 27–32.

DAWSON, L. A. & HILLIER, S. 2010. Measurement of soil characteristics for forensic applications. *Surface and Interface Analysis*, **42**, 363–377.

ECKARDT, R., KRUPICKA, E. & HOFMEISTER, W. 2012. Validation of powder x-ray diffraction following EN ISO/IEC 17025. *Journal of Forensic Sciences*, **57**, 722–737.

FITZPATRICK, R. W., RAVEN, M. D. & FORRESTER, S. T. 2009. A systematic approach to soil forensics: criminal case studies involving transference from crime scene to forensic evidence. *In*: RITZ, K., DAWSON, L. A. & MILLER, D. (eds) *Criminal and Environmental Soil Forensics*. Springer Science + Business Media, Berlin, 105–128.

HURLEY, C. 2012. gclus: Clustering graphics. R package version 1.3.1, http://CRAN.R-project.org/package=gclus

ICDD 2007. *The Powder Diffraction File 2 (PDF2) 2007 release*. International Center for Diffraction Data, Newton Square, PA.

ISACHSEN, Y., LANDING, E., LAUBER, J. M., RICKARD, L. V. & ROGERS, W. B. (eds) 2000. *Geology of New York: A Simplified Account*, 2nd edn. New York State Museum Educational Leaflet 28. New York State Museum/Geological Survey, Albany, NY.

ISPHORDING, W. 2004. The right way and the wrong way of presenting statistical and geological evidence in a court of law (a little knowledge is a dangerous thing!). *In*: PYE, K. & CROFT, D. J. (eds) *Forensic Geoscience: Principles, Techniques and Applications*. Geological Society, London, Special Publications, **232**, 281–288.

JENKINS, R. 1988. *X-Ray Fluorescence Spectrometry*. John Wiley and Sons, New York.

JENKINS, R. 1989. Experimental procedures. *In*: BISH, D. L. & POST, J. E. (eds) *Modern Powder Diffraction*. Reviews in Mineralogy. Mineralogical Society of America, Washington, DC, **20**, 46–71.

JENKINS, R. & SNYDER, R. L. 1996. *Introduction to X-ray Powder Diffractometry*. John Wiley and Sons, New York.

JUNGER, E. P. 1996. Assessing the unique characteristics of close-proximity soil samples: just how useful is soil evidence? *Journal of Forensic Sciences*, **41**, 27–34.

KABATA-PENDIAS, A. & MUKHERJEE, A. B. 2007. *Trace Elements from Soil to Human*. Springer, Berlin.

KAHLE, M., KLEBER, M. & JAHN, R. 2002. Review of XRD-based quantitative analyses of clay minerals in soils: the suitability of mineral intensity factors. *Geoderma*, **109**, 191–205.

KALNICKY, D. J. & SINGHVI, R. 2001. Field portable XRF analysis of environmental samples. *Journal of Hazardous Materials*, **83**, 93–122.

KLEEBERG, R., MONECKE, T. & HILLIER, S. 2008. Preferred orientation of mineral grains in sample mounts for quantitative XRD measurements: how random are powder samples? *Clays and Clay Minerals*, **56**, 404–415.

KONINKLIJKE PHILIPS ELECTRONICS 2001. *X-Ray Diffraction Sample Holders & Preparation*. Koninklijke Philips Electronics, Almeolo.

KUGLER, W. 2003. *X-Ray Diffraction Analysis in the Forensic Science: the Last Resort in Many Criminal Cases*. Advances in X-ray Analysis, **46**, International Centre for Diffraction Data, http://www.icdd.com/resources/axa/vol46/v46_01.pdf

MAECHLE, M., ROUSSEEUW, P., STRUYF, A., HUBERT, M. & HORNIK, K. 2012. Cluster: cluster analysis basics and extensions. R package version 1.14.2, http://cran.r-project.org/web/packages/cluster/

MARUMO, Y., NAGATSUKA, S. & OBA, Y. 1986. Clay mineralogical analysis using the <0.05-mm fraction for forensic science investigation – its application to volcanic ash soils and yellow-brown forest soils. *Journal of Forensic Sciences*, **31**, 92–105.

MELO, V. F., BARBAR, L. C., ZAMORA, P. G. P., SCHAEFER, C. E. & CORDEIRO, G. A. 2008. Chemical, physical and mineralogical characterization of soils from the Curitiba metropolitan region for forensic purpose. *Forensic Science International*, **179**, 123–143.

MOORE, D. M. & REYNOLDS, R. C. 1997. *X-Ray Diffraction and the Identification and Analysis of Clay Minerals*, 2nd edn. Oxford University Press, Oxford.

MURRAY, R. 2011. *Evidence from the Earth*, 2nd edn. Mountain Press, Missoula, MT.

MURRAY, R. C. & TEDROW, J. C. F. 1975. *Forensic Geology: Earth Sciences and Criminal Investigations*. Rutgers University Press, New Brunswick, NJ.

OTTNER, F., GIER, S., KUDERNA, M. & SCHWAIGHOFER, B. 2000. Results of an inter-laboratory comparison of methods for quantitative clay analysis. *Applied Clay Science*, **17**, 223–243.

PYE, K. 2007. *Geological and Soil Evidence: Forensic Applications*. CRC Press, Boca Raton, FL.

PYE, K. & BLOTT, S. J. 2004. Comparison of soils and sediments using major and minor trace element data. *In*: PYE, K. & CROFT, D. J. (eds) *Forensic Geoscience: Principles, Techniques and Applications*. Geological Society, London, Special Publications, **232**, 183–196.

PYE, K. & CROFT, D. 2007. Forensic analysis of soil and sediment traces by scanning electron microscopy and energy-dispersive X-ray analysis: an experimental investigation. *Forensic Science International*, **165**, 52–63.

PYE, K., BLOTT, S. J., CROFT, D. J. & CARTER, J. F. 2006. Forensic comparison of soil samples: assessment of small-scale variability in elemental composition, carbon and nitrogen isotope ratios, colour, and particle size distribution. *Forensic Science International*, **163**, 59–80.

R CORE TEAM 2012. *R: A Language and Environment for Statistical Computing*. R Foundation for Statistical Computing, Vienna, http://www.R-project.org/

RAWLINS, B. G., KEMP, S. J., HODGKINSON, E. H., RIDING, J. B., VANE, C. H., POULTON, C. & FREEBOROUGH, K. 2006. Potential and pitfalls in establishing the provenance of Earth-related samples in forensic investigations. *Journal of Forensic Sciences*, **51**, 832–845.

REYNOLDS, R. C. 1989. Principles and techniques of quantitative analysis of clay minerals by X-ray powder diffraction. *In*: PEVEAR, D. R. & MUMPTON, F. A. (eds) *Quantitative Mineral Analysis of Clays*. CMS Workshop Lectures, The Clay Minerals Society, Bloomington, Indiana, **1**, 4–37.

RUFFELL, A. & MCKINLEY, J. 2008. *Geoforensics*. Wiley-Blackwell, Chichester.

RUFFELL, A. & WILTSHIRE, P. 2004. Conjunctive use of quantitative and qualitative x-ray diffraction analysis of soils and rocks for forensic analysis. *Forensic Science International*, **145**, 13–23, doi: 10.1016/j.forsciint.2004.03.017.

ŚRODOŃ, J., DRITS, V. A., MCCARTY, D. K., HSIEH, J. C. C. & EBERL, D. D. 2001. Quantitative x-ray diffraction analysis of clay-bearing rocks from random preparations. *Clays and Clay Minerals*, **49**, 514–528.

ŠVARCOVÁ, S., KOČÍ, E., BEZDIČKA, P., HRADIL, D. & HRADILOVÁ, J. 2010. Evaluation of laboratory powder x-ray micro-diffraction for applications in the fields of cultural heritage and forensic science. *Analytical and Bioanalytical Chemistry*, **398**, 1061–1076.

TRIBUTH, H. 1991. Qualitative und quantitative Bestimmung der Tonminerale in Bodentonen. *In*: TRIBUTH, H. & LAGALY, G. (eds) *Identifizierung und Charahterisierung von Tonmineralen*. Beiträge zur Jahrestagung der DTTG, Giessen, 37–85.

US CENSUS BUREAU 2010. *Profile of General Population and Housing Characteristics: 2010: Buffalo–Niagara Falls Metropolitan area, New York*, http://factfinder2.census.gov/faces/tableservices/jsf/pages/productview.xhtml?pid=DEC_10_DP_DPDP1 (accessed 10 January 2013).

US CENSUS BUREAU 2012a. *Profile of General Population and Housing Characteristics: Cheektowaga CDP, New York*, Last Revised: Thursday, 06-Dec-2012 16:43:51 EST, http://quickfacts.census.gov/qfd/states/36/3615000.html (accessed 10 January 2013).

US CENSUS BUREAU 2012b. *Profile of General Population and Housing Characteristics: Tonawanda CDP, New York*, Last Revised: Thursday, 06-Dec-2012 16:43:51 EST, http://quickfacts.census.gov/qfd/states/36/3674183.html (accessed 10 January 2013).

US CENSUS BUREAU 2012c. *Table 3. Annual Estimates of the Resident Population for Incorporated Places in New York, April 1, 2010 to July 1, 2011* (SUB-EST2011-03-36), http://quickfacts.census.gov/qfd/states/36/36029lk.html (accessed 10 January 2013).

US EPA 1998. *Method 6200: Field Portable X-ray Fluorescence Spectrometry for the Determination of Elemental Concentrations in Soils and Sediments*, Revision 0, February 2007. United States Environmental Protection Agency, www.epa.gov/osw/hazard/testmethods/sw846/pdfs/6200.pdf

WEIR, A. H., ORMEROD, E. C. & EL MANSEY, I. M. I. 1975. Clay mineralogy of sediments of the Western Nile Delta. *Clay Minerals*, **10**, 369–386.

WICKHAM, H. 2009. *ggplot2: Elegant Graphics for Data Analysis*. Springer, New York, http://cran.r-project.org/web/packages/ggplot2/index.html

Automated forensic soil mineral analysis; testing the potential of lithotyping

DUNCAN PIRRIE[1]*, GAVYN K. ROLLINSON[2], MATTHEW R. POWER[3] & JULIA WEBB[4]

[1]*Helford Geoscience LLP, Menallack Farm, Treverva, Penryn, Cornwall TR10 9BP, UK*

[2]*Camborne School of Mines, College of Engineering, Mathematics & Physical Sciences, University of Exeter, Tremough Campus, Penryn TR10 9EZ, Cornwall, UK*

[3]*4669 West Lawn Drive, Burnaby, British Columbia, Canada VSC 3R2*

[4]*School of Natural and Social Sciences, University of Gloucestershire, Swindon Road, Cheltenham GL50 4AZ, UK*

*Corresponding author (e-mail: dpirrie@helfordgeoscience.co.uk)

Abstract: In the investigation of serious crimes, soil can be, in some cases, a very valuable class of trace evidence. The complexity of soil is part of the reason why it is useful as trace evidence but is also an inherent problem, as there are many different parameters in a soil sample that could potentially be characterized. The inorganic components of soils are dominated by minerals, along with anthropogenic particulate grains; thus, the analysis of soil mineralogy as the main technique for inorganic forensic soil characterization is recommended. Typical methods that allow the bulk mineralogy to be determined, such as X-ray diffraction (XRD), do not allow the texture of the particles to be characterized. However, automated scanning electron microscopy (SEM) provides both modal mineralogy and also allows particle textures to be characterized. A recent advance in this technique has been the ability to report the modal mineralogy of a sample as 'lithotypes', which are defined on the basis of a combination of mineralogy and other parameters, such as grain size and mineral associations. Defined lithotype groups may include monominerallic grains but also, importantly, allow the automated quantification of rock types and other anthropogenic materials. Based on a simulated forensic scenario, the use of lithotyping is evaluated as an aid in the analysis of soil samples. This technique provides additional discrimination when comparing different soil samples.

The analysis of the composition of soils and sediments has become relatively widely used in forensic investigations in serious crime cases (e.g. Ruffell & McKinley 2008). The inherent variability of soils, at a variety of different scales, and the potential ease of transfer from scene to suspect or victim makes soil, in particular, a very useful class of trace evidence (Fitzpatrick *et al.* 2009). However, the complexity of soil also results in an issue as to what parameters of a soil sample should be characterized as part of a forensic investigation. Typically, the overall bulk characteristics of a soil can be measured, such as colour, grain size, pH or conductivity (e.g. Guedes *et al.* 2009, 2011). Alternatively, individual components within a soil may be selected for analysis. Most soils comprise both an organic and an inorganic component (e.g. Dawson & Hillier 2010). The organic component may include spores, pollen, macroscopic plant debris and microorganisms, all of which can be forensically important (e.g. Brown 2006). The inorganic components of a soil will include naturally derived mineral grains from the underlying solid and drift geology, along with anthropogenic particulate grains reflecting current or past land use at the specific sampling location. When considering the inorganic component, some workers have advocated the use of bulk chemical analysis of soil (although it should be noted that such an approach has also been strongly questioned: e.g. Bull *et al.* 2008), the analysis of the chemistry of specific mineral grains, the identification of the overall soil mineralogy (Pirrie *et al.* 2004) or the examination of the surface texture of discrete mineralogical components present within the soil (e.g. the examination of quartz grain shape: Bull & Morgan 2006).

At present, there is no standard approach to the geoforensic analysis of soil samples and, typically, individual practitioners will utilize the facilities available to them, or those that they have greatest experience of, although a common approach is to start with a more generalized examination of a sample which then becomes more focused. Pirrie *et al.* (2009) advocated the analysis of the overall mineralogy of soil samples as the most robust parameter to measure the inorganic components of a soil. However, even with mineral analysis alone, there are different methods that might be used. Traditionally, mineral analysis in the geological sciences focused on the use of transmitted polarizing light microscopy, which with a trained operator allows both mineralogy and the sample texture to be quantified (see Bowen & Craven 2013 for a discussion on the value of transmitted light microscopy). Whilst still of considerable benefit in some cases, in a forensic context the sample size is commonly very small, and the material fine grained and, consequently, very difficult to identify using polarizing light microscopy. Another limitation with this method is that only translucent minerals can be identified; any opaque phases present will not be identified unless the sample is also examined using reflected light microscopy or scanning electron microscopy (SEM). There are two other relatively common methods to determine the overall bulk modal mineralogy: X-ray diffraction (XRD) and SEM with linked energy-dispersive spectrometers (EDS) (e.g. McVicar & Graves 1997; Ruffell & Wiltshire 2004). Both approaches have strengths and limitations. XRD identifies minerals on the basis of crystal structure. It is well suited to the analysis of clay minerals, although it should be noted that distinguishing between different clay mineral species requires the sample to be air dried, glycolated and heated in a stepwise approach. In addition, a recent validation of powder XRD found that minerals present in a known composite soil standard with a known abundance of up to about 9% were not identified (Eckardt et al. 2012). Bergliesen (2013) provides a thoughtful analysis of the strengths and limitations of XRD in a forensic context. As long as the XRD sample preparation and measurement parameters are clearly defined and adhered to, then this method does have some applications in forensic soil examination. Manual SEM-EDS analysis has also been widely used but it is difficult to generate quantitative datasets based on manual examination of a sample.

Automated SEM-EDS has the ability to characterize a large number of particles within a sample, based on the rapid acquisition of energy-dispersive spectra (Pirrie et al. 2004, 2009; Pirrie & Rollinson 2011). Mineral identification is based on the acquired chemical spectra, and limitations are that different minerals with the same, or very similar, chemistries cannot be distinguished from each other. If the modal mineralogy alone is reported, whether by XRD or automated SEM-EDS analysis, the texture of the sample is not known. For example, minerals may be present either as discrete particles composed of only a single mineral or they may be present within polyminerallic grains, such as natural rock fragments or as fragments of anthropogenic materials, such as aggregates. Such textural information can be very significant when testing the similarity, or otherwise, between different unknown samples. Previous work has shown how automated mineral analysis using QEMSCAN® SEM-EDS technology can characterize soil sample mineralogy (Pirrie et al. 2004, 2009).

In this paper, the potential forensic application of an additional function within the QEMSCAN® software that allows minerals to be grouped as 'lithotypes', which may be individual mineral grains, rock fragments or anthropogenic materials, is tested. The concept of lithotyping has recently been used in the oil and gas industry (Moscariello et al. 2010; Haberlah et al. 2012) to characterize the mineralogy and texture of well cuttings. The principle behind the technique is simple; individual particles are assigned to a compositional grouping based on expressions of their mineralogy along with other parameters, such as grain size, mineral associations, porosity and alteration. Thus, a coarse-grained rock fragment comprising quartz, alkali feldspar and plagioclase can be assigned as granite; a quartz sand with a calcite cement as a calcareous sandstone, and so on. Other particles within the sample may be monominerallic and can be assigned to a compositional grouping such as, for example, 'quartz grains'. In this way, the originally collected modal mineralogical data can be reassigned to lithotype groups, allowing the automated quantification of both mono- and polyminerallic grains. This is effectively the same process as carried out during optical microscopy but is automated.

Experimental design, soil sampling and analysis

To test the potential value of the use of automated mineralogy to not only characterize soils based on the overall bulk mineralogy but also on the basis of lithotypes, a simple forensic scenario was replicated. Our simulated scenario was the surface disposal of a murder victim in an area of woodland in Cornwall, UK. The underlying bedrock geology comprises coarse-grained granites of the Carnmenellis Granite and minor superficial deposits (Fig. 1a) (Leveridge et al. 1990). The selected location had a discrete parking area, surfaced with aggregate, reached along a series of minor roads some 5 km from the nearest significant town, Falmouth (Fig. 1b). From the car parking area, a relatively flat track led into an area of light woodland with abundant understorey vegetation (Fig. 2a, b). This location was selected as it was consistent with known offender profiling with respect to the disposal of murder victims (e.g. distance from town, nature of roads, secluded parking location, deposition on the

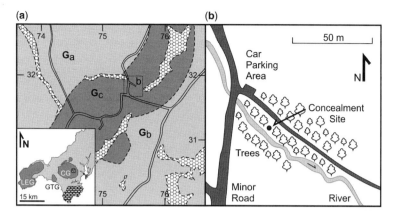

Fig. 1. Map showing the location of the study area (LEG, Lands End Granite; GTG, Godolphin-Tregonning Granite; CG, Carnmenellis Granite; F, Falmouth). (**a**) The underlying bedrock geology is entirely underlain by the Carnmenellis Granite and superficial deposits (adapted from Leveridge *et al.* 1990). G_a – coarse-grained granite abundant feldspar megacrysts >15 mm; G_b – coarse-grained granite sparse feldspar megacrysts >15 mm; G_c – coarse-grained granite abundant feldspar megacrysts <15 mm. (**b**) Detailed map of the study area showing the car parking area, track and concealment site.

flat or down-hill and not more than 30 m from a car parking area). For the purpose of the simulation, two 'offenders' wearing previously used footwear but new denim jeans drove to the car parking area. Both individuals then entered the woodland along the track. One individual was carrying a 50-litre rucksack wrapped in plastic and filled with ballast to replicate carrying a small victim. The two 'offenders' made their way along the track (Fig. 2c), and then one individual branched off into the undergrowth and hid the object, having covered it with surface vegetation (Fig. 2d). The two offenders then left the scene and went back to their respective home locations. The following day, their footwear and jeans were both 'seized', with the items being sealed individually within paper evidence bags.

Fig. 2. Photographs illustrating the study area. (**a**) The car parking area adjacent to the area of woodland. (**b**) The track leading towards the concealment site. (**c**) Offender B carrying an object into the woodland. (**d**) The final concealed package covered by vegetation.

On the day after the concealment of the item, the scene was examined by other members of our research group, who had not been present during the concealment. A search of the area was carried out and the concealed item was located. The area between the track and the concealment site was heavily vegetated (Fig. 2d), and there were no exposed surface soils in-amongst the vegetation. However, eight leaf litter samples were collected along the inferred offender approach path from the track to the concealment site. Fifteen soil samples were collected from the track (eight samples) and from the car parking area (seven samples). There were very common footwear tread marks along the track. The footwear tread of the offenders had not been described at the time of the scene visit; hence, it was not possible to specifically sample marks with a corresponding tread pattern. However, the footwear marks along the track were specifically targeted for sampling as they demonstrated that these areas had been recently contacted by footwear. Surface soil samples (to the depth of the observed tread marks) were collected from approximately 30 cm^2 areas by scraping the surface using a clean spatula, with the sample scrapped on to a piece of paper and then sealed within an evidence bag.

The surface of the car parking area was firmer than the surfaces along the track, although car tyre and diffuse footwear marks were observed. Consequently, samples collected in this area were collected in the same way as for the track but the sampling was of a shallower depth. The surface comprised soils, macroscopic plant debris and introduced geological materials, dominantly slates. There had been light rainfall both on the previous day and on the day the samples were collected, so the soil samples collected were damp. Therefore, the samples were gently dried (<50 °C) within the laboratory as soon as possible.

The footwear and clothing were examined at the University of Gloucestershire. The two footwear items comprised (a) a pair of black shoes (Fig. 3a, b) and (b) a pair of white trainers (Fig. 3c, d). There was abundant soil present on the soles of both items of footwear, although soil was more abundant on the soles of the trainers (the footwear worn by the 'offender' carrying the object for disposal: Fig. 3a, c). The sides of the uppers of the trainers also had quite abundant soil present, whilst the uppers of the black shoes were generally visually clean of soil other than a small area of soil on the outer surface of the left shoe (Fig. 3b, d). The amount of soil present on the footwear exhibits was generally greater than that usually encountered during forensic casework. The soles and uppers of both the left and right item of both pairs of footwear were washed separately using a dilute detergent solution, centrifuged and the soil recovered. Two pairs of jeans were also examined. One pair was very clean with no obvious areas of soil staining. The other pair of jeans had several discrete mud spots at the rear and bottom of the right-hand leg. Both pairs of jeans were prepared in the same way. First, the lower part of each leg was cut off the jeans approximately 20 cm above the hem. Secondly, the area from just above the knee to the cut edge of the jeans was also removed. Thus, from each leg of each pair of jeans, two samples were recovered. Each separate piece of fabric was then washed using a dilute detergent solution and the resultant sample centrifuged. Separate sample

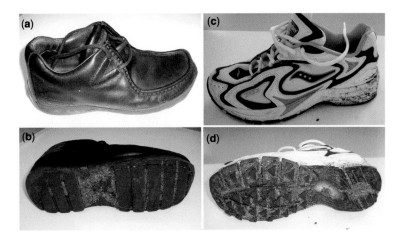

Fig. 3. Photographs showing footwear items (a) and (b) prior to the recovery of the soil trace evidence. (**a**) & (**b**) Uppers and sole of the right shoe from footwear item (a), a pair of black shoes. (**c**) & (**d**) Uppers and sole of the left shoe from footwear item (b), a pair of white patterned trainers.

aliquots were prepared from all of the footwear and clothing exhibits for (a) soil mineralogy and (b) palynology, although only the soil mineralogy is considered further in this paper. The soil samples collected from the access lane and car park area were subdivided with separate subsamples for palynology and mineralogy. The leaf litter samples were washed at the University of Gloucestershire, and the resultant debris was recovered and separate aliquots retained for palynology and mineralogy.

Mineralogical analysis

In total, 39 samples were prepared for mineral analysis. The samples were dried and then placed into clean 30 mm-diameter moulds. The samples were mixed with Epofix resin and then left in a pressure vessel overnight. The samples were then coded, and the moulds backfilled with resin and cured in an oven overnight. Samples were then cut and polished to a 1 μm finish and carbon-coated prior to analysis. Procedures for avoiding cross-contamination were followed throughout the preparation process. The samples were analysed at the University of Exeter, Cornwall Campus, QEMSCAN® facility. They were measured using the QEMSCAN® particle mineral analysis (PMA) mode using a 6 μm beam stepping interval (see Pirrie *et al.* 2004). Typically, in excess of 6000 mineral grains were measured in each sample and the resultant raw dataset was processed using the iDiscover v.4.2 software package. The data were processed and reported in two different ways, (a) modal mineralogy and (b) lithotypes. Initially, the dataset was reported as individual mineral categories (modal mineralogy). In this way the data are processed such that any area analysed with the same, or very similar, chemistry is grouped either as a specific mineral species (e.g. quartz) or as a chemical grouping (e.g. CaAl silicate). The resultant modal mineralogical data are comparable with a traditional point-counting technique widely used in the measurement of, for example, sandstone petrography. When the data are reported purely as modal mineralogy, it is not documented as to whether the minerals are occurring as discrete grains of a single composition (e.g. a grain composed entirely of quartz or feldspar) or whether the grains are, in fact, polyminerallic (e.g. either naturally occurring small rock fragments or as fragments

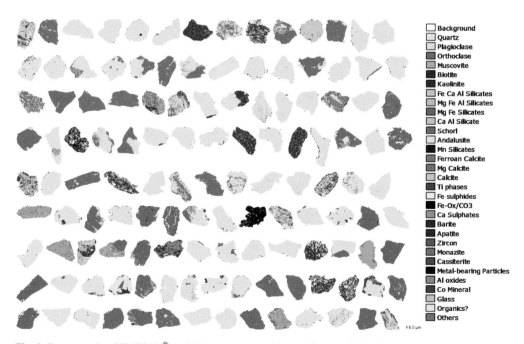

Fig. 4. Representative QEMSCAN® particle images arranged by area for one of the soil samples analysed. Note that the QEMSCAN® image shows that there are both particles composed of individual mineral grains (e.g. quartz, plagioclase, schorl) and composite particles composed of different mineral grains (i.e. rock fragments), such as grains composed of quartz and biotite with a well-defined fabric interpreted as metasedimentary rock fragments and coarse-grained quartz + orthoclase + plagioclase interpreted to be granite. Particles interpreted to be man-made also occur.

of aggregate). It is, however, possible to visualize this with the output of QEMSCAN® false-colour particle images (Fig. 4). It is also possible to quantify the mineral association data for the sample, which reflects the number of transitions between one mineral group and another between adjacent measured pixels (i.e. what is touching what).

The dataset was also reported as 'lithotypes'. A lithotype can either be a grain composed of a single mineral (e.g. a quartz lithotype) or polyminerallic grains can be identified and assigned a rock name based on, for example, the observed mineralogy or grain size (e.g. a granite lithotype). Lithotyping effectively allows the dataset to be re-reported as discrete mineral grains, rock fragments or anthropogenic components. Lithotyping was first developed as an application tool for automated mineralogy in the analysis of oilfield drill cuttings (e.g. Moscariello et al. 2010; Haberlah et al. 2012) but can be equally applicable in any study where the mineral particles present may be polyminerallic. The lithotype groupings are based on the standard geological definitions for rock types based on mineralogy, texture and grain or crystal size. As such, lithotyping is a means of generating automated quantitative petrographic data.

Results

The modal mineralogical data based on the QEMSCAN® analyses are presented first (a) as the overall sample modal mineralogy and then (b) grouped into lithotypes. The data from the concealment site are subdivided into three groupings: (a) soil from the car park area; (b) soil from the track; and (c) leaf litter washing samples collected along the offender approach path to the concealment site. The data from the two items of footwear are then considered. However, no more than 65 mineral grains were

Table 1. *Modal mineralogical data for soil samples collected from the car parking area**

Sample name	OF/37	OF/38	OF/39	OF/40	OF/41	OF/42	OF/43
No. of analysis points	280 157	172 664	131 225	402 341	296 281	511 828	813 423
No. of particles	7447	7024	7345	7353	7710	7583	6492
Quartz	29.49	18.01	27.06	30.86	28.45	32.39	34.11
Plagioclase	15.80	12.30	11.86	12.87	13.77	9.94	10.15
Orthoclase	20.92	17.69	12.00	17.57	17.30	16.44	18.68
Muscovite	11.03	12.98	9.80	10.11	9.32	8.57	9.10
Biotite	8.54	17.75	16.70	14.55	14.03	15.84	16.77
Kaolinite	4.48	4.77	3.50	1.93	2.58	1.35	0.85
FeCaAl silicates	0.29	0.51	0.82	0.44	0.73	0.81	0.31
MgFeAl silicates	4.17	6.90	11.35	4.83	5.27	4.33	3.61
MgFe silicates	0.89	0.77	0.73	0.85	2.16	0.81	0.93
CaAl silicates	0.93	0.87	1.38	0.67	1.22	1.44	0.41
Schorl	1.35	4.81	1.11	2.52	1.37	2.54	1.70
Andalusite	0.07	0.11	0.14	0.04	0.06	0.03	0.08
Mn silicates	0.01	0.03	0.09	0.03	0.04	0.01	0.02
Ferroan calcite	0.12	0.17	0.16	0.21	0.47	0.55	0.17
Mg calcite	0.00	0.00	0.00	0.00	0.01	0.00	0.01
Calcite	0.28	0.13	0.49	0.71	1.58	2.95	1.76
Ti phases	0.43	0.36	0.62	0.36	0.34	0.38	0.29
Fe sulphides	0.01	0.05	0.04	0.01	0.07	0.09	0.04
FeO_x/CO_3	0.65	1.35	1.56	1.03	0.64	0.80	0.62
Ca sulphates	0.03	0.01	0.05	0.02	0.03	0.03	0.01
Barite	0.00	0.01	0.01	0.00	0.00	0.00	0.00
Apatite	0.25	0.16	0.26	0.06	0.12	0.05	0.06
Zircon	0.06	0.10	0.09	0.05	0.09	0.06	0.03
Monazite	0.00	0.01	0.02	0.02	0.02	0.01	0.04
Cassiterite	0.01	0.01	0.00	0.04	0.01	0.14	0.05
Zn, Pb, Cu phases	0.00	0.01	0.00	0.01	0.01	0.01	0.00
Al oxides	0.11	0.08	0.06	0.11	0.08	0.08	0.08
Co mineral	0.00	0.00	0.00	0.00	0.00	0.00	0.00
Glass	0.02	0.02	0.03	0.05	0.19	0.30	0.09
Others	0.03	0.03	0.05	0.04	0.03	0.06	0.03

*Values shown as 0.00 may have the mineral present at an abundance of <0.01%.

identified in any of the washing samples recovered from the clothing samples. The low abundance of mineral grains on the clothing reflects the visual appearance of the clothing, which, other than several mud splash marks on one item, appeared clean, although Ruffell & Sandiford (2011) have suggested an improved method for soil recovery from clothing. The very low numbers of mineral particles located means that these samples are not considered further in this paper. Here, 'major' minerals are those forming >10% of the sample, 'minor' minerals are those between 1 and 10%, and 'trace' are <1%.

Modal mineralogy

The modal mineralogy of the soil samples collected from the car parking area is shown in Table 1. Between 6492 and 7710 individual mineral grains were characterized in the samples analysed. Major minerals present in these samples are quartz, plagioclase and orthoclase, along with major/minor biotite, muscovite and MgFeAl silicates. Other minor phases present are kaolinite and schorl (tourmaline), along with minor/trace MgFe silicates, CaAl silicates, calcite and FeO_x/CO_3. Trace minerals present in some or all of the samples measured are FeCaAl silicates, andalusite, Mn silicates, ferroan calcite, Mg calcite, Ti phases, Fe sulphides, Ca sulphates, barite, apatite, zircon, monazite, cassiterite, Zn, Pb, Cu phases, Al oxides, glass and 'others'.

The modal mineralogical data for the track are presented in Table 2. Eight soil samples were collected along the track between the car parking area and the location where the offender approach path entered the woodland. Within these samples between 6399 and 8485 discrete mineral grains were identified per sample. The samples are dominated by major quartz, plagioclase and orthoclase, and minor muscovite, biotite and MgFeAl silicates, along with minor/trace kaolinite, schorl (tourmaline) and FeO_x/CO_3. A wide range of trace mineral phases also occur in some, or all, of the samples

Table 2. *Modal mineralogical data for soil samples collected from the access track**

Sample name	OF/10	OF/30	OF/31	OF32	OF/33	OF/34	OF/35	OF/36
No. of analysis points	257 450	369 800	190 266	609 214	606 331	318 649	432 906	437 876
No. of particles	8485	6950	6399	6975	6897	6574	7108	6924
Quartz	37.67	39.63	38.16	41.88	42.56	40.01	38.60	32.79
Plagioclase	18.94	20.02	17.30	18.24	16.89	16.72	20.60	19.61
Orthoclase	14.04	21.87	17.55	24.90	22.05	22.98	21.73	23.81
Muscovite	9.42	8.46	10.31	7.59	8.85	7.68	9.22	10.48
Biotite	6.98	3.28	6.43	2.98	3.88	4.63	4.29	4.99
Kaolinite	2.08	1.37	1.71	0.77	0.79	1.29	1.57	2.53
FeCaAl silicates	0.21	0.10	0.16	0.14	0.09	0.11	0.08	0.15
MgFeAl silicates	6.10	1.66	2.81	0.79	1.34	2.11	1.27	1.35
MgFe silicates	0.65	0.20	0.29	0.40	0.22	0.49	0.45	0.56
CaAl silicates	0.53	0.30	0.85	0.24	0.28	0.32	0.37	0.52
Schorl	0.55	1.19	2.48	0.86	2.10	1.65	0.54	1.29
Andalusite	0.03	0.17	0.05	0.13	0.09	0.08	0.17	0.21
Mn silicates	0.08	0.01	0.04	0.00	0.01	0.02	0.01	0.01
Ferroan calcite	0.02	0.03	0.06	0.09	0.03	0.14	0.03	0.06
Mg calcite	0.00	0.00	0.00	0.00	0.00	0.00	0.00	0.00
Calcite	0.09	0.04	0.30	0.10	0.07	0.10	0.09	0.15
Ti phases	0.55	0.32	0.36	0.21	0.31	0.67	0.18	0.60
Fe sulphides	0.01	0.00	0.01	0.00	0.00	0.00	0.00	0.01
FeO_x/CO_3	1.58	0.57	0.87	0.33	0.29	0.54	0.38	0.53
Ca sulphates	0.01	0.00	0.02	0.00	0.00	0.00	0.01	0.02
Barite	0.00	0.00	0.00	0.00	0.00	0.00	0.00	0.00
Apatite	0.20	0.44	0.07	0.23	0.05	0.19	0.23	0.19
Zircon	0.07	0.13	0.07	0.05	0.03	0.15	0.08	0.06
Monazite	0.11	0.03	0.02	0.01	0.03	0.04	0.01	0.03
Cassiterite	0.01	0.14	0.01	0.01	0.00	0.00	0.00	0.01
Zn, Pb, Cu phases	0.00	0.00	0.00	0.00	0.00	0.01	0.00	0.00
Al oxides	0.03	0.02	0.03	0.02	0.01	0.03	0.06	0.03
Co mineral	0.00	0.00	0.00	0.00	0.00	0.00	0.00	0.00
Glass	0.02	0.01	0.02	0.01	0.01	0.02	0.01	0.02
Others	0.01	0.01	0.02	0.01	0.00	0.01	0.01	0.02

*Values shown as 0.00 may have the mineral present at an abundance of <0.01%.

Table 3. *Modal mineralogical data for the washing samples recovered from the leaf litter samples**

Sample name	OF/11	OF/12	OF/13	OF/14	OF/15	OF/16	OF/17	OF/18
No. of analysis points	14 500	4372	1068	3321	2992	2477	386	51 236
No. of particles	644	253	80	170	194	163	31	1619
Quartz	43.84	45.14	43.83	28.33	33.80	40.95	43.42	27.23
Plagioclase	14.71	14.80	5.71	6.94	13.60	13.60	0.77	16.20
Orthoclase	14.29	14.23	8.53	13.00	9.64	13.25	8.88	29.38
Muscovite	13.55	10.13	10.71	25.99	10.96	10.53	11.95	9.44
Biotite	6.60	2.91	10.62	3.22	6.00	4.55	1.29	5.18
Kaolinite	0.95	1.13	0.78	2.71	1.11	3.05	0.00	0.93
FeCaAl silicates	0.11	0.21	1.02	0.77	0.28	0.62	0.00	0.19
MgFeAl silicates	1.56	2.41	1.79	1.31	3.53	3.90	0.86	4.57
MgFe silicates	0.54	0.23	1.79	1.05	0.95	1.51	0.00	0.29
CaAl silicates	0.36	0.87	0.67	0.83	2.84	0.64	0.00	0.56
Schorl	0.36	0.64	0.84	0.00	0.12	1.17	0.00	2.01
Andalusite	0.18	0.00	0.00	0.00	0.00	0.00	0.00	0.02
Mn silicates	0.09	0.09	0.38	1.11	0.26	0.16	0.00	0.02
Ferroan calcite	0.16	0.03	0.00	0.03	0.26	0.00	0.00	0.82
Mg calcite	0.00	0.07	0.00	0.00	0.03	0.00	0.00	0.00
Calcite	0.68	2.91	0.00	1.09	4.85	0.81	30.55	0.16
Ti phases	0.24	1.10	0.27	0.13	0.39	0.89	0.00	0.68
Fe sulphides	0.00	0.00	0.00	0.00	0.00	0.00	0.00	0.01
FeO_x/CO_3	1.41	2.22	13.07	6.72	7.45	3.84	0.81	1.05
Ca sulphates	0.01	0.02	0.00	0.02	0.00	0.00	0.00	0.75
Barite	0.00	0.00	0.00	0.00	0.00	0.00	0.00	0.00
Apatite	0.00	0.00	0.00	5.92	0.00	0.00	0.00	0.01
Zircon	0.11	0.04	0.00	0.10	0.34	0.14	0.90	0.05
Monazite	0.00	0.00	0.00	0.00	0.00	0.00	0.00	0.00
Cassiterite	0.00	0.00	0.00	0.00	0.00	0.00	0.00	0.00
Zn, Pb, Cu phases	0.00	0.00	0.00	0.00	0.12	0.00	0.00	0.00
Al oxides	0.01	0.00	0.00	0.09	0.43	0.16	0.00	0.27
Co mineral	0.00	0.00	0.00	0.00	0.00	0.00	0.00	0.00
Glass	0.21	0.02	0.00	0.03	0.10	0.12	0.00	0.00
Others	0.03	0.81	0.00	0.60	2.95	0.12	0.56	0.18

*Values shown as 0.00 may have the mineral present at an abundance of <0.01%.

including FeCaAl silicates, MgFe silicates, CaAl silicates, andalusite, Mn silicates, ferroan calcite, calcite, Ti phases, Fe sulphides, Ca sulphates, apatite, zircon, monazite, cassiterite, Zn, Pb, Cu phases, Al oxides and glass.

Eight samples of leaf litter were collected along the offender approach path from the track to the concealment site. Along this route there were no exposed soils. The leaf litter samples were washed to recover mineral grains and pollen. The number of mineral grains present in the samples analysed is very variable, ranging between 31 and 1619 discrete mineral grains. In most quantitative mineralogical studies a minimum of 300 grains is usually considered to be required for a data set to be considered to be statistically robust. In this case, only two samples had >300 mineral grains. It is, however, interesting to note that the modal mineralogical data for some of the samples with <300 grains are not dissimilar to the results for the samples with >300 grains. Although all of the data are presented in Table 3, only those data for samples OF/11 and OF/18 are considered further. These two samples comprise major quartz, plagioclase and orthoclase, along with major/minor muscovite and minor biotite, MgFeAl silicates and FeO_x/CO_3. Schorl (tourmaline) and MgFe silicates are present as minor/trace phases in these two samples, along with trace kaolinite, FeCaAl silicates, CaAl silicates, andalusite, Mn silicates, ferroan calcite, calcite, Ti phases, Fe sulphides, Ca sulphates, apatite, zircon, Al oxides, glass and 'others'.

Soil samples were recovered from both the soles and uppers of both the left and right shoe of the two items of footwear. The modal mineralogical data for these eight samples are shown in Table 4. Between 5961 and 6522 mineral grains were identified in the four samples from the soles of the items of footwear, whilst between 1003 and 1810 were recovered from the washing samples recovered from the uppers of the two items of footwear, despite them

Table 4. Modal mineralogical data for the samples recovered from the two items of footwear*

Sample	Sole left shoe A	Uppers left shoe A	Sole right shoe A	Uppers right shoe A	Sole left trainer B	Uppers left trainer B	Sole right trainer B	Upper right trainer B
Sample name	OF/1/1	OF/1/2	OF/2/1	OF/2/2	OF/4/1	OF/4/2	OF5/1	OF/5/2
No. of analysis points	142 670	23 109	221 063	20 786	302 574	39 001	473 132	23 185
No. of particles	6522	1003	6215	1412	6278	1810	5961	1104
Quartz	34.36	21.28	35.15	26.96	36.67	33.45	34.54	27.45
Plagioclase	16.43	22.06	16.28	15.35	19.60	17.06	17.39	10.96
Orthoclase	12.26	12.47	20.18	10.04	20.11	15.03	23.44	8.45
Muscovite	13.06	13.75	13.55	16.15	10.24	13.45	9.27	9.38
Biotite	9.13	14.22	6.24	12.38	5.45	7.86	6.82	11.00
Kaolinite	2.27	2.78	2.18	2.81	2.68	2.93	1.97	1.90
FeCaAl silicates	0.33	0.64	0.20	0.43	0.19	0.24	0.18	0.16
MgFeAl silicates	4.75	3.26	2.10	5.36	1.29	2.61	1.81	2.63
MgFe silicates	0.59	1.90	0.56	2.37	0.60	0.85	0.82	0.93
CaAl silicates	1.51	1.74	0.68	1.61	0.60	0.88	0.30	0.68
Schorl	1.37	0.37	0.49	0.57	1.05	0.53	1.48	0.14
Andalusite	0.17	0.16	0.08	0.03	0.12	0.83	0.44	0.10
Mn silicates	0.01	0.01	0.02	0.00	0.01	0.07	0.00	0.00
Ferroan calcite	0.34	0.35	0.33	0.11	0.06	0.09	0.07	0.11
Mg calcite	0.00	0.07	0.00	0.00	0.00	0.22	0.00	0.30
Calcite	1.02	1.12	0.37	0.18	0.13	0.38	0.41	2.31
Ti phases	0.52	0.82	0.59	0.54	0.29	1.20	0.25	5.17
Fe sulphides	0.01	0.04	0.01	0.15	0.01	0.02	0.03	0.03
FeO$_x$/CO$_3$	1.33	1.62	0.75	3.03	0.49	1.39	0.36	1.11
Ca sulphates	0.04	0.16	0.02	0.08	0.03	0.01	0.01	0.04
Barite	0.00	0.19	0.00	0.00	0.00	0.38	0.00	15.94
Apatite	0.13	0.02	0.08	0.31	0.16	0.19	0.18	0.24
Zircon	0.08	0.35	0.05	0.41	0.09	0.04	0.04	0.34
Monazite	0.01	0.00	0.01	0.00	0.02	0.00	0.02	0.03
Cassiterite	0.07	0.03	0.00	0.02	0.00	0.00	0.08	0.02
Zn, Pb, Cu phases	0.01	0.14	0.00	0.49	0.01	0.03	0.01	0.14
Al oxides	0.13	0.23	0.04	0.23	0.03	0.14	0.04	0.30
Co mineral	0.00	0.00	0.00	0.00	0.00	0.00	0.00	0.00
Glass	0.03	0.09	0.02	0.12	0.05	0.05	0.02	0.08
Others	0.04	0.13	0.02	0.26	0.01	0.04	0.01	0.06

*Values shown as 0.00 may have the mineral present at an abundance of <0.01%.

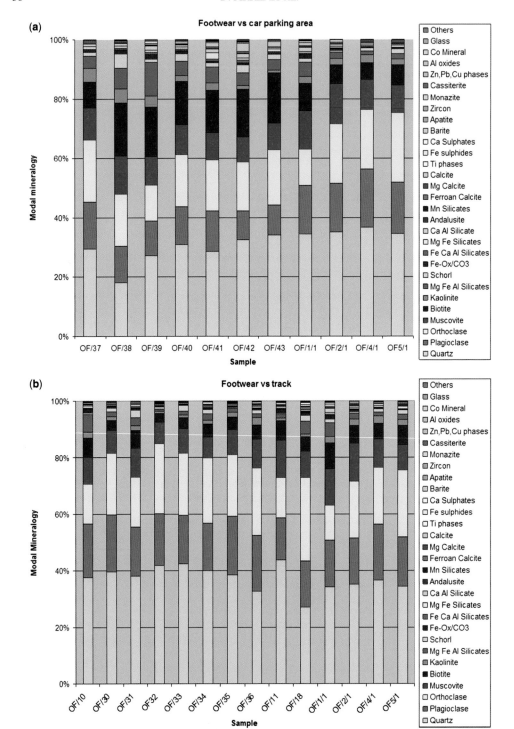

Fig. 5. Comparison of the overall modal mineralogy of: (**a**) the soil samples collected from the car parking area (OF/37, OF/38, OF/39, OF/40, OF/41, OF/42 and OF/43); and (**b**) soil samples from the access track (OF/10, OF/30, OF/31, OF/32, OF/33, OF/34, OF/35 and OF/36) and leaf litter (OF/11 and OF/18) with the mineralogy of the soil samples recovered from the soles of both items of footwear (OF/1/1, OF/2/1, OF/4/1 and OF/5/1).

Table 5. *Definitions for the lithotype groupings used in this study**

Lithotype grouping	Definition
Quartz grains	Area % of quartz >75%
Orthoclase grains	Area % of orthoclase >75%
Tourmaline grains	Area % of schorl >50%
Plagioclase grains	Area % of plagioclase >75%
Muscovite grains	Area % of muscovite >50%
Biotite grains	Area % of biotite >50%
Calcite grains	Area % of calcite >50%
Granite grains	Area % of feldspar >10% and size >40% or area % of feldspar >5% and area % of mica >5% and area % of feldspar + area % of mica >5% and area % of feldspar + area % of mica >75% or area % of quartz >5% and area % of mica >5% and area % of quartz + area % of mica >75% or area % of feldspar >75% or area % of quartz >5% and area % of feldspar >10%
Metasediment	Area % of quartz >20% and area % of mica >10% or area % of quartz >50% or area % of kaolinite >10%
Chlorite	Area % of MgFeAl silicates >5% and area % of quartz >5% and area % of MgFeAl silicates + area % of quartz >50% or area % of MgFeAl silicates >75%
Kaolinite	Area % of kaolinite >50%
Tourmalinized granite	Area % of schorl >5% and area % of quartz >5% and area % of quartz and area % of schorl >50% or area % of schorl and area % of biotite >5% and area % of biotite and area % of schorl >50%
Cassiterite-bearing grains	Area % of cassiterite >0.5%
Al silicate grains	Area % of andalusite >50%
Mafic grains	Area % of MgFe silicate >50% or area % of plagioclase >5% and area % of MgFe silicate >5% and area % of plagioclase and area % of MgFe silicate 0.50%
Glass	Area % of glass >75%
Ti minerals	Area % of Ti phases >50%
Barium phases	Area % of barite >25%
Apatite	Area % of apatite >50%
Zircon/monazite	Area % of zircon >50% or area % of monazite >50%
Fe oxides and sulphides	Area % of $FeOx/CO_3$ >50 or area % of Fe sulphides >50%
Gypsum	Area % of Ca sulphate >50%
Calcareous sandstone	Area % of quartz >4% and area % of calcite >4%
CaAl silicate (glass/slag)	Area % of CaAl silicate >50% or area % of FeCaAl silicate >50%
Metal particles	Area % of Al oxide >50% or area % metal-bearing particles >50% or area 5 Co mineral >50%
Aggregates of grains	Area % silicates >5%

*The expression of the definitions are as used in the iDiscover v 4.2 software.

appearing visually quite clean during examination. The samples are composed of major quartz, plagioclase and orthoclase, along with major/minor muscovite and biotite. Minor phases present in these samples are kaolinite and MgFeAl silicates, along with minor/trace MgFe silicates, CaAl silicates, schorl, calcite, Ti phases and FeO_x/CO_3. Trace phases present in all, or some, of these samples are FeCaAl silicates, andalusite, Mn silicates, ferroan calcite, Mg calcite, Fe sulphides, Cu sulphates, apatite, zircon, monazite, cassiterite, ZnPbCu phases, Al oxides, glass and 'others'. An anomaly that is discussed in greater detail below is that the soil sample recovered from the uppers of the right trainer is markedly different to the other footwear samples in that it also contained 15.94% of spherical Ba–Sb particles. These particles are interpreted to be anthropogenic.

The modal mineralogical data for the soil samples collected from the soles of the two items of footwear are compared with the soil samples from (1) the car parking area (Fig. 5a) and (2) the track and the leaf litter samples (Fig. 5b). Based on the modal mineralogical data the soil samples from footwear item B (samples OF/4/1 and OF/5/1) are directly comparable with the modal mineralogical data for the soil samples collected along the access track, and would support a hypothesis that the soils present on footwear item B were derived through contact with the exposed soils present along the track. However, although the data for the right shoe from footwear item A (sample OF/2/1) are comparable with the soil samples present from the access track, the data for the left shoe of footwear item A (sample OF/1/1) are more consistent with the data from the car parking area. Thus, the mineralogical data could be used to suggest that the individual wearing footwear item A did not enter the track leading to the concealment site. Where more than one individual is involved in a body deposition, it is not an unusual defence proposition that one or other of the individuals might have been present in an area but not directly involved in the victim deposition.

Thus, based on the modal mineralogical data alone, whilst the data are consistent with the hypothesis that footwear item B had contacted soils along the track leading to the body deposition site, the data are not entirely conclusive regarding whether or not footwear item A was only in contact with the soils in the car parking area, or whether the footwear also contacted exposed soils along the track.

Lithotype data

To test this hypothesis further the QEMSCAN® mineralogical data were reprocessed and exported as lithotype groupings. The lithotype groupings used are shown in Table 5, and include: (a) grains that are composed of a single mineral type (e.g. plagioclase grains); (b) naturally occurring rock types (e.g. granite grains or tourmalinized granite); and (c) anthropogenic particle types (e.g. CaAl silicates (glass/slag)). The lithotype groupings are rigorously defined based on mineralogy and other parameters, and, once defined, all data can be output directly within these lithotype 'classes' and compared. Within a QEMSCAN® particle mineral analysis, false-colour maps of all particles are generated and these display the textures of the particles being measured in each sample. Effectively, the lithotype data enable the textural data displayed in the QEMSCAN® particle images to be quantified.

The lithotype data are presented in Tables 6–8, and are shown graphically in Figure 6a, b. In Figure 6a, b only the major/minor lithotype groups are plotted. As can be seen in Figure 6a, when the data for both items of footwear are compared with the data from the car parking area, although there are some similarities with some of the samples, the majority of the samples from the car parking area are distinctly different to the samples from

Table 6. *Lithotype data for the soil samples collected from the car parking area**

	OF/37	OF/38	OF/39	OF/40	OF/41	OF/42	OF/43
Quartz grains	25.66	13.53	18.48	27.57	25.47	28.21	31.38
Orthoclase grains	18.80	13.08	6.77	14.70	14.10	13.55	15.41
Tourmaline grains	0.71	3.88	0.22	1.91	0.80	2.05	1.20
Plagioclase grains	10.71	8.15	6.25	10.30	9.33	6.39	7.13
Muscovite grains	6.14	8.17	3.56	5.41	3.37	3.70	4.22
Biotite grains	4.21	12.78	6.31	11.77	9.32	13.33	18.13
Calcite grains	0.24	0.17	0.31	0.74	1.67	3.14	1.67
Granite grains	19.96	20.52	29.64	16.27	20.03	17.56	13.84
Metasediment	3.54	4.87	4.84	1.42	3.02	1.50	1.03
Chlorite	0.87	1.81	5.13	1.03	1.12	1.75	0.40
Kaolinite	2.69	3.13	2.13	1.14	1.53	0.81	0.52
Tourmalinized granite	0.42	0.52	0.86	1.47	0.48	0.94	1.36
Cassiterite-bearing grains	0.00	0.04	0.02	0.20	0.09	0.89	0.15
Al silicate grains	0.01	0.05	0.11	0.01	0.03	0.00	0.05
Mafic grains	1.55	0.70	0.45	1.09	2.12	0.28	0.92
Glass	0.00	0.00	0.00	0.00	0.08	0.00	0.00
Ti minerals	0.17	0.07	0.19	0.03	0.04	0.05	0.01
Barium phases	0.00	0.01	0.00	0.00	0.00	0.00	0.00
Apatite	0.21	0.12	0.21	0.02	0.10	0.01	0.00
Zircon/monazite	0.02	0.01	0.00	0.02	0.01	0.01	0.02
Fe oxides and sulphides	0.06	0.48	0.23	0.16	0.07	0.14	0.16
Gypsum	0.03	0.01	0.00	0.00	0.00	0.01	0.00
Calcareous sandstone	0.32	0.30	0.84	0.62	0.82	1.21	0.65
CaAl silicate (glass/slag?)	0.03	0.05	0.07	0.09	0.05	0.09	0.14
Metal particles	0.01	0.00	0.00	0.01	0.00	0.00	0.00
Aggregates of grains	0.12	0.34	1.06	0.47	0.40	0.22	0.10
Undifferentiated	3.52	7.22	12.34	3.57	5.93	4.18	1.49

*Values shown as 0.00 may have the lithotype present at an abundance of <0.01%.

the footwear. Thus, if one was testing the hypothesis that the soil on both footwear items was derived through contact with the exposed soils/surfaces in the car parking area, then the lithotype data for both sets of footwear would not support this hypothesis. If, however, the degree of similarity between the lithotype data for the footwear and the soil samples collected from the track leading towards the deposition site are considered (Fig. 6b), then the lithotype data would support a hypothesis that the soil on both items of footwear was consistent with contact between the footwear and the exposed soils along the track leading to the deposition site.

Discussion

Once a sample has been measured using automated mineralogy, the raw data can be output in a variety of different formats. In this study, the data presented allow the overall modal mineralogy and the lithotype data to be compared, which, whilst based on the modal mineralogy, also identifies whether mineral particles are present as discrete mineral grains or as polyminerallic grains (e.g. rock fragments). Thus, the lithotype data combine the mineralogical and textural data in exactly the same way as a geologist defines a rock type. Whilst there should be a broad correspondence between the two datasets, the lithotype data should allow the similarity or otherwise between two samples to be assessed more robustly than that based entirely on the modal mineralogical data. Indeed, this has been what has been done during forensic casework where, in addition to the overall modal mineralogy, the textural data contained within the QEMSCAN® particle images are also considered (see Fig. 4). The advantage of lithotyping is that this allows the mineralogical and textural data to be quantified.

Table 7. *Lithotype data for soil samples collected from the access track and also the leaf litter samples that comprised >300 grains**

	OF/10	OF/11	OF/18	OF/30	OF/31	OF/32	OF/33	OF/34	OF/35	OF/36
Quartz grains	30.43	42.20	22.54	35.90	34.24	40.52	41.56	38.18	36.28	30.24
Orthoclase grains	10.01	10.30	28.31	19.76	13.88	22.75	18.61	21.74	18.96	21.17
Tourmaline grains	0.34	0.12	2.39	0.83	2.33	0.79	1.84	1.34	0.34	0.87
Plagioclase grains	14.39	14.26	12.12	15.71	12.36	14.19	12.11	13.31	16.08	15.45
Muscovite grains	4.29	11.46	5.53	4.68	6.28	5.06	5.59	4.12	5.93	6.16
Biotite grains	1.86	5.40	2.37	1.22	3.16	1.74	2.37	2.68	2.74	3.08
Calcite grains	0.01	0.81	0.59	0.03	0.23	0.14	0.06	0.19	0.08	0.11
Granite grains	27.39	10.41	14.14	17.08	20.72	12.31	15.30	12.35	15.11	15.98
Metasediment	1.50	0.59	0.74	0.58	1.48	0.31	0.41	0.77	0.71	1.13
Chlorite	2.30	0.26	4.63	0.48	0.89	0.35	0.27	0.82	0.86	0.22
Kaolinite	1.24	0.75	0.36	0.60	0.63	0.17	0.27	0.63	0.84	1.91
Tourmalinized granite	0.17	0.10	0.09	0.44	0.08	0.06	0.03	0.60	0.25	0.20
Cassiterite-bearing grains	0.01	0.00	0.00	0.51	0.01	0.01	0.00	0.00	0.00	0.00
Al silicate grains	0.01	0.15	0.00	0.16	0.04	0.09	0.08	0.08	0.07	0.03
Mafic grains	0.65	0.65	0.20	0.16	0.24	0.48	0.23	0.55	0.69	0.89
Glass	0.00	0.18	0.00	0.00	0.00	0.00	0.00	0.00	0.00	0.00
Ti minerals	0.07	0.00	0.30	0.06	0.11	0.08	0.15	0.34	0.07	0.29
Barium phases	0.00	0.00	0.00	0.00	0.00	0.00	0.00	0.00	0.00	0.00
Apatite	0.18	0.00	0.00	0.37	0.06	0.19	0.04	0.16	0.19	0.13
Zircon/monazite	0.10	0.00	0.00	0.04	0.02	0.02	0.02	0.09	0.03	0.02
Fe oxides and sulphides	0.30	0.71	0.14	0.05	0.03	0.05	0.02	0.05	0.03	0.05
Gypsum	0.00	0.00	0.63	0.00	0.03	0.00	0.00	0.00	0.00	0.02
Calcareous sandstone	0.11	0.06	0.14	0.07	0.24	0.09	0.06	0.08	0.06	0.11
CaAl silicate (glass/slag?)	0.01	0.06	0.58	0.05	0.02	0.01	0.01	0.01	0.01	0.02
Metal particles	0.00	0.00	0.00	0.00	0.00	0.00	0.00	0.00	0.00	0.01
Aggregates of grains	0.37	0.00	0.25	0.02	0.08	0.02	0.07	0.02	0.01	0.06
Undifferentiated	4.26	1.51	3.95	1.20	2.85	0.56	0.89	1.87	0.65	1.86

*Values shown as 0.00 may have the lithotype present at an abundance of <0.01%.

Table 8. *Lithotype data for the samples recovered from the two items of footwear**

	OF/1/1	OF/1/2	OF/2/1	OF/2/2	OF/4/1	OF/4/2	OF/5/1	OF/5/2
Quartz grains	30.49	18.03	32.96	21.92	34.96	30.98	33.16	27.96
Orthoclase grains	9.40	9.10	16.98	4.71	17.74	10.55	22.41	5.82
Tourmaline grains	0.97	0.00	0.19	0.30	0.76	0.18	0.88	0.00
Plagioclase grains	12.28	15.88	12.52	10.12	17.68	10.42	13.58	9.07
Muscovite grains	7.91	8.01	10.31	9.84	6.44	10.06	6.38	5.94
Biotite grains	4.70	12.17	3.89	8.02	4.22	4.99	6.06	9.86
Calcite grains	1.19	1.08	0.57	0.21	0.14	0.52	0.43	2.98
Granite grains	19.65	17.83	15.37	25.44	11.87	20.51	10.77	13.64
Metasediment	2.77	3.63	1.47	3.47	1.35	1.96	0.81	1.08
Chlorite	1.41	1.05	0.45	0.96	0.24	1.39	0.59	1.12
Kaolinite	0.95	0.86	1.18	1.65	1.92	1.88	1.19	1.45
Tourmalinized granite	0.21	0.13	0.26	0.24	0.05	0.24	0.60	0.00
Cassiterite-bearing grains	0.00	0.20	0.00	0.22	0.00	0.00	0.31	0.20
Al silicate grains	0.06	0.12	0.04	0.00	0.07	0.74	0.47	0.00
Mafic grains	0.60	4.57	1.19	2.25	0.57	0.86	0.88	1.17
Glass	0.00	0.06	0.00	0.06	0.01	0.00	0.00	0.01
Ti Minerals	0.19	0.50	0.13	0.25	0.10	0.65	0.07	0.81
Barium phases	0.00	0.14	0.00	0.00	0.00	0.23	0.00	14.66
Apatite	0.10	0.00	0.05	0.28	0.11	0.18	0.07	0.23
Zircon/monazite	0.02	0.06	0.00	0.09	0.07	0.00	0.02	0.05
Fe oxides and sulphides	0.18	0.45	0.09	0.43	0.12	0.27	0.02	0.33
Gypsum	0.00	0.05	0.01	0.13	0.04	0.06	0.01	0.00
Calcareous sandstone	0.68	0.73	0.26	0.17	0.08	0.37	0.08	0.19
CaAl silicate (glass/slag?)	0.03	0.03	0.06	0.23	0.35	0.08	0.05	0.19
Metal particles	0.00	0.12	0.01	0.38	0.00	0.11	0.00	0.25
Aggregates of grains	0.38	0.35	0.01	0.61	0.01	0.00	0.00	0.00
Undifferentiated	5.85	4.85	2.00	8.03	1.11	2.78	1.18	2.97

*Values shown as 0.00 may have the lithotype present at an abundance of <0.01%.

In the case study presented in this paper, the modal mineralogical data alone supported the hypothesis that footwear item B had come into contact with the exposed soils along the track leading to the deposition site. However, the modal mineralogy data alone for footwear item A indicated that one of the pair of shoes had contacted the exposed soils, yet the data for the other item were ambiguous. In court, the data could be used to argue that the footwear could not be proven to have contacted the exposed soils, and that the data may have suggested that the footwear was only in contact with the exposed surfaces in the car park area. However, the lithotype data not only indicate that both items of footwear have contacted the exposed soils from the track leading to the deposition site but also indicate that the soils were not derived from contact with the exposed surfaces in the car park area. Thus, if this had been a real forensic case, then the lithotyping data would have been crucial in terms of testing alternative scenarios put forward as a defence.

Traditional mineralogical techniques such as polarizing light microscopy provide mineralogical data in a textural context. This is, however, difficult to quantify, and, more significantly, from a forensic context, it is rare to have a sample of sufficient size to allow optical microscopy. Automated mineralogy with the ability to 'lithotype' the data set allows mineralogy and texture to be quantified. This is not possible based on other commonly used mineral analysis methods such as XRD. Similarly, whilst manual SEM of either sectioned and polished mineral grains, or the analysis of grain surface textures, can provide valuable data from a forensic context, it is difficult to rigorously quantify data from manual SEM.

Potential limitations to the application of lithotyping are that whilst data collection during automated mineralogy is operator independent, the processing of the data is not (Pirrie & Rollinson 2011). With lithotyping, the individual processing the raw data defines the mineralogical groupings used, as shown for this study in Table 5. Thus, the groups defined will, in part, be based on the experience and skill of the individual processing the data, and would need to be adapted on a case-by-case basis. From a forensic viewpoint, a dataset in which the lithotype groupings are defined too widely could potentially provide an inaccurate

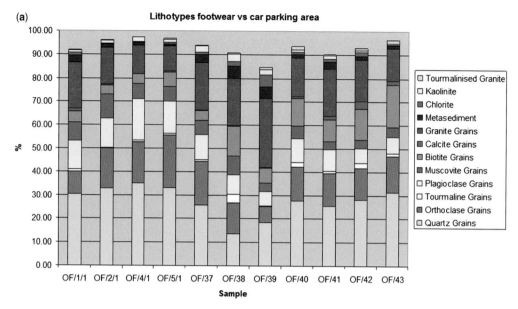

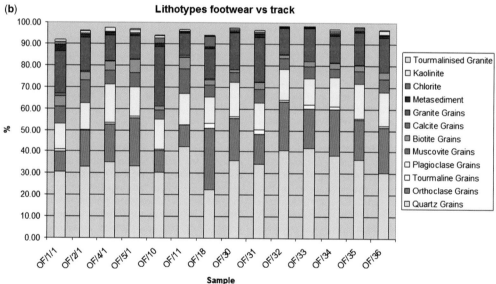

Fig. 6. Comparison of the relative abundance of the major/minor lithotypes present in: (**a**) the soil samples collected from the car parking area (OF/37, OF/38, OF/39, OF/40, OF/41, OF/42 and OF/43); and (**b**) soil samples from the access track (OF/10, OF/30, OF/31, OF/32, OF/33, OF/34, OF/35 and OF/36) and leaf litter (OF/11 and OF/18) with the soil samples recovered from the soles of both items of footwear (OF/1/1, OF/2/1, OF/4/1 and OF/5/1).

degree of similarity between different samples. In contrast, a lithotyping study in which the groupings were very tightly defined would provide a robust test of the level of similarity between different samples. If lithotyping were to become widely adopted in criminal soil forensics, clearly defined protocols for the lithotype groupings could be defined and adhered to. However, no two studies are identical as a result of the wide variability of soils and, as such, the lithotype groupings would still have to be adapted on a case-by-case basis.

Finally, an additional advantage of an automated mineralogy approach to soil forensics, whether with the reporting of the results as overall modal

Fig. 7. QEMSCAN® particle images for a spherical Ba–Sb phase present on the uppers of footwear item (b). The source of these particles is unknown.

mineralogy or as lithotypes, is that all particle types present are characterized. In this study, a surprising result was that in the sample recovered by washing the uppers of the right white trainer (footwear item B), 15.94% of the particles present reported to the mineral category 'barite'. However, these particles, which are characteristically spherical in shape (Fig. 7), are a Ba–Sb phase and are interpreted as anthropogenic particulates. Similar particle types were not recovered in the soil samples from the car parking area, access lane or concealment site, although two spherical Ba–Sb particles were present in the washing sample from the uppers of the left white trainer (footwear item B). Whilst Sb–Ba particles are considered to be consistent with particles formed during the discharge of a firearm (ASTM 2010), spherical particles of Ba–Sb are known from, for example, the use of fireworks (Grima *et al.* 2012) and are also known to occur from motor vehicle disk brakes, although in this case the particles are not typically spherical in shape (Garofano *et al.* 1999). The source of the Ba–Sb particles in this study is not known. Although the shoes were pre-worn prior to this field trial, the owner denies: (a) discharging a firearm whilst wearing the footwear or being close to a firearm when being discharged; (b) being in close proximity to fireworks whilst wearing the footwear; or (c) having come into contact with motor vehicle brake shoes. It is, however, interesting to note that the area used for the concealment of the item is occasionally used by individuals for shooting, and also a sculptor who works using metal frequently stores items in, and around, the car parking area. It should, however, be noted that no spherical Ba–Sb particles were recovered in any of the soil samples analysed and there were no Ba–Sb particles present on the other item of footwear.

Conclusions

One of the key reasons why soil is an important and useful class of trace evidence is because of its complexity and variability (e.g. Fitzpatrick *et al.* 2009). Its complexity also brings with it an inherent issue in that, whilst ideally in a forensic context one would wish to examine several different parameters within the soil, there is no standard protocol as to what aspects of a soil should be examined. Whilst numerous papers have advocated the use of grain size, colour and bulk chemistry as a means of discriminating between different soils, all of these attributes have inherent problems when viewed from a forensic context. For instance, it is widely recognized that there may be differential transfer and retention of different particle sizes on, for example, different types of clothing, such that a direct comparison between the grain size of soil on clothing v. the grain size of a soil at a scene may provide a false negative response (although see Morgan & Bull 2007). Colour can at best be used to exclude the possibility of a similar origin of two questioned soils of different colours but a similarity in colour does not provide sufficient evidence to indicate an association between two soils. Bulk chemistry is not only destructive but may also provide either false positive or false negative interpretations. Characterizing the mineralogy of soils is the most important way in our opinion to discriminate between samples based on the inorganic components present. If through automated mineralogy this can be combined with particle texture and mineralogical associations, then it allows even the typically very small samples encountered in forensic casework to be fully characterized geologically. This allows a greater degree of precision when evaluating whether or not an unknown sample could have been derived through contact with soils at a known location. However, lithotyping is potentially also very significant as a means of providing critical provenance and geolocation data where a soil sample is being examined to try to identify the potential location(s) from where that soil may have been derived. The data in this study are consistent with the dominant soils in the area being derived from a granitic bedrock geology. In particular, the underlying geology is composed of the coarse-grained granite comprising alkali feldspar, plagioclase, quartz, biotite white mica and tourmaline (Leveridge et al. 1990). The reason why the soil samples present in the car parking area were distinctive when compared with the track is because this surface had been dressed with metasedimentary rock fragments, predominantly slates, as a hardcore, and this is reflected in the lithotype data set. It is this introduced material that made the car parking area distinctive from the track and the concealment site.

We are grateful for the ongoing support and interest in our research from Dr A. Butcher (FEI Company). P. Frost and G. Rollinson (CSM) both kindly donated clothing and footwear for analysis, and were willing volunteers during the clandestine disposal of the object in the woodland (without being caught). B. Snook is thanked for assistance in preparing the figures and H. Campbell commented on early drafts of the manuscript. We are grateful for the comments of the referees.

References

ASTM 2010. *E-1588-10. Standard Guide for Gunshot Residue Analysis by Scanning Electron Microscopy/ Energy Dispersive X-ray Spectrometry*. ASTM International, West Conshohocken, PA.

BERGLIESEN, E. T. 2013. X-ray diffraction and field portable X-ray fluorescence analysis and screening of soils: project design. *In*: PIRRIE, D., RUFFELL, A. & DAWSON, L. A. (eds) *Environmental and Criminal Geoforensics*. Geological Society, London, Special Publications, **384**, first published online July 16, 2013, http://dx.doi.org/10.1144/SP384.14

BOWEN, A. M. & CAVEN, E. A. 2013. Forensic provenance investigations of soil and sediment samples. *In*: PIRRIE, D., RUFFELL, A. & DAWSON, L. A. (eds) *Environmental and Criminal Geoforensics*. Geological Society, London, Special Publications, **384**, first published online May 16, 2013, http://dx.doi.org/10.1144/SP384.4

BROWN, A. G. 2006. The use of forensic botany and geology in war crimes investigations in NE Bosnia. *Forensic Science International*, **163**, 204–210.

BULL, P. A. & MORGAN, R. M. 2006. Sediment fingerprints: a forensic technique using quartz sand grains. *Science and Justice*, **46**, 64–81.

BULL, P. A., MORGAN, R. M. & FREUDIGER-BONZON, J. 2008. A critique of the present use of some geochemical techniques in geoforensic analysis. *Forensic Science International*, **178**, e35–e40.

DAWSON, L. A. & HILLIER, S. 2010. Measurement of soil characteristics for forensic applications. *Surface and Interface Analysis*, **42**, 363–377.

ECKARDT, R., KRUPICKA, E. & HOFMEISTER, W. 2012. Validation of powder X-ray diffraction following EN ISO/IEC 17025. *Journal of Forensic Science*, **57**, 722–737.

FITZPATRICK, R. W., RAVEN, M. D. & FORRESTER, S. T. 2009. A systematic approach to soil forensics: criminal case studies involving transference from crime scene to forensic evidence. *In*: RITZ, K., DAWSON, L. & MILLER, D. (eds) *Criminal and Environmental Soil Forensics*. Springer, Heidelberg, 105–127.

GAROFANO, L., CAPRA, M., FERRARI, F., BIZZARO, G. P., DITULLIO, D., DELL'OLIO, M. & GHITTI, A. 1999. Gunshot residue further studies on particles of environmental and occupational origin. *Forensic Science International*, **103**, 1–21.

GRIMA, M., BUTLER, M., HANSON, R. & MOHAMEDEN, A. 2012. Firework displays as sources of particles similar to gunshot residue. *Science and Justice*, **52**, 49–57.

GUEDES, A., RIBEIRO, H., VALENTIM, B. & NORONHA, F. 2009. Quantitative colour analysis of beach and dune sediments for forensic applications: a Portuguese example. *Forensic Science International*, **190**, 42–51.

GUEDES, A., RIBEIRO, H., VALENTIM, B., RODRIGUES, A., SANT'OVAIA, H., ABREU, I. & NORONHA, F. 2011. Characterisation of soils from the Algarve region (Portugal): a multidisciplinary approach for forensic applications. *Science and Justice*, **51**, 77–82.

HABERLAH, D., BOTHA, P. W. S. K., DOBRZINSKI, N., BUTCHER, A. R. & KALDI, J. G. 2012. Petrological reconstruction of the subsurface based on PDC drill cuttings: an advanced rock typing approach. *In*: BROEKMANS, M. A. T. M. (ed.) *Proceedings of the 10th International Congress for Applied Mineralogy (ICAM)*. Springer, Heidelberg, 275–283.

LEVERIDGE, B. E., HOLDER, M. T. & GOODE, A. J. J. 1990. *Geology of the Country Around Falmouth*. Geological Memoirs & Sheet Explanations (England & Wales), **352**. British Geological Survey, Keyworth, Nottingham.

MCVICAR, M. J. & GRAVES, W. J. 1997. The forensic comparison of soils by automated scanning electron microscopy. *Journal of the Canadian Society of Forensic Scientists*, **30**, 241–261.

MORGAN, R. M. & BULL, P. A. 2007. The use of grain size distribution analysis of sediments and soils in forensic enquiry. *Science and Justice*, **47**, 125–135.

MOSCARIELLO, A., BURNS, S., WALKER, D., POWER, M. R., KITSON, W. S. & SLIWINSKI, J. 2010. Sequence stratigraphy and reservoir characterisation of barren fluvial sequences using rock-typing analyses of core and cuttings. *AAPG GEO 2010 Middle East Geoscience Conference & Exhibition. Innovative Geoscience Solutions – Meeting Hydrocarbon Demand in Changing Times*, 7–10 March, Manama, Bahrain. http://www.searchanddiscovery.com/abstracts/html/2010/geo_bahrain/abstracts/Moscariello.html

PIRRIE, D. & ROLLINSON, G. K. 2011. Unlocking the applications of automated mineral analysis. *Geology Today*, **27**, 226–235.

PIRRIE, D., BUTCHER, A. R., POWER, M. R., GOTTLIEB, P. & MILLER, G. L. 2004. Rapid quantitative mineral and phase analysis using automated scanning electron microscopy (QemSCAN); potential applications in forensic geoscience. *In*: PYE, K. & CROFT, D. J. (eds) *Forensic Geoscience: Principles, Techniques and Applications*. Geological Society, London, Special Publications, **232**, 123–136.

PIRRIE, D., POWER, M. R., ROLLINSON, G. K., WILTSHIRE, P. E. J., NEWBERRY, J. & CAMPBELL, H. E. 2009. Automated SEM-EDS (QEMSCAN®) mineral analysis in forensic soil investigations: testing instrumental reproducibility. *In*: RITZ, K., DAWSON, L. & MILLER, D. (eds) *Criminal and Environmental Soil Forensics*. Springer, Heidelberg, 411–430.

RUFFELL, A. & WILTSHIRE, P. 2004. Conjunctive use of quantitative and qualitative X-ray diffraction analysis of soils and rocks for forensic analysis. *Forensic Science International*, **145**, 13–23.

RUFFELL, A. & MCKINLEY, J. 2008. *Geoforensics*. Wiley, New York.

RUFFELL, A. & SANDIFORD, A. 2011. Maximising trace soil evidence: An improved recovery method developed during investigation of a $26 million bank robbery. *Forensic Science International*, **209**, e1–e7.

The potential application of magnetic susceptibility as a technique for soil forensic examinations

ALEXANDRA GUEDES[1]*, RAYMOND C. MURRAY[2], HELENA RIBEIRO[1],
HELENA SANT'OVAIA[1], BRUNO VALENTIM[1], ANDREIA RODRIGUES[1],
SARA LEAL[1] & FERNANDO NORONHA[1]

[1]Centro de Geologia e Departamento de Geociências, Ambiente e Ordenamento
do Território, Faculdade de Ciências, Universidade do Porto,
Rua do Campo Alegre, 687, 4169-007 Porto, Portugal

[2]University of Montana, Missoula, MT, USA

*Corresponding author (e-mail: aguedes@fc.up.pt)

Abstract: Magnetic susceptibility measurement methodology, its reproducibility and accuracy of analysis were tested on soil samples to evaluate its potential use in forensic applications. It was observed that magnetic susceptibility can enable discrimination between soil samples, measured values are reproducible over time and the analysis can be carried out on small samples. However, in order for this method to be used in a forensic soil investigation, it is important to always adopt the same analytical protocol during analysis.

The evidential value of soil lies in the diversity of this material, making the power to discriminate among soils equally large (Murray & Tedrow 1992; Murray 2004). An important issue in a geoforensic investigation is the demonstration of whether or not variations in the soil properties result in dissimilarity between samples, and a large number of analytical methods have been applied to characterize soils, to test whether or not sites can be excluded. Magnetic parameters such as magnetic susceptibility are important in characterizing materials, and detectable quantities of magnetic minerals are almost always found in soils. The magnetic susceptibility of a soil is the sum of all of the contributions from the minerals present, and varies owing to differences in the concentration and composition of those minerals. In a forensic investigation, it is important to not only obtain samples that are representative and comparable, but also develop standardized protocols and frameworks of best practice to better evaluate soil properties in a representative manner.

Several experiments on the magnetic susceptibility of soils have been conducted recently, mostly related to pollution monitoring and mapping. In some of the works, the accuracy and precision of this analytical method on soils have been evaluated (Schibler et al. 2002; Boyko et al. 2004; Maier et al. 2006; D'Emilio et al. 2007); however, these reported the use of magnetic susceptibility probes and in situ analysis.

Although there are only a few reported applications of magnetic susceptibility in forensic investigations (e.g. Chen et al. 2009), the importance of the use of this technique for forensic applications has been reported in work relating to forensic archaeology, detecting clandestine graves and forensic searches for buried targets. Most of the research and applications deal with the survey application of portable magnetometers, and the advantages and limitations of the technique are reported (Fenning & Donnelly 2004; Hunter & Cox 2005; Bergslien 2012; Pringle et al. 2012).

This study is based on research and experimentation based on soil samples for forensic applications. The magnetic susceptibility measurement methodology, reproducibility and accuracy in analysis were tested on soil samples collected from different locations. The magnetic susceptibility measurement of soils recovered from a crime scene, at room temperature, is non-destructive; therefore, the same material is available for further analysis with any other technique. Additionally, it does not require sample preparation, and can be used as a simple and rapid method that is operable with small sample sizes; thus, it is an advantageous method for the study of soils being used as trace evidence in forensic investigations.

Low-field magnetic susceptibility

Low-field magnetic susceptibility is a measure of the magnetic response of a material to an external magnetic field. The volume susceptibility k, measured in dimensionless units, is defined as the ratio of the material magnetization J (per unit volume) to the weak external magnetic field:

$$H : J = kH$$

Alternatively, the specific or mass susceptibility χ, measured in units of $m^3 \, kg^{-1}$, is defined as the ratio of the material magnetization J (per unit mass) to the weak external magnetic field:

$$H : J = \chi H$$

On the atomic scale, magnetism arises from the uncompensated spin moment of the outermost electrons orbiting around a nucleus of iron. For the more common minerals, the spin moments of electrons cancel each other out. In other situations, the uncompensated spins may line up and their individual magnetic moments can be added to obtain a macroscopic magnetism. The susceptibility of each mineral depends on their behaviour and, therefore, it is important to distinguish diamagnetism, paramagnetism and ferromagnetism.

Diamagnetism

The diamagnetic response to the application of a magnetic field, H, is the acquisition of a small induced magnetization, J, opposite to the applied field. The magnetization depends linearly on the applied field and reduces to zero on removal of the field. Application of the magnetic field alters the orbital motion of electrons to produce the small magnetization antiparallel to the applied magnetic field. Magnetic susceptibility, χ, for a diamagnetic material is negative and independent of temperature.

Paramagnetism

Paramagnetic solids contain atoms with atomic magnetic moments (but no interaction between adjacent atomic moments) and acquire induced magnetization, J, parallel to the applied field, H. The magnetization is linearly dependent on the magnetic field. As with diamagnetic materials, magnetization reduces to zero when the magnetizing field is removed.

Ferromagnetism

Ferromagnetic solids have atoms with magnetic moments, but unlike the paramagnetic case, adjacent atomic moments interact strongly. The effect of interaction is to produce magnetizations in ferromagnetic solids that can be orders of magnitude larger than for paramagnetic solids in the same magnetizing field. For a given ferromagnetic material and temperature there is a maximum magnetization referred to as saturation magnetization, J_s, and increasing the magnetic field, H, beyond the level needed to reach J_s will not result in increased magnetization. The main property of ferromagnetic solids is their ability to record the direction of an applied magnetic field.

The low-field magnetic susceptibility, which can be rapidly measured, can be used as a proxy indication of the magnetic grain concentration, and is related to the size of these grains. The main minerals found in soils, such as carbonates, feldspars and quartz, are diamagnetic. Among the Fe-containing minerals, biotite, tourmaline, siderite and pyrite are paramagnetic, whereas magnetite, hematite and pyrrhotite are ferromagnetic. Magnetic susceptibility measurements in soils are sensitive if paramagnetic and ferromagnetic minerals are present. These minerals can be inherited or newly formed by weathering and breakdown of parent rocks during soil formation, which is affected by bacterial activity, surface processes and anthropogenic activities (Dekkers 1997). The anthropogenic contribution to magnetic parameters can also be an important factor in soils. Anthropogenic particles originating during high-temperature combustion of fossil fuels, dusts and fly ashes from various industries and from vehicle emissions create a significant enhancement of soil magnetic susceptibility (Flanders 1994).

Material and methods

Sample collection and handling

Seventeen soil samples were collected along several Portuguese sites (Fig. 1). Beach and dune sediments surrounded by different geological settings (limestones, granites and mica schists) and soil samples from different underlying parent rocks (carbonates, granites and mica schists) were selected (Table 1). At each site, samples were manually collected from the surface soil (<5 cm depth; Fig. 2) with a plastic spade (carefully cleaned after each sample), put into a plastic bag and then labelled. After the samples were transported to the laboratory and dried in an oven at 40 °C, they were divided into two sub-samples, one retained as a duplicate and stored in a plastic box kept in a cool place, and the other used for magnetic susceptibility analyses. The different samples were homogenized and five sub-samples removed with a spatula and weighed for magnetic susceptibility assays. Measurements were performed: (a) directly on two dried, unsieved bulk samples, one weighing 1 g and the other weighing 0.5 g; (b) on 1 g of dried, and sieved <150 μm fraction using a <150 μm sieve; (c) on 1 g of dried, sieved <150 μm fraction and ground

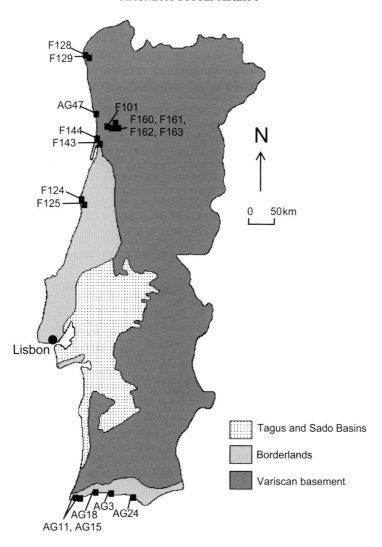

Fig. 1. Simplified geological map of Portugal and sampling locations.

Table 1. *Studied samples, type of material, their location in Portugal, and underlying geology*

Samples	Material	Location	Lithology of country rocks
AG3, AG11, AG15, AG18 and AG24	Mineral soil	Algarve	Limestones and carbonate rocks
F101, F162 and F163	Mineral soil	Gondomar	Mica schists
F160	Mineral soil	Gondomar	Two-mica granites
F161	Organic soil	Gondomar	Two-mica granites
AG47	Beach sand	Azurara	Biotite granites
F128	Beach sand	Afife	Two-mica granites
F144	Beach sand	Espinho	Mica schist
F125	Beach sand	Quiaios	Limestones
F124	Dune sand	Quiaios	Limestones
F129	Dune sand	Afife	Two-mica granites
F143	Dune sand	Espinho	Mica schists

Fig. 2. (a) Photograph of the dune area surrounding the location where the sample F124 was collected; (b) photograph showing the collection of surface sand for forensic research.

to around 4 μm – grinding was performed using an agate pestle; and (d) on 1 g of unsieved bulk sample inside a transparent plastic bag.

Magnetic susceptibility analysis

The magnetic susceptibility was measured by applying an external magnetic field of 300 A m^{-1} to the sample, using a Kappabridge model KLY-4S Agico balance equipped with the Sumean software. Before measurement the equipment was calibrated, and it was regularly calibrated throughout the measurement period owing to the repeated nature of the measurements carried out.

The accuracy of the Kappabridge was checked using a calibration standard with a bulk susceptibility of 136.7 × 10^{-3} SI provided by Agico. Additionally, the diamagnetism of the (plastic) sample holder was removed. The bulk susceptibility of the empty holder was automatically measured three times and its mean value and standard error of the average calculated. If the measurements are inconsistent, for example, if the holder bulk susceptibility does not lie within the interval between -20×10^{-6} SI and $+5 \times 0^{-6}$ SI or the standard error is $>0.1 \times 10^{-6}$ SI, then the equipment indicates that the procedure should be repeated. Taking this procedure into account, the bulk susceptibility of the empty holder was measured and the value was automatically corrected on the sample measurement. Measures were performed on each sample after homogenization and the magnetic susceptibility was calculated. Magnetic susceptibility was calculated in m^3 kg^{-1}.

In order to evaluate the magnetic susceptibility utility for forensic purposes, tests were performed to establish: (a) if the technique allowed the discrimination between natural geological samples (with similar and different geological and geographical origins); (b) the measurement reproducibility within samples and over time; and (c) the variation within a single sample as a result of sample quantity (1 and 0.5 g), size fraction (dried, unsieved bulk sample; dried sieved <150 μm; dried sieved ground <150 μm fractions) and the presentation method (the sample directly and the sample inside a transparent plastic bag). Descriptive statistical analysis was performed and the different experiments were evaluated using the coefficient of variation (CV%).

Results and discussion

Magnetic susceptibility variation between geological samples

The first test performed aimed to ascertain the potential of this analytical technique in a forensic investigation to differentiate between different samples. Samples from similar and different geological and geographical origins were compared (Table 1). Variations of magnetic susceptibility between samples were observed, even within samples from similar geological and geographical origins. The lower values were obtained on sediment samples (beach and dune sands) and the higher values on soils (Table 2). Even though representative magnetic susceptibility values could not be determined for samples with low magnetic susceptibility presenting high coefficient of variation (CV%; Table 2: values in italics), this feature is none the less important because it can be used as a discriminatory characteristic.

In general the magnetic susceptibility variation within samples was low, indicating good instrument reproducibility (Table 2), although variability increased in samples with low magnetic susceptibilities. The different magnetic susceptibility values obtained indicate that magnetic susceptibility is useful for sample discrimination, and as a soil property to be used in a forensic context.

MS variation owing to sample quantity

An important point in any methodological protocol developed for forensic applications is the amount of sample needed for analysis. In a crime scene investigation, it is not usually difficult to obtain relatively large control samples. However, evidential questioned soil samples obtained from spades, shoes or clothes are typically very small. Therefore, it is essential for magnetic susceptibility to be measured using small samples. The mean, standard deviation and coefficient of variation values for magnetic susceptibility measured on 1 g and 0.5 g from different soil sample types are summarized in Table 2. Apart from the samples with weaker magnetic susceptibilities, variations within sample were low for both the 0.5 and 1 g samples, indicating a good reproducibility (Table 2: CV% overall). However, samples AG3, AG15, AG18, AG24, F129 and F163 presented quite different average magnetic susceptibility values when measured on 1 and 0.5 g samples (high overall variability). In samples AG15, AG18 and AG24, the magnetic susceptibility values measured in 0.5 g of the sample were approximately two-fold higher, and in sample F163 three-fold higher than those measured using a 1 g sample.

The magnetic susceptibility measurements performed in the Kappabridge are obtained using a magnetic field acting along a direction in which the sample is placed. If the sample is homogeneous in terms of the presence of diamagnetic, paramagnetic and ferromagnetic minerals, then the measurements of the same sample show small variability. However, if the sample is heterogeneous, owing to the type of sediment and magnetic mineralogy, the magnetic susceptibility of the sample may be different according to the minerals that are magnetized along the magnetic field. In this case, the homogenization of the sample should be carried out manually, and the magnetic susceptibility obtained in the samples of 0.5 g corresponds to a more accurate value because it reflects the magnetization of the whole sample along the field.

MS variation between samples over time

Testing the precision of any equipment is essential in order to ascertain its potential application in a forensic context. Consequently the magnetic susceptibility was measured repeatedly on the same samples over time. For samples with magnetic properties, measurement reproducibility was observed over time, in both the 1 and 0.5 g samples (Table 2).

MS variation between sample size fractions

A key point in forensic soil investigation is the use of representative and comparable samples. In this context, various authors have tested different soil particle size ranges in the evaluation of different instrumental methods for forensic applications, such as spectrophotometers for colour analysis (Croft & Pye 2004; Guedes et al. 2009), laser diffraction for particle size distribution (Blott et al. 2004) and ICP-MS for geochemical analysis (Pye & Blott 2004). Consequently the measurement of the magnetic susceptibility of different size particle ranges from soil samples was carried out in order

Table 2. Variation in magnetic susceptibility values ($\times 10^{-8}\ m\ kg^{-1}$) from 17 sample types

	Measures 1 g (n = 60)					Measures 0.5 g (n = 60)					Overall
	t_1	t_2	t_3	Total		t_1	t_2	t_3	Total		
		$x^* \pm H^\dagger$		$x \pm H$	CV%[‡]		$x \pm H$		$x \pm H$	CV%	CV%
AG3	20.2 ± 10.9	11.5 ± 5.9	4.9 ± 13.4	12.2 ± 12.1	99.4	1.2 ± 1.6	2.1 ± 0.8	2.4 ± 1.6	1.9 ± 1.4	76.2	142.4
AG11	307.1 ± 0.7	307.6 ± 0.9	314.9 ± 0.8	309.9 ± 3.7	1.2	305.8 ± 0.8	303.9 ± 0.8	310.3 ± 1.2	306.7 ± 2.9	0.9	1.2
AG15	360.3 ± 1.1	358.2 ± 1.6	359.6 ± 0.9	359.4 ± 1.5	0.4	610.8 ± 6.3	603.1 ± 4.9	614.4 ± 6	609.4 ± 7.4	1.2	25.9
AG18	16.1 ± 0.6	18.5 ± 0.9	15.7 ± 0.6	16.7 ± 1.4	8.5	37.5 ± 2.3	34.5 ± 3.4	36.7 ± 2.4	36.2 ± 2.9	8.3	37.9
AG24	817.9 ± 5.1	813.6 ± 5.8	819.9 ± 5.9	817.2 ± 6.1	0.8	1191.1 ± 8.7	1202.7 ± 10.7	1199.5 ± 13.1	1197.7 ± 11.9	0.9	18.9
AG47	131.8 ± 1.5	131.6 ± 1.2	129.7 ± 1.7	131.0 ± 1.7	1.3	155.2 ± 1.4	150.8 ± 1.6	152.9 ± 2.1	153.0 ± 2.5	1.6	7.9
F101	114.8 ± 0.6	113.7 ± 0.4	113.7 ± 0.6	114.1 ± 0.7	0.6	124.8 ± 3.2	119.4 ± 2.7	122.6 ± 2.9	122.2 ± 3.7	3.0	4.1
F124	1.0 ± 4.2	−0.2 ± 3.0	1.4 ± 0.4	0.7 ± 3.0	414.2	−2.4 ± 0.5	0.2 ± 3.6	2.7 ± 1.3	0.2 ± 3.0	1522.6	651.3
F125	0.9 ± 3.9	−0.4 ± 0.4	−0.3 ± 1.8	0.1 ± 2.5	3631.7	1.9 ± 5.7	0.9 ± 0.2	0.7 ± 3.1	1.1 ± 3.7	329.2	534.3
F128	6.2 ± 0.5	6.3 ± 0.4	7.5 ± 0.6	6.7 ± 0.8	11.9	5.7 ± 0.7	6.5 ± 1.2	6.0 ± 1.4	6.1 ± 1.2	19.3	16.3
F129	15.9 ± 0.6	15.3 ± 0.5	16.2 ± 0.9	15.8 ± 0.8	5.0	10.1 ± 1.3	9.4 ± 0.6	8.7 ± 0.5	9.4 ± 1.0	10.8	26.5
F143	15.4 ± 1.6	13.6 ± 0.9	14.5 ± 1.5	14.5 ± 1.5	10.6	14.5 ± 1.0	14.4 ± 1.2	14.9 ± 1.2	14.6 ± 1.1	7.8	9.3
F144	1.3 ± 1.1	−0.4 ± 2.1	−0.8 ± 2.2	0.04 ± 2.1	4687.6	3.4 ± 1.2	1.4 ± 2.9	1.9 ± 1.2	2.2 ± 2.1	94.9	207.7
F160	1.9 ± 0.3	0.9 ± 0.4	1.0 ± 0.6	1.3 ± 0.7	50.8	1.7 ± 1.1	0.07 ± 2.6	2.4 ± 0.8	1.4 ± 1.9	140.2	107.4
F161	62.6 ± 0.9	64.6 ± 2.1	64.6 ± 1.7	63.9 ± 1.8	2.9	66.6 ± 1.5	69.2 ± 0.9	69.8 ± 1.6	68.5 ± 1.9	2.8	4.5
F162	85.3 ± 0.5	84.8 ± 0.4	86.1 ± 0.7	85.4 ± 0.8	0.9	98.6 ± 1.0	94.4 ± 1.0	98.9 ± 5.3	97.3 ± 3.7	3.8	7.2
F163	32.7 ± 0.7	33.3 ± 0.7	33.3 ± 0.7	33.1 ± 0.7	2.2	102.7 ± 0.9	103.4 ± 1.1	104.0 ± 1.7	103.4 ± 1.4	1.3	51.7

Measurements were performed on 1 and 0.5 g dried, unsieved bulk from each sample (tested 20 times) over three consecutive months (t_1, t_2 and t_3).
*Average.
[†]Standard deviation.
[‡]Coefficient of variation; highest values are shown in italic font.

Table 3. MS variation with particle size fractions for six soil types (measures performed on 1 g)

		AG11	AG15	AG18	AG24	AG47	F101
DUBS	Average	319.3	344.4	18.8	785.1	127.5	163.1
$n = 10$	Std dev*	0.2	1.0	0.4	4.6	0.9	2.2
	CV%[†]	0.1	0.3	2.2	0.6	0.7	1.3
DSF < 150 µm,	Average	387.6	559.2	1583.1	1896.9	441.4	118.5
$n = 10$	Std dev	0.3	0.4	13.7	17.2	5.7	0.4
	CV%	0.1	0.1	0.8	0.9	1.4	0.3
DSFG <150 µm,	Average	384.7	569.6	1696.1	1932.0	496.5	120.3
$n = 10$	Std dev	0.9	1.9	4.0	20.0	3.1	0.2
	CV%	0.2	0.3	0.2	1.0	0.6	0.1
Overall (CV%)							
DUBS[‡] v. DSF < 150 µm[§]		9.4	24.8	157.5	44.4	64.0	14.0
DUBS v. DSFG < 150 µm[‖]		9.1	25.8	129.5	47.2	69.4	12.3
DSF < 150 µm v. DSFG < 150 µm		0.4	1.0	3.6	1.3	6.2	0.8

*Standard deviation.
[†]Coefficient of variation.
[‡]DUBS, Dried, unsieved bulk sample.
[§]DSF, Dried, sieved <150 µm fraction.
[‖]DSFG, Dried, sieved < 150 µm fraction, ground.

to ascertain the most appropriate size fraction for analysis.

In Table 3, the magnetic susceptibility values obtained in 1 g samples with different particle size ranges from six soil samples are summarized. The values obtained for the dried, unsieved bulk samples were very different from the values obtained for the other two tested size fractions (Table 3: CV% overall). Lower magnetic susceptibility values were observed when the measurements were performed on bulk samples (Table 3). This can be explained by variations in the level of magnetic susceptibility in soils depending, in part, on the particle size distribution, which in turn can affect the sample mineralogy and therefore the relative abundance of the magnetic minerals present (Schibler et al. 2002).

Variation in within-sample replicate measurements was low (CV% <2.2). This was also observed for variations between the magnetic susceptibility values measured on the dried, sieved <150 µm fraction, and on the dried, sieved and ground <150 µm sub-samples (CV% <1.4 and 1.0, respectively; Table 3). Since grinding the sample is a destructive methodology and there were no major differences in the measured magnetic susceptibility values between the dried and ground samples, it is suggested that grinding the samples is not required and should not be carried out. Similarly it is recommended that the bulk

Table 4. Variation in magnetic susceptibility values measured directly on 1 g of the sample inside a transparent plastic bag and on 1 g of the sample of six different soil sample types

		AG11	AG15	AG18	AG24	AG47	F101
Bag,* $n = 10$	Average	124.8	141.5	8.3	318.4	55.2	65.7
	Std dev[†]	0.3	0.5	0.1	2.3	0.5	3.2
	CV%[‡]	0.3	0.3	1.3	0.7	1.0	4.9
No bag, $n = 10$	Average	319.3	344.4	18.8	785.1	127.5	163.1
	Std dev	0.2	1.0	0.4	4.6	0.9	2.2
	CV%	0.1	0.3	2.2	0.6	0.7	1.3
Overall	Std dev	91.2	95.1	4.9	218.8	33.9	45.7
	CV%	34.6	33.2	31.1	33.6	31.7	33.8

*Sample measured inside a plastic bag.
[†]Standard deviation.
[‡]Coefficient of variation.

MS variation between sample presentation methods

In order to validate a technique for future forensic analysis, it is important to conduct experiments concerning the way samples are prepared to evaluate the most suitable method. In this test the magnetic susceptibility measurements were performed directly on the sample and were compared with analyses of the sample inside a transparent plastic bag (Table 4).

Overall variations between the two methods were observed. It is clear that the transparent plastic bag does interfere with the sample magnetic signal; however, some investigations may require no sample handling or contact, in which case the analysis is able to meet this requirement. In contrast, when it comes to an environmental crime investigation, the measurements should be performed directly on the sample in order to accurately determine the magnetic susceptibility.

Conclusions

Magnetic susceptibility measurement methodology, reproducibility and accuracy in analysis were tested on soil samples to evaluate the application of this approach for forensic investigations. It has been shown that magnetic susceptibility can enable discrimination between different soil samples since variations in magnetic susceptibility between samples are observed, even when samples were collected from areas with a similar geological and geographical setting. However, representative magnetic susceptibility values could not be determined for samples that contained low levels of magnetic minerals, although it should be noted that, in samples with low magnetic susceptibility, it could also be used as a discriminatory characteristic.

Apart from the samples with low magnetic susceptibility values, measurement reproducibility was observed over time, in both 1 and 0.5 g samples. Measurement reproducibility was also observed for the measurements on dried, sieved <150 µm fraction, samples and on dried, sieved <150 µm and ground samples. Transparent plastic sample bags can interfere with the sample magnetic signal and therefore the measurements should be performed, when possible, directly on the sample in order to accurately determine the magnetic susceptibility. However, differences in the measured magnetic susceptibility values show the importance of always adopting the same analytical protocol. Overall soil magnetic susceptibility based on the proposed measurement methodology is appropriate for forensic investigations, although further research on the potential applications of magnetic susceptibility should be carried out before its application to operational forensic case work can be fully approved.

This work has been financially supported by the project PTDC/CTE-GEX/67442/2006 (FCT-Portugal). We are grateful to the two anonymous reviewers and the Editor D. Pirrie for improving this paper.

References

BERGSLIEN, E.. 2012. *An Introduction to Forensic Geoscience*. Wiley-Blackwell, Chichester.

BLOTT, S. J., CROFT, D. J., PYE, K., SAYE, S. E. & WILSON, H. E. 2004. Particle size analysis by laser diffraction. *In*: PYE, K. & CROFT, D. J. (eds) *Forensic Geoscience: Principles, Techniques and Applications*. Geological Society, London, Special Publications, **232**, 63–73.

BOYKO, T., SCHOLGER, R. & STANJEK, H. 2004. Topsoil magnetic susceptibility mapping as a tool for pollution monitoring: repeatability of in situ measurements. *Journal of Applied Geophysics*, **55**, 249–259.

CHEN, M., YU, L., NIU, X. & CHEN, B. 2009. Application of environmental magnetism on crime detection in a highway traffic accident from Yangzhou to Guazhou, Jiangsu Province, China. *Forensic Science International*, **187**, 29–33.

CROFT, D. J. & PYE, K. 2004. Colour theory and evaluation of an instrumental method of measurement using geological samples for forensic applications. *In*: PYE, K. & CROFT, D. J. (eds) *Forensic Geoscience: Principles, Techniques and Applications*. Geological Society, London, Special Publications, **232**, 49–62.

DEKKERS, M. J. 1997. Environmental magnetism. *Geologie en Mijnbouw*, **76**, 163–182.

D'EMILIO, M., CHIANESE, D., COPPOLA, R., MACCHIATO, M. & RAGOSTA, M. 2007. Magnetic susceptibility measurements as proxy method to monitor soil pollution: development of experimental protocols for field surveys. *Environmental Monitoring and Assessment*, **125**, 137–46.

FENNING, P. J. & DONNELLY, L. J. 2004. Geophysical techniques for forensic investigation. *In*: PYE, K. & CROFT, D. J. (eds) *Forensic Geoscience: Principles, Techniques and Applications*. Geological Society, London, Special Publications, **232**, 11–20.

FLANDERS, P. J. 1994. Collection, measurements and analysis of airborne magnetic particulates from pollution in the environment. *Journal of Applied Physics*, **75**, 5931–5936.

GUEDES, A., RIBEIRO, H., VALENTIM, B. & NORONHA, F. 2009. Quantitative colour analysis of beach and dune sediments for forensic applications: a Portuguese example. *Forensic Science International*, **190**, 42–51.

HUNTER, J. & COX, M. 2005. *Forensic Archaeology: Advances in Theory and Practice*. Routledge, Abingdon.

MAIER, G., SCHOLGER, R. & SCHÖN, J. 2006. The influence of soil moisture on magnetic susceptibility measurements. *Journal of Applied Geophysics*, **59**, 162–75.

MURRAY, R. C. 2004. *Evidence from the Earth*. Mountain Press, Missoula, MT.

MURRAY, R. C. & TEDROW, J. 1992. *Forensic Geology*. Prentice Hall, Englewood Cliffs, NJ.

PRINGLE, J. K., HOLLAND, C., SZKORNIK, K. & HARRISON, M. 2012. Establishing forensic search methodologies and geophysical surveying for the detection of clandestine graves in coastal beach environments. *Forensic Science International*, **219**, e29–e36.

PYE, K. & BLOTT, S. J. 2004. Comparison of soils and sediments using major and trace element data. *In*: PYE, K. & CROFT, D. J. (eds) *Forensic Geoscience: Principles, Techniques and Applications*. Geological Society, London, Special Publications, **232**, 183–196.

SCHIBLER, L., BOYKO, T., FERDYN, M., GAJDA, B., HÖLL, S., JORDANOVA, N. & RÖSLER, W. 2002. Topsoil magnetic susceptibility mapping: data reproducibility and compatibility, measurement strategy. *Studies in Geophysics and Geodesy*, **46**, 43–57.

Analysis of geological trace evidence in a case of criminal damage to graves

ROSA MARIA DI MAGGIO[1]* & LEONARDO NUCCETELLI[2]

[1]*Geoscienze Forensi, Viale Mediterraneo, 77, Rome 00122, Italy*
[2]*Servizio Polizia Scientifica, Direzione Centrale Anticrimine, Via Tuscolana, Roma, 1548, Italy*
*Corresponding author (e-mail: rosamaria.dimaggio@gmail.com)

Abstract: Limestone and travertine headstones of Jewish graves in the Verano Monumental Cemetery in Rome were discovered severely damaged. Police investigators suspected a group of gardeners who worked illegally in the cemetery. Tools, and other items belonging to the gardeners, were seized by the police. Traces of a white substance, and soil, were analysed using a variety of methods. The trace evidence from the graves was very similar to that retrieved from the gardeners' tools. Evidence from several independent analytical methods strongly suggested that the gardeners' tools had been involved in the offence.

In a judicial investigation, the evidence provided by geological materials, such as samples of rock, stone or the inorganic fraction of soil, can be strongly supportive of the prosecution's case. There is little doubt that excellent results can be achieved with microtrace evidence (e.g. Ruffell & McKinley 2008). The forensic geologist is able, in many instances, to obtain compatible results from geological and pedological materials by combining a variety of analytical methods (Murray 2004; Di Maggio *et al.* 2009). In this paper, a case study is presented that exemplifies the use of geological trace evidence in a criminal investigation.

The case

The day before a Jewish religious celebration, Jewish tombs in the Verano Monumental Cemetery in Rome were severely damaged. Headstones, made from limestone marls and travertine, had been broken and some graves had been opened. It was a high visibility case because it was the first time that anti-Semitic-related acts had taken place in the Rome Monumental Cemetery. Police investigators suspected that a group of gardeners, who were carrying out unofficial gardening and repair work in the cemetery, might have been involved. The gardeners claimed they had only used their tools to restore some partition walls in the graveyard with mortar a few days before the damage. The walls were unrefined and made from tufaceous bricks, with dark mortar filling the gaps; some gaps had been roughly filled with an easily removable, and brittle, white material. The police carried out a judicial site survey of the crime scene and seized gardening equipment belonging to the gardeners under investigation. They seized three picks and two iron bars from a box inside the cemetery. On one of the picks and on a bar (Fig. 1) there were microtraces of a white material and soil. These were mainly located on the head of the tools.

Aims

The aims of the investigation were to compare (a) soil, (b) samples of the damaged headstones and (c) the white material from the walls with (d) the trace evidence present on the tools. If the results of the inorganic analyses showed a greater concordance between the material on the tools and the headstones than between the material on the tools and the material from the wall, then the testimony of the gardeners could be challenged.

Methods

Geologists collected samples of soil from inside and outside the Jewish section of the cemetery. The soil was collected along the narrow and short paths among the damaged tombs. The samples were collected from different locations along each path. They also collected samples from damaged headstones, and white material from the partition walls in order to compare them with the soil traces and the white trace evidence on the tools. Samples from the headstone were taken where they showed fresh fractures, which could represent possible points of impact. The white marks on the tools were very small (approximately 1 mm^2) and limited in number. Because of this, every care was taken to avoid dispersing the microtraces of the material.

From: PIRRIE, D., RUFFELL, A. & DAWSON, L. A. (eds) 2013. *Environmental and Criminal Geoforensics.*
Geological Society, London, Special Publications, **384**, 75–79.
First published online May 16, 2013, http://dx.doi.org/10.1144/SP384.2
© The Geological Society of London 2013. Publishing disclaimer: www.geolsoc.org.uk/pub_ethics

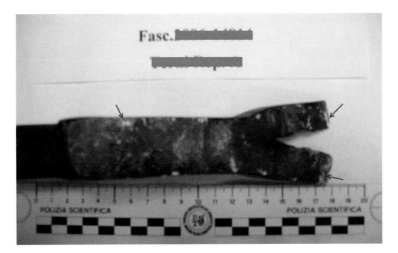

Fig. 1. One of the tools seized from the unofficial gardeners' box inside the cemetery. The red arrows show the white micromarks.

Control soil samples were dried at 60°C, weighed into tubes and disaggregated in an ultrasonic bath. These were wet-sieved using sieves with mesh sizes spaced at 1φ intervals between 2000 and 4 μm. Samples were viewed with stereoscopic microscopy (within a magnification range of $10\times-50\times$), and the colour of the clay fractions was assessed using Munsell colour charts. Thin sections of the sand particles, embedded in resin, were viewed using polarizing light microscopy. The whole soil samples were subjected to X-ray diffraction (XRD). XRD analyses were performed using a Philips PW1800 diffractometer with Cu–Kα radiation generated at 40 kV and 40 mA. Each of the XRD analyses was performed with an angular value range of $5°-80°$, at a step size of 0.01 and at a time per step of 0.9 s.

The samples from the headstones, the white material from the walls and the white marks on the tools were subjected to non-destructive techniques, including stereoscopic microscopy, X-ray diffraction and elemental analysis using SEM–EDX (scanning electron microscopy–energy-dispersive X-ray spectroscopy). Each of the XRD analysis charts was carried out with an angular value range of $5°-80°$, at a step size of 0.01 and at a time per step of 0.5 s. SEM–EDX analyses were performed to detect the elemental composition of the marks on the tools, the headstone samples and the material from the walls. SEM–EDX analysis involved using a Tescan Vega microscope, at variable pressure, and EDAX (EDX) microanalysis. Each of the EDX analyses was done with a hundred counts.

There are a very large number of papers dealing with the examination of rocks and industrial materials, such as bricks and concrete, using SEM–EDX and X-ray diffraction (e.g. Pye & Krinsley 1984; Evans & Tokar 2000; Schiavon 2002). However, useful overviews of these techniques applied to forensic investigation are provided by Pye (2004), Ruffell & Wiltshire (2004) and Kugler (2003), which suggest how the data can be useful only if they are suitably placed in the context of the investigation.

Results

Figure 2 shows a comparison between the X-ray diffractograms for the soil samples. The detailed stereoscopic microscope examination of the white marks on the tools revealed that they were doughy and compact but easily removable. They also featured micrometric stripes on the surface (Fig. 3). The position of the marks on the pick and the bar, and the orientation of the stripes, were consistent with an up and down movement of the tools. Figure 4 shows a comparison between the X-ray diffractograms of the headstone, the white marks and the white material from the walls.

Soils

The soil analysis showed a strong similarity between the morphological, physical and mineralogical features of the samples collected in the cemetery and the ones found on the tools. X-ray diffraction showed that all of the samples consisted of the same crystalline phases, namely quartz, plagioclase, calcite, analcime, mica and kaolinite.

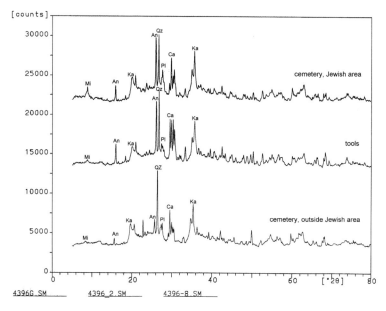

Fig. 2. The comparison between the X-ray diffractograms of the whole soil sampled at the cemetery, in the Jewish area, soil on the tools and a soil sampled outside the Jewish area. All of the samples consist of quartz, plagioclase, calcite, analcime, mica and kaolinite. The critical peaks are labelled with the initials of the minerals: Ca, calcite; Qz, quartz; Pl, plagioclase; An, analcime; Mi, mica; Ka, kaolinite.

Headstones

The headstones, made from limestone marl, gave diffraction patterns typical of pure calcite. The only other peak identified was attributed to a small quantity of quartz. The headstone made from travertine consisted of calcite. Quantitative analysis showed the elements in the headstone were dominated by calcium with smaller quantities of aluminium and silicon. The percentage of these

Fig. 3. White marks on the pick, observed by means of a stereoscopic microscope. Magnification of ×24.

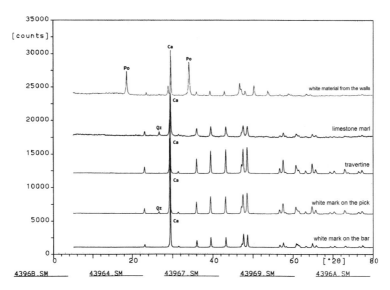

Fig. 4. X-ray diffractograms of limestone marl, travertine, marks on the pick and the bar, and the white material from the walls. The comparison between all of the spectra shows the similarity between the headstones and the white marks. The critical peaks are labelled with the initials of the main minerals: Ca, calcite; Po, portlandite; Qz, quartz.

elements in each sample was: limestone marl, Ca 87.0–Si 9.4–Al 3.5; and travertine, Ca 93.7–Si 3.9–Al 2.3.

Pick

The pick gave similar results to those of the headstones made from limestone marl: calcite and a small quantity of quartz, and elements with percentage values of Ca 88.2–Si 8.3–Al 3.4.

Bar

The marks on the bar consisted of the same crystalline phase as that of the gravestone made from travertine: calcite. The elements showed similar percentage values to those of the travertine: Ca 92.9–Si 4.0–Al 3.1.

Wall

The white material collected from the partition walls consisted of calcite and portlandite. Portlandite, a calcium hydroxide $Ca(OH)_2$, is the major bonding agent in cement and concrete. Portlandite is formed during the hydration process of lime (CaO), and will later react with CO_2 from the air to form calcite by carbonation (Callebaut et al. 2000). The mortar should contain quartz and/or other minerals as inert material in the white material from the walls but none of these minerals were found. The XRD data suggested that the white material could be a hydrated lime paste. The diffraction pattern showed close intensities of the peaks of calcite and portlandite (Allen et al. 2000). Even if the carbonation process is quite slow, Hughes & Swann (1998) have shown that some limes, even when taken as fresh samples, contain remarkably little portlandite with respect to calcite. The quantitative analysis showed the element in the material from the wall was only calcium. Further, the SEM analysis revealed the presence of well-formed hexagonal plates about 10 μm across that correspond to portlandite formed during the hydration of the lime. The dimensions of portlandite crystals were compatible with a hydrated lime paste that was about 3 months old (Colantuono et al. 2011).

Discussion

The pick and the bar were found inside the cemetery, and this accounted for the presence of graveyard soil on the tools. Thus, in this instance, the soils were not useful for comparison purposes as the soil could have been innocently present on the tools. However, XRD analysis showed the material from the walls was different from the white marks on the tools. Furthermore, the peaks of calcite had similar intensities in the headstone samples and in the marks, while the peak of calcite in the white material from the wall did not fully correspond with those of the headstone or the marks.

The quantitative data suggested a similarity between the travertine and the mark on the bar, and between the limestone marl and the mark on the pick. The silicon and the aluminium in the limestone marl linked up with the presence of small quantities of quartz and aluminium silicate hydrate. The presence of small quantities of silicon and aluminium in the travertine were consistent with the inclusion of impurities during its formation. The material from the partition walls consisted of calcium but no aluminium and/or silicon were found. The total absence of these elements in this sample supported the XRD data and suggested, again, that the material from the wall could be a hydrated lime. Furthermore, the dimensions of the well-shaped hexagonal crystals of portlandite suggested that the hydrated lime paste had been used about 3 months before. This time did not correspond with the gardeners' claim. The SEM-EDX data provided further evidence of the similarity between the samples from the gravestone with the marks on the tools, and the difference between the marks on the tools and the material from the wall where they were stated as having been used.

Conclusion

The analysis of the material from the walls revealed that it was not a mortar but probably a hydrated lime paste (made from portlandite and calcite), which had been improperly used to restore the partition walls. Indeed, the mineralogical and chemical data did not show the presence of sand or pozzolana, which are used as inert material in mortar. However, analysis using stereo-microscopy, X-ray diffraction and SEM-EDX showed that, although the white material from the walls was different, the headstones contained the same minerals and elements as the marks on the tools. These results suggested that the white marks on the tools were derived from the damaged headstones rather than from the hydrated lime of the wall mortar.

The authors wish to thank Dr R. Murray and Dr D. Pirrie for suggesting and helping them in writing this article, and Dr A. D'Ippolito, Judge Prosecutor of the Criminal Court of Law of Rome, for permission to publish the detailed scientific data in this case.

References

ALLEN, G. C., HALLAM, K. R., RADONJIC, M., ELTON, N. J., FAREY, M. & ASHURST, J. 2000. X-ray diffraction and environmental electron microscopy of hydrate limes. *In*: BARTOS, P., GROOT, C. & HUGHES, J. J. (eds) *Historic Mortars: Characteristics and Tests*. Proceedings of the International RILEM Workshop. RILEM, Bagneux, France.

CALLEBAUT, K., ELSEN, J., VAN BALEN, K. & VIAENE, W. 2000. Historical and scientific study of hydraulic mortars from the 19th century. *In*: BARTOS, P., GROOT, C. & HUGHES, J. J. (eds) *Historic Mortars: Characteristics and Tests*. Proceedings of the International RILEM Workshop. RILEM, Bagneux, France.

COLANTUONO, A., DEL VECCHIO, S., MARINO, O., MASCOLO, G., MASCOLO, M. C. & VITALE, A. 2011. Evoluzione del grassello in calce con il tempo di stagionatura. *Forum Italiano Calce News*, **2**, 1–7.

DI MAGGIO, R. M., MAIO, M. & NUCCETELLI, L. 2009. Geologia Forense. *In*: PICOZZI, M. & INTINI, A. (eds) *Scienze Forensi: Teoria e Prassi dell'Investigazione Scientifica*. UTET, Torino, 255–266.

EVANS, J. E. & TOKAR, F. J. 2000. Use of SEM/EDX and X-ray diffraction analyses for sand transport studies, Lake Eire, Ohio. *Journal of Coastal Research*, **16**, 926–933.

HUGHES, D. & SWANN, S. 1998. Hydraulic limes – a preliminary investigation. *Lime News*, **6**, 41–52.

KUGLER, W. 2003. X-ray diffraction analysis in the forensic science: the last resort in many criminal cases. *Advances in X-ray Analysis*, **46**, 1–16.

MURRAY, R. C. 2004. *Evidence from the Earth*. Mountain Press, Missoula, MT.

PYE, K. 2004. Forensic examination of rocks, sediments, soils and dust using scanning electron microscopy and X-ray chemical microanalysis. *In*: PYE, K. & CROFT, D. J. (eds) *Forensic Geoscience. Principles, Techniques and Application*. Geological Society, London, Special Publications, **232**, 103–122.

PYE, K. & KRINSLEY, D. H. 1984. Petrographic examination of sedimentary rocks in the SEM using backscattered electron detectors. *Journal of Sedimentary Research*, **54**, 877–888.

RUFFELL, A. & MCKINLEY, J. 2008. *Geoforensics*. John Wiley & Sons, Chichester.

RUFFELL, A. & WILTSHIRE, P. 2004. Conjunctive use of quantitative and qualitative X-ray diffraction analysis of soil and rocks for forensic analysis. *Forensic Science International*, **145**, 13–23.

SCHIAVON, N. 2002. Biodeterioration of calcareous and granitic building stones in urban environment. *In*: SIEGESMUND, S., WEISS, T. & VOLLBRECHT, A. (eds) *Natural Stone, Weathering Phenomena, Conservation Strategies and Case Studies*. Geological Society, London, Special Publications, **205**, 195–205.

Using soil mineral signatures to confirm sources of industrial contaminant trespass

WAYNE C. ISPHORDING

School of Continuing Studies, Tulane University, Biloxi, MS 39531, USA and University of South Alabama (retired), Mobile, AL 36688, USA (e-mail: isphordingw@bellsouth.net)

Abstract: An accompanying consequence of new industrial operations, or the expansion and enlargement of existing facilities, is contaminant encroachment from these operations onto adjacent residential properties. This often results in a favourable jury decision for the plaintiffs, providing the case can be proven. While smokestack emissions or wind-blown debris from some manufacturing operations can be readily identified in samples obtained from nearby properties (e.g. coal fly ash and mine wastes), obtaining proof that a site, or individuals, have been impacted by emissions from other processes may pose a challenge for the expert witness. As an example of the former, a mineral phase universally present in electric arc furnace smokestack discharges by mini steel mills (now widespread throughout the United States) is magnetite (FeO · Fe$_2$O$_3$). When this mineral is present as non-detrital anthropogenic grains in dust and soil samples (and further identified by a manganese content three or more orders of magnitude greater than in detrital magnetite) its presence is conclusive evidence that airborne contamination from an anthropogenic source has taken place.

A problem arises, however, when contaminants are emitted as a mixed gas–particulate phase whose particles are <2.5 μm in size. Emissions from paper mills may fall into this category, and invariably will contain hydrogen sulphide (H$_2$S), sulphur dioxide (SO$_2$) and sulphur trioxide (SO$_3$) as gas phases. These rapidly combine with water vapour to form sulphuric acid. Dispersal can occur over a wide area as a toxic plume, and can cause a variety of acutely or chronically harmful medical problems, especially in young children and the elderly. Although acknowledged as present in stack emissions by the paper companies, evidence of the vapour phase is often difficult to find on surrounding residential properties (although its corrosive effects are invariably present). An associated particulate phase (calcite), which is used as a whitener in paper production, however, is emitted from a facility's lime kiln and can often be used to show that contamination from a paper-manufacturing source has taken place.

Soil and dust mineralogy has proven to be a valuable tool in situations involving both criminal and civil investigations. Many examples can be cited (see Murray 2004; Isphording 2007; Ruffell & McKinley 2008) wherein mineral evidence was crucial in solving cases involving, for example, mineral and mining fraud, hit-and-run accidents, burglary and rape felonies, establishing the presence of individuals at a specific crime scene, or confirming that a victim was moved from a site where an act of violence was committed to a different location. Similarly, mineral evidence has been used to clearly establish that residences, properties or individuals in proximity to industrial sites have been impacted, damaged or physically harmed by off-site migration of surface, subsurface or airborne emissions from industrial operations (Isphording 2007). The latter has become an ever-more common challenge for the forensic mineralogist because of the inevitable encroachment that has taken place by construction of residential properties adjacent to land previously zoned as 'industrial' (e.g. housing development v. airports, manufacturing or other commercial operations) or the opposite, wherein zoning changes have allowed industries to be established on land immediately adjacent to areas formerly restricted to residential development. Regardless as to whether the expert witness has been called upon by plaintiffs or the defence, knowledge of factors controlling the provenance, morphology, weathering behaviour and chemistry of soil minerals becomes critical.

Two case studies are described that illustrate these issues. The first describes emissions from a steel mill operation and the second describes identification of emissions from a large paper mill.

Case 1: airborne smokestack emissions from a steel 'mini-mill' operation

World-wide production of steel is derived from two principal sources, integrated mills and mini-mills. The former are very large tonnage operations (>2 000 000 t a^{-1}) that have continued to produced high-quality steel from their initiation in the mid-nineteenth century up to the present day. These operations use a charge of iron ore, coal (coke) and limestone fed to a blast furnace, and can produce a wide variety of finished products, including pipe, bars, steel plate and wire. Because they require vast tonnages of raw ore, coal and limestone, they are located in close proximity to port facilities to permit the easy acquisition of the needed raw materials. Mini-mills, in contrast, are smaller operations largely dating from the late 1960s, which utilize scrap steel melted in electric arc furnaces (Figs 1 & 2). Finished products were originally restricted to bar stock but since the late 1980s new processes have allowed production of strip casting and slab production. These mills typically produce several hundred thousand tons per year (t a^{-1}), and can be located where labour costs are lower and in proximity to actual sales markets. One such facility was established in the early 1970s near a residential site in Alabama (USA) and, shortly after operations commenced, became the subject of considerable acrimony with adjacent property owners.

Property owner complaints and objections

Litigation between residential property owners and the steel mill operators stemmed from two major problems: (1) noise pollution; and (2) the frequent rain of dust from the mill operations. Noise pollution originated from a variety of sources (trucks, trains and other heavy equipment operating in the mill) but the principal complaint was associated with the stacking of the finished slabs. These massive sheets weigh nearly 50 t and were produced 24 h/day. As they were dropped in the storage yard, a marked 'seismic shock' was felt over a distance of greater than 1 km and was accompanied by a loud, audible 'clang' as sheet was stacked upon sheet. Detailed measurements of these phenomena were obtained, and recorded, by the plaintiff's experts for production at trial.

Fig. 1. Magnetite generated in an electric arc furnace from fusion of scrap metal is both morphologically and geochemically distinct from natural forms.

The second complaint, that of airborne dust generation, originated from debris carried by prevailing winds for distances of 10 km, or greater, from the manufacturing site. The source of the dust was fine-grained material associated with different stages of the manufacturing process but the chief source was electric arc furnace dust that passed through the bag house and was emitted from the smokestack. This material not only settled throughout the adjacent area, causing accumulation on automobiles and other surfaces, but also resulted in discoloration of painted surfaces on homes in the affected area. While this constitutes an apparent aesthetic nuisance, the principal problem of the dust was its very small particle size coupled with its particulate chemistry. The dust particles were typically in the fine- to medium-silt range (5–50 μm) but a significant portion were present in the easily respirable <2.5 μm size. This, combined with the presence of a number of potentially toxic heavy metals that were present in the dust (e.g. mercury, lead, zinc, manganese, chromium and zinc), further rendered it as a potential health problem. Information available from regulatory agencies disclosed that the annual release of bag house dust was in excess of 600 000 kg a^{-1}. This material is classified by the US Environmental Protection Agency (EPA) as a 'hazardous waste' and, as such, must be removed from the site and transported to a hazardous waste-receiving facility. Evidence that the dust was derived from furnace emissions was clearly present in the quantities of magnetite found in samples collected from numerous residences. While the proportions of the magnetic fraction of dust for eight 'background' samples ranged from 0.004 to 0.129%, and averaged 0.08%, those proportions from 28 residences near the facility were one or two orders of magnitude greater, ranging from slightly greater than 0.1 to over 26%, and averaged 6.10%. Actual bag house samples were almost wholly magnetic in character and averaged 90% magnetic material. Although the proportions found on the plaintiff's properties are clearly 'anomalous' and were undoubtedly derived from mill dust, the burden was on the plaintiff's expert to provided clear evidence that the dust was from the mill and not from some other source, such as fly ash derived from wind-borne debris from a power plant or, simply, indigenous detrital magnetite.

Geological evidence confirming the source of the dust

Microscopic analysis of dust samples from the bag house and from the plaintiff's properties revealed that magnetite (FeO · Fe$_2$O$_3$) was ubiquitous in all samples, and was also a major constituent in the

Fig. 2. Steel manufactured in electric arc furnaces in the United States, in terms of total volume, has now largely supplanted quantities produced by older, integrated mill operations.

electric arc furnace fly ash dust collected from the bag house. The magnetite from the steel furnace was morphologically distinct from samples of coal-derived fly ash obtained from the nearby power plant, and, when viewed using a petrographical microscope, was similar to classic 'string magnetite' that has been described by McCrone & Delly (1973) and is shown in Figure 3. Magnetite observed from coal fly ash from electric power stations, in contrast, is invariably spherical in form (Fig. 4) and easily allowed particulate material from the power plant to be eliminated as a source of dust contamination (see Tishmack & Burns 2004; Goodarzi 2005; Kutchko & Kim 2006).

Proof countering the argument that the magnetite was actually an indigenous soil mineral component was not only provided from numerous heavy mineral analyses of soils carried out in the surrounding region in past years by the expert witnesses but was also shown from published literature, which demonstrated that magnetite is a very uncommon mineral in the heavy mineral fraction of sediments throughout the coastal plain of eastern and SE United States (see Hsu 1960; Pilkey 1963; Isphording 2007; Milliken 2007; Pirkle *et al.* 2007). By thus eliminating the argument that magnetite found on the plaintiff's properties was not a 'natural' constituent, the witness was thereby able to trace all such dust, unequivocally, to airborne emission from the steel mill's bag house. A favourable verdict for substantial property damages was subsequently rendered by the jury for the various plaintiffs in the case.

Case 2: airborne emissions from a paper mill

More challenging for the expert witness was providing evidence that airborne contaminants have similarly impacted residential areas adjacent to paper mills. Nearly 3000 paper producers are present in the United States, a number of which are classified as 'Kraft mills' (also known as 'Kraft pulping' or 'Sulphate process' mills). The Kraft process was developed in Germany in 1879 and, by the early

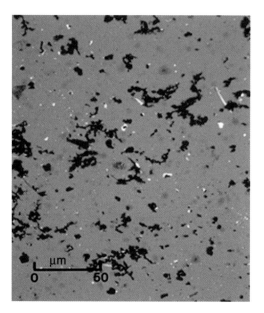

Fig. 3. Classic 'string magnetite' produced in a steel mill furnace (McCrone & Delly 1973).

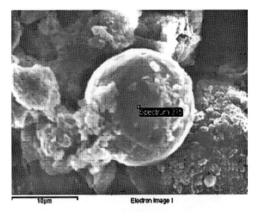

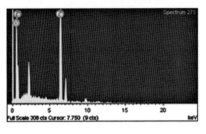

Fig. 4. SEM (scanning electron microscope) photomicrograph of spherical magnetite, produced by burning coal in a power plant, along with a representative spot EDS (energy-dispersive spectrometry) analysis.

1940s, became the dominant method for producing wood pulp. Operationally, it involves a process whereby wood chips are fed into high-pressure digester vessels, and then converted into wood pulp by treatment with a mixture of sodium hydroxide and sodium sulphide. The method is largely carried out in a closed system and produces a wide variety of high-quality paper products. By-products are also generated, some of which are economically useful (e.g. turpentine). Others are less desirable and include hydrogen sulphide, methyl mercaptan, dimethyl sulphide and other volatile sulphur compounds that are malodorous and give the mills their distinctive 'aroma'. Past processes used elemental chlorine in the bleaching process, which generated organic waste products such as chlorinated dioxin. Dioxins are highly toxic, and are now regulated internationally and by the US EPA at the 'part per quadrillion' level.

Although ideally designed as a 'closed system', the Kraft process varies from site to site and, locally, may operate with a reduced efficiency. One such example is discussed from a mill in Alabama where adjacent property owners were less than enthusiastic about the mill's presence.

Property owner complaints and objections

Litigation between mill operators and property owners originally arose because of complaints that vapours and particulate emissions were becoming a corrosion nuisance on metal objects on adjacent properties and, especially, on the painted surfaces of automobiles. Subsequently, medical problems were identified, especially in children, whose suspected cause was believed related to emission of specific mill-generated volatiles. The primary mill emissions implicated in these problems included sulphur dioxide (SO_2), sulphur trioxide (SO_3) and hydrogen sulphide (H_2S). Each of these compounds when released to the atmosphere reacts with water vapour to form sulphuric acid (H_2SO_4). Although admitting that 'some releases' of hydrogen sulphide and other volatiles occurred from units involved in the wastewater treatment system, the company minimized any 'distant environmental effects' on the neighboring populace and contended that they were operating in accordance with limits promulgated by the State. The Materials Safety Data sheet (MSD) that accompanies all containers of manufactured sulphuric acid in the United States, however, is quite explicit and carries the warning:

> DANGER! CORROSIVE LIQUID AND MIST CAUSE SEVERE BURNS TO ALL BODY TISSUE. MAY BE FATAL IF SWALLOWED OR CONTACTED WITH SKIN. HARMFUL IF INHALED. AFFECTS TEETH. CANCER HAZARD. STRONG INORGANIC ACID MISTS CONTAINING SULFURIC ACID CAN CAUSE CANCER. RISK DEPENDS ON DURATION AND LEVEL OF EXPOSURE.

Complaints of corrosion of metals on fences, automobiles and air conditioners were dismissed by the company as 'normal usage deterioration', in spite of the fact that: (1) many air conditioning units were failing in less than one year; (2) similar deterioration was observed after a short time on exposed gates, fences and other exposed metal items; and (3) the company maintained on its premises a 'car wash' and encouraged its use by employees. Medical complaints were, similarly, dismissed as 'speculative', despite doctor's suggestions that the only sure relief for asthma patients, and those with acute and chronic bronchial and pulmonary conditions, was to 'move away!'.

Forensic mineralogy challenge for the expert witness

A complicating factor for the expert witness was that the volatile components were being emitted from the stacks as a mixed gas–particulate phase whose particles were generally <2.5 μm in size (PM < 2.5). Although acknowledged as present in stack emissions from many paper mills, the presence of such particles on adjacent properties is often difficult to identify. This is because the emitted particles themselves can become transformed into a gaseous or liquid phase by reaction with atmospheric water vapour. To convince a jury that such emissions are, in fact, impacting the surrounding community the expert witness relied on: (1) historical annual wind-rose diagrams produced by the National Weather Service and compiled over a 20 year period; these clearly refuted the mill's contention that wind directions do not, typically, blow from the mill towards the residential area and are, invariably, 'east–west,' rather than 'north–south' (the data clearly showed 'north–south' winds over 25% of the time); and (2) the witness pointed to the widespread presence of $CaCO_3$ (calcite) that was present on automobiles and other exposed surfaces throughout the adjacent residential area. This material was extensively present throughout the mill itself and originated from the facility's lime kiln. $CaCO_3$ is commonly used as a whitener for paper products and also has other uses in paper manufacture. Its presence, therefore, was irrefutable evidence that not only was this material being carried by winds on to the properties but any gaseous phase was similarly being transported to the residential area.

This case did not actually result in a trial. The company, following review of the plaintiff's evidence and other pre-trial depositions, interrogatories, etc., agreed on a sealed settlement that was deemed equitable by the company, the plaintiffs and their attorneys.

Conclusions

Two examples are presented to illustrate the straightforward manner by which trespass of emissions from steel mills and paper mills can be efficiently, and unequivocally, identified on properties adjacent to manufacturing operations. No sophisticated instrumentation (other than a petrographical microscope) is generally needed and the identification of non-indigenous particulates can be carried out, usually in a matter of minutes.

The presence of magnetite serves as 'fingerprint' mineral phase that is invariably traceable to a steel-manufacturing site, and its identification should pose little in the way of a technical challenge for a forensic mineralogist. If further evidence is desired, chemical analysis of other airborne dust (fly ash) materials associated with magnetite is also useful. Because steel is an alloy product, certain chemical elements will be present in the dust that reflect the original additives that were present in the scrap metal. Manganese is always present in elevated amounts and results from the universal use of this element as a de-sulphurizing additive in steel (proportions are typically one or two orders of magnitude lower in 'natural' magnetite). Other metals that may be present in anomalous quantities in steel-derived fly ash dust include arsenic, vanadium, nickel, cobalt, chromium, zirconium, lead and zinc. Elevated lead and zinc (and arsenic) owe their presence to paint present on automobiles that commonly are used as scrap metal sources in mini-mills.

Potentially harmful emissions from paper mills are largely caused by the generation and release of sulphur dioxide (SO_2), sulphur trioxide (SO_3) and hydrogen sulphide (H_2S). All of these readily combine with water vapour in the atmosphere and are quickly converted to sulphuric acid (H_2SO_4), a highly corrosive compound. As such, breathing and exposure to this volatile phase can produce serious medical problems. Because these are vapour phases, detection of their actual presence requires sophisticated air-monitoring equipment. Indirect evidence, however, can be observed either from observation of extensive corrosion of metal surfaces (automobiles, chain link fences and metal lawn furniture), premature failure of common household air-conditioning units, or corrosion of exposed wiring, etc. Because many mills utilize calcium carbonate as 'whiteners' for paper (and for other purposes), those with lime kilns afford the investigator a rapid means of identifying airborne trespass. Calcium carbonate (calcite) emitted from the lime kiln stack and carried by wind (along with the harmful sulphuric acid vapour phases) can be easily identified by its characteristic optical properties when viewed using a petrographical microscope. Its presence on automobiles, lawn furniture and other exposed surfaces, in the absence of any indigenous source (e.g. a local quarrying operation), also serves to confirm industrial trespass.

Special thanks are extended to Dr George C. Flowers, Tulane University, Coordinated Instrumentation Facility, New Orleans, LA, for the SEM/EDS analysis included in this paper.

References

GOODARZI, F. 2005. Morphology and chemistry of fine particles emitted from a Canadian coal-fired power plant. *Fuel*, **85**, 273–280.

HSU, K. J. 1960. Texture and mineralogy of the recent sands of the Gulf Coast. *Journal of Sedimentary Petrology*, **30**, 380–403.

ISPHORDING, W. C. 2007. Forensic use of heavy minerals in civil and criminal investigations. *In*: MANGE, M. &

WRIGHT, D. (eds) *Heavy Minerals in Use*. Developments in Sedimentology. Elsevier, Amsterdam, **58**, 963–982.

KUTCHKO, B. G. & KIM, A. N. 2006. Fly ash characterization by SEM-EDS. *Fuel*, **85**, 2537–2544.

MCCRONE, W. C. & DELLY, J. G. 1973. *The Particle Atlas*. Ann Arbor Science, Ann Arbor, MI.

MILLIKEN, K. L. 2007. Provenance and diagenesis of heavy minerals, Cenozoic units of the Northwestern Gulf of Mexico sedimentary basin. *In*: MANGE, M. & WRIGHT, D. (eds) *Heavy Minerals in Use*. Developments in Sedimentology. Elsevier, Amsterdam, **58**, 247–262.

MURRAY, R. C. 2004. *Evidence from the Earth, Forensic Geology and Criminal Investigation*. Mountain Press Publishing, Missoula, MT.

PILKEY, O. H. 1963. Heavy minerals of the U.S. South Atlantic continental shelf slope. *Geological Society of America Bulletin*, **76**, 641–648.

PIRKLE, F. L., PIRKLE, W. A. & PIRKLE, E. C. 2007. Heavy mineral sands of the Atlantic and Gulf Coastal Plains, USA. *In*: MANGE, M. & WRIGHT, D. (eds) *Heavy Minerals in Use*. Developments in Sedimentology. Elsevier, Amsterdam, **58**, 1145–1232.

RUFFELL, A. & MCKINLEY, J. 2008. *Geoforensics*. Wiley, New York.

TISHMACK, J. K. & BURNS, P. E. 2004. The chemistry and mineralogy of coal and coal combustion products. *In*: GIERÉ, R. & STILLE, P. (eds) *Energy, Waste, and the Environment: A Geochemical Perspective*. Geological Society, London, Special Publications, **236**, 223–246.

Geological and palynological characterization of a river beach in Portugal for forensic purposes

ÁUREA CARVALHO[1], HELENA RIBEIRO[1], ALEXANDRA GUEDES[2],
ILDA ABREU[3] & FERNANDO NORONHA[1]*

[1]*Centro de Geologia da Universidade do Porto (CGUP), Faculdade de Ciências, Universidade do Porto, Rua do Campo Alegre 687, 4169-007 Porto, Portugal*

[2]*CGUP e Departamento de Geociências, Ambiente e Ordenamento do Território, Faculdade de Ciências, Universidade do Porto, Rua do Campo Alegre s/n, 4169-007 Porto, Portugal*

[3]*CGUP e Departamento de Biologia, Faculdade de Ciências, Universidade do Porto, Rua do Campo Alegre s/n, 4169-007 Porto, Portugal*

*Corresponding author (e-mail: fmnoronh@fc.up.pt)

Abstract: Currently, it is important to use a combination of independent analytical techniques in order to reliably characterize between different samples of geological trace evidence in criminal investigations. Areínho beach, in the town of Vila Nova de Gaia (northern Portugal), is located on the southern bank of the Douro River. Although isolated, it is a very busy area during the summer and other seasons due to its popular water sports. The geology of the area is complex, with numerous different outcrops. In addition, there is a considerable variety of vegetation in the region, and many materials are transported by the Douro River and deposited in the area.

In order to characterize this beach for forensic purposes (both criminal and environmental), sediment samples were collected and analysed using a combination of geological techniques, namely colour, particle size distribution and mineralogy. These geological techniques were combined with a palynological assessment of the sediment samples. In addition, the surrounding vegetation was characterized and seasonal vegetation surveys were carried out.

In this study, the results obtained revealed homogeneous data throughout the profile in all the properties analysed. This research demonstrated that the combination of different, independent techniques used to characterize sediments from this river beach has the potential to contribute substantially to future forensic investigations.

Geology and palynology can be useful in forensic investigations, and are currently used by forensic laboratories in many countries, providing important contributions to the evidence-based decision making in courts. These two independent fields have a complementary nature and can provide information that is useful in forensic investigations, especially for the comparison of geological and biological material occurring as evidence that can be used to associate a victim and/or suspect with a crime scene (Erdtman 1969; Murray & Tedrow 1986). Forensic geology includes the analysis of soils and other geologically related materials, which typically consist of both organic and inorganic materials. Forensic soil analysis commonly involves comparison of both bulk properties and certain soil components, including colour, grain size distribution, mineralogy and bulk chemistry. Palynology deals with the study of palynomorphs, and can contribute to forensic soil analysis by providing temporal and spatial information, as many plants produce large quantities of pollen that are dispersed and can adhere to a variety of surfaces, including those with natural, artificial and human origins. Pollen assemblages differ from region to region (Faegri & Iversen 1989; Reille 1992), making it possible to compare an unknown sample containing pollen with pollen data obtained from a potential source region. So, forensic pollen analysis can provide information about whether a crime occurred at the location where the victim was found and/or associate a suspect with a crime scene (Milne *et al.* 2004).

Practices and procedures similar to those used in geological and palynological research are used in forensic investigations. However, it is important to optimize and adapt protocols to ensure that they are appropriate for forensic scenarios. Their use is essential for providing evidence that can stand up to scrutiny in a court of law.

The Centro de Geologia da Universidade do Porto (CGUP) is currently active in the development and implementation of geological and palynological methods that are appropriate for forensic science, paving the way for additional research in this area in Portugal (Guedes *et al.* 2009, 2011). The primary objectives of this work are to develop scientific protocols that are useful in forensic investigations, to collect data that provide a basis for interpretation of forensic soil analysis and to complement existing projects in CGUP by further characterizing the geology of the Portuguese territory. While many geological materials, including soils, are currently used as evidence in criminal cases in Portugal, additional work is still needed to determine which properties of geological materials are most relevant for forensic comparisons, what analytical techniques, methods and protocols are most appropriate for the analysis of soils in criminal cases, and to build databases that can support forensic investigations involving soil and other geological materials.

In this context, we aimed to characterize a portion of a river beach that is surrounded by a wooded area. Of specific interest are variations in geological and botanical materials along a transect perpendicular to the river bank. The data presented here are the results of a combined geological and palynological characterization of soil samples, specifically soil colour, particle size distribution, mineralogy and pollen content.

The location selected for sampling is a river beach located in northern Portugal, on the southern bank of the Douro River in Oliveira do Douro, Vila Nova de Gaia, near Porto (the second largest city in Portugal). This beach has the standard features typical of most river beaches and it is located in a geological environment that is distinct from that in other locations in the Porto area, marked by the nearby occurrence of several different lithologies. The surrounding vegetation is also diverse. The beach is in a secluded location near an area that is a popular destination for water sports, especially during the summer bathing season. This makes it a potential location for the commission of suicides, homicides or for the occurrence of accidental deaths.

Materials and methods

Sample collection and handling

The sampling was performed in the spring of 2009. Samples were collected from eight locations, each 15 m apart, along a transect perpendicular to the river bank. At each sampling location, three samples were collected from the corners of a triangular array, 1 m apart. This sampling procedure was designed to produce representative samples in order to reliably compare the accuracy of different instrumental techniques (Fig. 1). The samples were manually collected from a depth of less than 5 cm using a plastic spade that was thoroughly cleaned between each sample. In the laboratory, the samples were dried at 40°C, homogenized and subsequently split for several analyses. A plant survey of the river beach was also performed during the sampling period.

The surrounding geology is characterized by several distinct lithologies, the most important of which are the Porto two-mica granite, migmatites, gneisses, schists and metagreywackes of the Complexo Xisto-grauváquico. An attempt was made to assess the contributions of each of these parent rocks to the mineralogy of the beach sediments.

Fig. 1. Sampling profile of the Oliveira do Douro river beach.

Colour analysis

Colour analysis was carried out on dried unsieved bulk samples. The samples were homogenized and placed in a standard glass Petri dish, and four colour measurements were taken for each sample. Guedes et al. (2009) demonstrated that this methodology enables greater discrimination between sediment samples than measurements performed on other fractions or with other pre-treatment methods. Colour measurements were performed using a Konica Minolta CM-2600d spectrophotometer programmed with the following settings: measurement area of 0.8 mm diameter; specular component included; CIE Standard Illuminant D65 representing average daylight with a correlated colour temperature of approximately 6504 K, including the ultraviolet wavelength region; and CIE 1964 Standard Observer (10° Observer). The spectrophotometer was calibrated prior to and during the sample measurements to take into account the influence of daylight variation. Negative calibration was performed by directing the apparatus-measuring port into the air while positive calibration was performed with an international standard white calibration plate. The recorded colour parameters correspond to the uniform colour space CIELAB (or CIE $L^*a^*b^*$, where L^* is lightness, a^* represents the red/green axis and b^* represents the yellow/blue axis) and were directly computed by the spectrophotometer using the SpectraMagic NX software. The results were analysed statistically by SPSS Statistics 18, and the reproducibility within and between samples was evaluated using the coefficient of variation (CV).

Particle size distribution

Samples were subjected to dry sieving using a column of 10 sieves, resulting in 11 different size fractions for each sample: >32 mm; 32–16 mm; 16–8 mm; 8–4 mm; 4–2 mm; 2–1 mm; 1–0.5 mm; 0.5–0.25 mm; 0.25–0.125 mm; 0.125–0.063 mm; and <0.063 mm. The bulk samples were weighed prior to sieving and the individual size fractions were weighed after sieving. The data were analysed statistically using Microsoft Office Excel 2007 and GRADISTAT software (Blott & Pye 2001).

Heavy mineral analysis

Heavy mineral separation was carried out on the 0.5–0.25 mm size fraction using bromoform ($d = 2.89$). The heavy minerals collected were washed and dried in an oven. They were then weighed and the magnetic minerals were separated using a hand-held magnet. Further magnetic separation of heavy minerals was done using a Frantz laboratory magnetic separator. Paramagnetic and diamagnetic fractions were separated by using various amperages: 0.2, 0.4, 0.6, 0.8 and 1 A. The resulting fractions were weighed. Each fraction was examined using a binocular petrographical microscope, and the minerals were examined to determine their type and abundance. When necessary, complementary scanning electron microscopy and energy dispersive spectroscopy (SEM–EDS) analyses were performed.

Pollen analysis

Ten grammes of sediment were removed from each sample, and submitted to a series of sequential chemical procedures in order to remove as much mineral material and extraneous organic matter as possible, leaving the chemically resistant organic matter that includes pollen and other palynomorphs (Horrocks 2004). To remove any fine adhering organic material from the pollen surface, and to remove the protoplasm from the pollen interior in order to render the exines more visible by light

Table 1. *Means and coefficients of variation of $L^*a^*b^*$ parameters measured for the samples collected along the sampling profile*

Samples		L^* (D_{65})		a^* (D_{65})		b^* (D_{65})	
		Mean	CV	Mean	CV	Mean	CV
A1	A	58.1	1.0	2.5	6.3	11.1	2.1
	B	58.0		2.8		11.3	
	C	57.0		2.6		10.9	
A2	A	56.7	0.4	3.1	4.1	11.1	1.4
	B	57.0		3.3		11.3	
	C	57.2		3.1		11.3	
A3	A	58.1	2.0	2.6	7.1	10.1	4.7
	B	56.7		2.9		9.9	
	C	59.1		3.0		10.8	
A4	A	59.9	1.0	3.2	4.7	11.5	2.2
	B	59.1		2.9		11.2	
	C	60.4		3.2		11.7	
A5	A	59.4	0.2	2.7	5.0	10.9	2.7
	B	59.5		3.0		11.4	
	C	59.3		2.7		10.8	
A6	A	59.6	1.6	2.9	5.8	11.5	4.9
	B	58.5		2.6		10.6	
	C	57.8		2.7		10.6	
A7	A	58.0	0.5	2.7	2.4	10.5	0.4
	B	58.2		2.7		10.4	
	C	58.5		2.8		10.5	
A8	A	56.4	0.7	2.6	2.6	9.9	1.8
	B	55.7		2.5		9.7	
	C	56.4		2.6		10.0	
CV inter-sample		2.2		8.1		5.3	

microscopy, the samples were put through a procedure known as acetolysis (Erdtman 1969). Subsequently, the pollen residue was mounted onto microscope slides in glycerol jelly, and the pollen types identified and quantified by performing 10 equidistant horizontal traverses of the preparation using a light microscope, with a magnification of ×400. Pollen counts for each identified taxon were converted into percentages of total counts and divided into classes according to their representation percentage (higher than 10%, 5–10%, 1–5% and lower than 1%).

The pollen grains were identified on the basis of their shape, number of apertures, size and ornamentation with the help of reference preparations and published literature (Reille 1992, 1995, 1998; Horrocks 2004). In order to obtain a composite pollen spectrum, the results from the three samples collected at each location along the collection traverse were combined.

Results and discussion

Colour analysis

A statistical analysis of the colour measurements carried out on the triplicate samples collected at each sampling location revealed similar colour parameters of the soil samples over distances on the order of 1 m (Table 1). For future work, colour analysis of a single sample from each sampling location should be sufficient. It was also determined that there is very minor variation in the colour parameters across the entire transect, which was sampled at the Oliveira do Douro beach (the CV was always less than 8.1%). These results are consistent with visual observations made during sampling because no perceptible differences in the colour of the soil from the different sampling locations was observed. These results indicate the utility of performing an initial visual examination of a scene prior to deciding on the number of samples and the locations to be sampled during a forensic investigation.

The samples analysed produced colour values close to the red ($2.5 < a < 3.3$) and yellow ($9.7 < b < 11.7$) colour values, and a lightness value ($55.7 < L < 60.4$) close to the paler portion of the $L^*a^*b^*$ system colour sphere. Compared to results obtained for marine sands from northern Portugal by Guedes et al. (2009), the river-bank samples had lower L^* values, and higher a^* and b^* values. The lower L^* values are probably due to the fact that we are comparing river sands with marine sands and the latter typically have higher brightness values. The differences between the

Table 2. *Statistical data related to the particle size distribution of the samples collected along the sampling profile (Folk & Ward 1957 method)*

Samples	Mean	Sorting	Skewness	Kurtosis
A1A	Very coarse sand	Poorly sorted	Symmetrical	Platykurtic
A1B	Very coarse sand	Poorly sorted	Symmetrical	Mesokurtic
A1C	Very coarse sand	Poorly sorted	Symmetrical	Platykurtic
A2A	Very coarse sand	Poorly sorted	Symmetrical	Platykurtic
A2B	Very coarse sand	Poorly sorted	Coarse skewed	Platykurtic
A2C	Very coarse sand	Poorly sorted	Symmetrical	Platykurtic
A3A	Very coarse sand	Poorly sorted	Symmetrical	Platykurtic
A3B	Very coarse sand	Poorly sorted	Coarse skewed	Platykurtic
A3C	Very coarse sand	Poorly sorted	Symmetrical	Platykurtic
A4A	Very coarse sand	Poorly sorted	Symmetrical	Mesokurtic
A4B	Very coarse sand	Poorly sorted	Symmetrical	Leptokurtic
A4C	Very coarse sand	Poorly sorted	Symmetrical	Platykurtic
A5A	Very coarse sand	Poorly sorted	Symmetrical	Mesokurtic
A5B	Very coarse sand	Poorly sorted	Symmetrical	Platykurtic
A5C	Very coarse sand	Poorly sorted	Symmetrical	Platykurtic
A6A	Very coarse sand	Poorly sorted	Symmetrical	Platykurtic
A6B	Very coarse sand	Poorly sorted	Symmetrical	Mesokurtic
A6C	Very coarse sand	Moderately sorted	Coarse skewed	Platykurtic
A7A	Very coarse sand	Poorly sorted	Symmetrical	Platykurtic
A7B	Very coarse sand	Poorly sorted	Symmetrical	Platykurtic
A7C	Very coarse sand	Moderately sorted	Symmetrical	Mesokurtic
A8A	Very coarse sand	Poorly sorted	Fine skewed	Leptokurtic
A8B	Very coarse sand	Poorly sorted	Fine skewed	Leptokurtic
A8C	Very coarse sand	Poorly sorted	Symmetrical	Mesokurtic

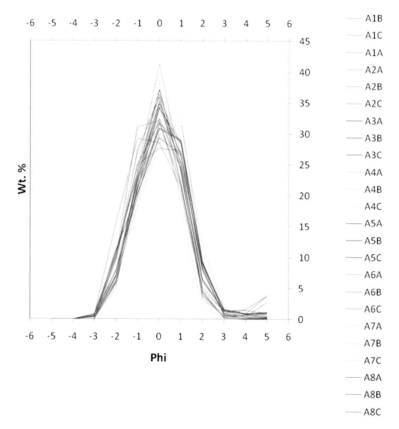

Fig. 2. Frequency curves of particle size distribution from the samples collected in the sampling profile.

$a*$ and $b*$ parameters of the river and marine sands are probably due to differences in their parent rocks.

Particle size distribution

Particle size descriptive parameters showed a mean size distribution between [0.56; −0.10φ]. The sediment sorting varies between [0.97; 1.45φ], the skewness varies between [−0.16; 0.17φ] and the kurtosis varies between [0.74; 1.27φ]. The particle size distributions observed for the triplicate samples collected at each sampling location were observed to be homogeneous. There were also relatively few differences for samples taken from different locations along the sampling profile. The percentage of gravel varies between 24.8 and 45.4%, the percentage of sand varies between 54.6 and 74.3%, and the percentage of mud between 0.1 and 3.8%. There appeared to be a trend in the particle size data, with a tendency to find coarser sediments near the river bank and progressively finer sediments at greater distances from the river bank. The sediments taken from the eighth sampling point exhibited a prominent increase in the percentage of mud. This was interpreted as being due to the contributions to the sediment from agricultural areas bordering the beach (Table 2, Figs 2 & 3). In general, all of the sediment samples were characterized as very coarse, poorly sorted sands (Table 2). Since these data are very similar along all the entire profile, they are useful forensic parameters for characterizing this particular river beach.

Heavy mineral analysis

The results of the heavy mineral analyses indicate that the heavy mineral assemblages did not vary significantly along the sampled profile. The primary heavy minerals identified were magnetite (in the magnetic fraction), ilmenite (in the 0.2 A fraction), biotite (in the 0.4 and 0.6 A fractions), staurolite (in the 0.6 A fraction), tourmaline (in the 0.8 A fraction) and andalusite (in the diamagnetic fraction) (Fig. 4). The occurrence of biotite is a useful discriminating parameter. Biotite is not a very resistant mineral, which suggests that a source rock

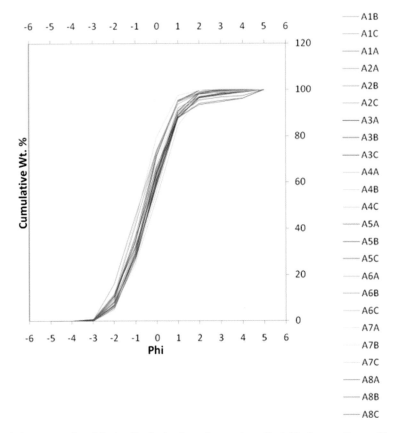

Fig. 3. Cumulative curves of particle size distribution from the samples collected in the sampling profile.

containing biotite is likely to occur in close proximity to the sampling site.

Pollen analysis

The plant survey performed in the sampling site determined that a large diversity of annual weeds, and *Platanus* spp., *Populus* spp. and *Metrosideros* spp. trees, occur in the immediate vicinity of the sampling site (Table 3). Pollen analysis of the samples collected revealed that the pollen assemblages from every sampling location primarily contain pollen types from arboreal vegetation. The majority of arboreal pollen types are not present in the immediate vicinity of the beach but, rather, occur in the surrounding areas within a 5 km

Fig. 4. (a) Biotite observed in the heavy mineral assemblage of the river beach sand samples in the 0.4 and 0.6 A fractions. (b) Magnetite observed in the heavy mineral assemblage in the magnetic fraction.

Table 3. *Spring plant survey from the Areinho river beach in Oliveira do Douro, Vila Nova de Gaia town (north of Portugal)*

\multicolumn{4}{c}{Plants}			
Family	Species	Family	Species
Asteraceae	*Arctotheca calendula*	Oxalidaceae	*Oxalis pes-caprae*
	Cotula coronopifolia		
	Sonchus oleraceus		
Boraginaceae	*Echium plantagineum*	Plantaginaceae	*Plantago lanceolata*
Brassicaceae	*Capsella bursa-pastoris*	Platanaceae	*Platanus* spp.
	Cisymbrium officinale		
	Raphanus raphanistrum		
Caryophylaceae	*Cerastium diffusum*	Poaceae	*Bromus diandrus*
	Silene gallica		*Poa pratensis*
	Stellaria media		*Poa trivialis*
			Vulpia muralis
Fabaceae	*Lupinus angustifolius*	Polygonaceae	*Rumex bucephalophorus*
	Medicago Arabica		
	Melilotus spp.		
	Ornithopus compressus		
	Ornithopus pinnatus		
	Trifolium nigrescens		
Geraniaceae	*Erodium moschatum*	Ranunculaceae	*Ranunculus muricatus*
Hypolepidaceae	*Pteridium aquilinum*	Rubiaceae	*Galium aparine*
Lamiaceae	*Stachys arvensis*	Salicaceae	*Populus* spp.
Liliaceae	*Allium triquetrum*	Solanaceae	*Solanum sublobatum*
Myrtaceae	*Metrosideros* spp.	Urticaceae	*Urtica dioica*
			Urtica membranacea

radius. This illustrates the importance of assessing not only the vegetation at a crime scene but also surveying the vegetation in the surrounding areas (Table 4). Pollen taxa consistent with the annual weeds observed at the sampling site were also identified in samples from the entire profile; however, most were present at trace levels (below 1%). Exceptions to this included the annual weeds *Plantago*, Asteraceae and Poaceae, which were found in the samples in substantial quantities, with the latter two families representing the abundant pollen types from each sampling location. This fact is probably due to the anemophilous nature of these plants and, as a result, their production of large quantities of pollen. This demonstrates the importance of distinguishing between anemophilous and entomophilous plant species, as knowledge of pollination mechanisms and pollen production makes it possible to account for the apparent over- or under-representation of certain plant taxa in the pollen spectra of sediment samples. Sixty-one different pollen types were identified along the profile, and 23 of them were present in all of the samples. Despite the pollen diversity, a similar dominant pollen assemblage was observed along the entire profile, differing only in their relative abundances.

A greater diversity of pollen was found immediately next to the river, with greater amounts of pollen types including Poaceae, *Pinus*, *Alnus*, Brassicaceae, *Salix*, Lamiaceae, *Castanea* and Aceraceae being observed at the river bank, and progressively smaller amounts observed at greater distances from the river. These results suggest the possibility that certain pollen types are being carried by the river and deposited on the river bank, including pollen from *Pinus*, *Alnus*, *Salix*, *Castanea* and *Aceraceae*, as these taxa are not represented at the sampling location. Increasing amounts of *Platanus*, Asteraceae tubuliflorae, Caryophylaceae and *Chenopodium* pollen were observed at progressively greater distances from the shore, suggesting that this pollen was mainly derived from the surrounding vegetation.

However, other pollen types dispersed along the beach were not found to have any pattern associated with their variation, including pollen from *Cupressus*, *Quercus*, Oleaceae, Ericaceae, Urticaceae, Myrtaceaea and Asteraceae liguliflorae. Some of these taxa that are present in the surrounding vegetation, namely *Cupressus*, *Quercus*, Oleaceae and Ericaceae, are anemophilous and, therefore, could have been transported by wind and randomly dispersed along the sampling profile. In the current study, the pollen assemblages along the profile generally contained the same pollen types but with different relative abundances. A similar trend was also reported by Guedes *et al.* (2011), who found that pollen assemblages from several beaches in

Table 4. Pollen types observed in the samples collected along the sampling profile

Classes*	A1		A2		A3		A4		A5		A6		A7		A8	
>10%	Poaceae	32.1	Poaceae	18.3	Poaceae	31.9	Poaceae	27.7	Pinus	19.0	Platanus	27.2	Platanus	41.2	Platanus	42.1
			Pinus	15.4	Asteraceae Tubuliflorae	12.5	Pinus	17.7	Poaceae	18.1	Poaceae	15.8			Poaceae	15.6
									Platanus	10.7	Pinus	15.7				
5–10%	Ericaceae	7.7	Myrtaceae	5.2	Platanus	8.3	Asteraceae	8.6	Poaceae	7.4	Asteraceae	9.2	Poaceae	8.0	Ericaceae	9.7
	Pinus	7.6	Platanus	5.2	Pinus	8.2	Tubuliflorae		Asteraceae	6.2	Tubuliflorae		Asteraceae	7.5	Asteraceae	5.4
	Alnus	6.4	Ranunculus	5.2			Platanus	5.8	Tubuliflorae				Tubuliflorae		Tubuliflorae	
			Alnus	5.1					Quercus	6.2			Pinus	5.3		
1–5%	Asteraceae	4.9	Urticaceae	4.9	Plantago	4.9	Urticaceae	4.4	Cupressus	4.2	Quercus	5.0	Urticaceae	4.4	Quercus	4.1
	Liguliflorae		Ericaceae	4.0	Alnus	3.4	Quercus	3.8	Ericaceae	3.1	Urticaceae	3.8	Ericaceae	3.7	Urticaceae	3.0
	Brassicaceae	3.9	Acacia	3.2	Quercus	2.9	Cupressus	2.9	Castanea	2.2	Cupressus	2.8	Quercus	3.0	Asteraceae	1.9
	Cupressus	3.3	Asteraceae	2.9	Urticaceae	2.4	Myrtaceae	2.3	Myrtaceae	1.9	Myrtaceae	2.7	Cupressus	1.6	Liguliflorae	
	Quercus	3.0	Liguliflorae		Oleaceae	2.3	Ericaceae	2.1	Oleaceae	1.8	Ericaceae	2.4	Plantago	1.6	Oleaceae	1.6
	Urticaceae	2.8	Salix	2.8	Acacia	2.1	Alnus	1.5	Alnus	1.3	Oleaceae	1.3	Myrtaceae	1.5	Cupressus	1.5
	Myrtaceae	2.6	Aceraceae	2.5	Castanea	1.6	Fabaceae	1.5	Brassicaceae	1.3	Plantago	1.0	Alnus	1.2	Pinus	1.5
	Salix	1.7	Brassicaceae	2.3	Cupressus	1.6	Asteraceae	1.3					Caryophyllaceae	1.1	Chenopodium	1.2
	Lamiaceae	1.4	Cupressus	1.7	Myrtaceae	1.6	Liguliflorae									
	Castanea	1.3	Oleaceae	1.7	Brassicaceae	1.4	Oleaceae	1.3								
	Plantago	1.2	Castanea	1.2	Ericaceae	1.2	Acacia	1.1								
	Oleaceae	1.0														
	Platanus	1.0														

*Percentage of representation (pollen counts of each identified taxon were converted into % of total counts). Only percentages above 1% were considered.

the Algarve region generally contained the same pollen types but with different relative abundances, allowing the different sites to be distinguished.

Conclusions

Our results revealed that the colour, grain size distribution and heavy mineral assemblages are similar across the entire sampling profile along the river beach. Although the relative abundance of different pollen types changed along the sampled profile, the dominant pollen types were also similar in the eight sampled points. The data reported here offer a complete characterization of this river beach in terms of its colour, grain size distribution, heavy mineral and pollen assemblages. These data could be used to compare sand from the Areinho fluvial beach with sand from other locations. The data from this study will be incorporated into a geographical information systems (GIS) database that could be used as an investigative aid in forensic investigations involving sand, soil or sediment samples.

Á. Carvalho benefits from a PhD scholarship (SFRH/BD/61460/2009) and H. Ribeiro from a post-doctoral scholarship (SFRH/BDP/43604/2008), both funded by Fundação da Ciência e Tecnologia (QREN-POPH-Type 4.1-Advanced Training, subsidized by the European Social Fund and national funds MCTES). The authors also acknowledge funding from POCI 2010 (Fundação para a Ciênciae e a Tecnologia).

References

BLOTT, S. J. & PYE, K. 2001. Gradistat: a grain size distribution and statistics package for the analysis of unconsolidated sediments. *Earth Surface Processes and Landforms*, **26**, 1237–1248.

ERDTMAN, G. 1969. *Handbook of Palynology. An Introduction to the Study of Pollen Grains and Spores.* Hafner, New York.

FAEGRI, K. & IVERSEN, J. 1989. *Textbook of Pollen Analysis*, 4th edn. Blackwell Scientific, Oxford.

FOLK, R. L. & WARD, W. C. 1957. Brazos River bar: a study in the significance of grain size parameters. *Journal of Sedimentary Petrology*, **27**, 3–26.

GUEDES, A., RIBEIRO, H., VALENTIM, B. & NORONHA, F. 2009. Quantitative colour analysis of beach and dune sediments for forensic applications: a Portuguese example. *Forensic Science International*, **190**, 42–51.

GUEDES, A., RIBEIRO, H., VALENTIM, B., RODRIGUES, A., SANT'OVAIA, H., ABREU, I. & NORONHA, F. 2011. Characterization of soils from the Algarve region (Portugal): a multidisciplinary approach for forensic applications. *Science and Justice*, **51**, 77–82.

HORROCKS, M. 2004. Sub-sampling and preparing forensic samples for pollen analysis. *Journal of Forensic Sciences*, **49**, 1024–1027.

MILNE, L., BRYANT, V. & MILDENHALL, D. 2004. Forensic palynology. *In*: COYLE, H. M. (ed.) *Forensic Botany. Principles and Applications to Criminal Casework.* CRC Press, Boca Raton, FL.

MURRAY, R. & TEDROW, J. C. F. 1986. *Forensic Geology: Earth Sciences and Criminal Investigation.* Rutgers University Press, New York.

REILLE, M. 1992. *Pollen et spores d'Europe et d'Afrique du Nord.* Éditions du Laboratoire de Botanique Historique et Palynologie, Marseille.

REILLE, M. 1995. *Pollen et spores d'Europe et d'Afrique du Nord (Vol. Supplément 1).* Éditions du Laboratoire de Botanique Historique et Palynologie, Marseille.

REILLE, M. 1998. *Pollen et spores d'Europe et d'Afrique du Nord (Vol. Supplément 2).* Éditions du Laboratoire de Botanique Historique et Palynologie, Marseille.

Integration of different sediment characteristics to discriminate between sources of coastal sediments

ALEXANDRA GUEDES[1]*, RAYMOND C. MURRAY[2], HELENA RIBEIRO[1], ANDREIA RODRIGUES[1], BRUNO VALENTIM[1], HELENA SANT'OVAIA[1] & FERNANDO NORONHA[1]

[1]*Centro de Geologia e Departamento de Geociências, Ambiente e Ordenamento do Território, Faculdade de Ciências, Universidade do Porto, Porto, Portugal*

[2]*106 Ironwood Place, University of Montana, Missoula, MT 59803, USA*

*Corresponding author (e-mail: aguedes@fc.up.pt)

Abstract: In order to investigate the effects of geological setting and the surrounding lithology on coastal sediment samples, four properties were analysed in a series of samples collected from different localities: colour determined by spectrophotometry; particle size distribution determined by laser granulometry; chemical composition determined by ICP-MS (inductively coupled plasma mass spectrometry); and low-field magnetic susceptibility determined using a susceptibility meter. A hierarchical cluster analysis was applied to ascertain the capacity of the different properties for discrimination between samples from the different geological settings.

The study reveals that colour analysis only allowed discrimination between different geographical areas; particle size distribution allowed separation between dune and beach samples; chemical composition allowed discrimination between both different geological settings and also between dune and beach sands; and, finally, the combination of magnetic susceptibility with colour and particle size distribution allowed for clear discrimination between geological settings, and between beach and sand dunes.

The studied samples are part of a larger and growing collection of data, and a more extensive evaluation will continue to be made for a more definitive assessment of the use of different characteristics for discrimination. So far, the results give confidence that it is meaningful to use these characteristics, especially when examined in combination, to distinguish between different locations.

Soils are an important class of trace evidence owing to the almost unlimited number and types of materials they contain, and the large number of observations and analyses that can be carried out on them (Murray & Tedrow 1992; Murray 2011). Also important in forensic investigations is the comparison of questioned samples with samples of known origin referenced in a database, and the quantification of the degree of similarity observed between samples and its significance (Morgan & Bull 2007; Pye & Blott 2009).

In Portugal, a soils database for forensic purposes is being developed, with the referenced soils analysed and studied using different techniques. At the present time, this database has more than 200 geographically well-referenced samples from coastal Portugal. The samples used in this study, in addition to completing the reference database under development, are from distinct geological settings and, therefore, can be used to ascertain whether it is possible to discriminate between them based only on their geological attributes. If they can be distinguished, then samples that are geographically well referenced in the database will provide relevant information for the selection of search areas for forensic investigations in the future. The database will provide valuable contextual information on the characteristics of the samples collected (Dawson & Hillier 2010; Guedes *et al.* 2011) that are representative and that can be compared. The importance of integrating data from a range of standard procedures and analytical techniques, such as described in the publications by Pye & Blott (2004, 2009), Pye *et al.* (2006*a*, *b*, 2007), Morgan & Bull (2007), Pye (2007), Manrong *et al.* (2009) and Dawson & Hillier (2010) amongst others, is very well established, especially the combination of independent techniques such as geological and biological properties. The combination of a range of datasets underpins the interest in the development of databases that contain the most common soils, which could then be applied as reference materials for comparison with samples related to crime scenes. In fact, nowadays, the importance of soil databases for forensic purposes is widely considered, although the amount of work involved, together with the lack of human resources available, has resulted in the fact that only now are they starting to be developed, most of them in the UK (Saye & Pye 2004; Macaulay Land Use Research Institute 2009), the USA (Menchaca et al. 2007), where geoforensics is relatively well advanced, and in Canada (Dalpé et al. 2010). In this study, and in order to investigate the effects of geological setting and the surrounding lithology on the different geological properties, several samples of beach and dune sand were collected from coastal areas in the NW of Portugal and the following properties analysed: colour determined by spectrophotometry; particle size distribution determined by laser granulometry; chemical composition determined by ICP-MS (inductively coupled plasma mass spectrometry); and low-field magnetic susceptibility determined using a susceptibility meter. These data will be added to the Portuguese soils forensic database, which is under construction.

Materials and methods

Sample collection and handling

Samples, of beach and/or dune sand, were collected from five coastal areas of NW Portugal (Quiaios, Figueira da Foz, Cabedelo, Espinho and Afife), each of them surrounded by different bedrock geology: limestones and other carbonate rocks from the Mesozoic and Cenozoic borderland (Quiaios, dune (F124) and beach (F125) samples; Figueira da Foz, dune sample (F126); Cabedelo, spit sample on the Mondego river mouth (F127)); metasedimentary rocks from the Variscan basement (Espinho, dune (F143) and beach (F144) samples); and granites, also from the Variscan basement (Afife, beach (F128) and dune (F129) samples) (Fig. 1).

In order to quantify the small-scale local variability, three samples were taken (A, B and C in Table 1) at each point following the procedure described in previous studies (Saye & Pye 2004; Pye 2007). A total of 24 surface samples were collected over a 3 week period during 2008. At each site 200–400 g per sample of sediment were manually collected from the surface (<5 cm depth), using a plastic spade (carefully cleaned after each sample), and placed within a plastic bag.

The samples were dried at 40°C and divided into two subsamples: one retained as a duplicate and stored; and the other used for colour, particle size distribution, elemental composition and magnetic susceptibility analyses. The subsamples for analysis were dry-sieved for particle size distribution (<1 mm size fraction), chemical composition (<150 μm size fraction) and magnetic susceptibility analyses (<150 μm size fraction).

Colour analysis

Colour measurements were performed on bulk samples (Guedes et al. 2009) using a Konica Minolta CM-2600d spectrophotometer programmed with the following settings: measurement area of 0.8 mm of diameter; specular component included; CIE Standard Illuminant D65 representing average daylight with a correlated colour temperature of approximately 6504 K, including the ultraviolet wavelength region; and CIE 1964 Standard Observer (10° Observer). The spectrophotometer was set to take three sequential measurements, giving the means of the colour co-ordinates obtained. The spectrophotometer was calibrated prior to and during the sample measurements to take into account the influence of daylight variation. In addition, before measurement, the spectrophotometer was calibrated according to the manufacturer's instructions, and it was regularly calibrated throughout the measurement period due to the repeated nature of the measurements being carried out. Negative calibration was performed by directing the apparatus measuring port into the air while positive calibration was performed with an international standard white calibration plate. Sample material was homogenized and presented in a standard glass Petri dish, and an average of five–eight measurements were taken for each of the three samples collected at each site. The colour parameters recorded correspond to the uniform colour space CIELAB (or CIE $L*a*b*$) and were directly computed by the spectrophotometer using the SpectraMagic NX software.

Particle size distribution

The <1 mm dry-sieved fraction was automatically analysed using a Coulter LS130 Laser Beam Granulometer, coupled with a liquid module, and the particle size distribution of the <900 μm size fraction determined. The accuracy and precision of this method is regularly tested.

Chemical composition

Major, trace and rare earth elements (REE) were determined on the ground standardized <150 μm size fraction. This size fraction has been used extensively in previous forensic investigations (Pye & Blott 2009). However, a composite sample from each site was used as not enough of the <150 μm fraction was obtained for geochemical analysis from just one individual sample from each site. Therefore, the results provide an average value for

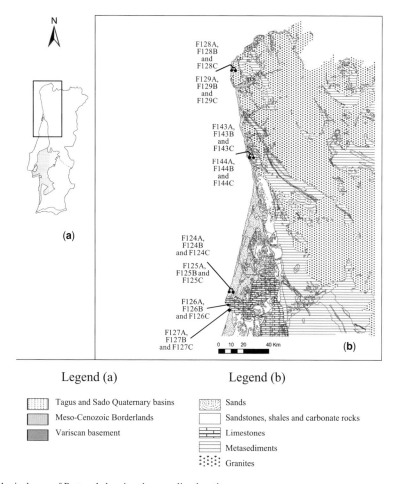

Fig. 1. Geological map of Portugal showing the sampling locations.

the area sampled. After sieving and homogenization, a subsample of approximately 1 g in mass was removed with a spatula and ground, using an agate pestle, for analysis. Since studies of repeatability and reproducibility with ICP-MS have already been carried out for forensic purposes by Pye et al. (2006b), and a standardized procedure has been developed, ICP-MS was selected to determine the elemental composition of the sediments using acid digestion Ultratrace ICP-MS analysis at ACME labs (Canada).

Magnetic susceptibility

Magnetic susceptibility determination was performed with a Kappabridge, model KLY-4S, Agico balance equipped with the Sumean software. Measurement was performed on 1 g of each dry-sieved <150 μm fraction, and the magnetic susceptibility (χ) calculated in m^3/kg. Magnetic susceptibility is directly proportional to the quantity and grain size of ferromagnetic materials in the sample. The accuracy and precision of this analytical method is regularly tested.

Statistical analysis

Measurement precision was evaluated only in terms of colour analysis, with reproducibility within-sample $L^*a^*b^*$ colour parameters evaluated using the coefficient of variation (CV) (where L^* represents lightness; a^* represents the red/green axis; and b^* represents the yellow/blue axis). Between-sample reproducibility was evaluated in terms of colour, particle size distribution and magnetic susceptibility analyses. Since $L^*a^*b^*$ measures present a normal distribution (evaluated by the Shapiro–Wilk test p-value <0.01), an analysis of

Table 1. Descriptive statistics and ANOVA test of $L^*a^*b^*$ indices, particle size and magnetic susceptibility descriptive parameters measured at each of the three samples collected at each site from different geological settings: Carbonate surroundings: Figueira da Foz, dune sample; Quiaios, beach and dune samples; Cabedelo, spit sample

		Colour													Particle size distribution (μm)									Magnetic susceptibility ($\times 10^{-8}$ m³/kg)	
		Measures						ANOVA																	
		L^*		a^*		b^*				L^*		a^*		b^*											
	Samples	Mean	CV%[1]	Mean	CV%	Mean	CV%	df[2]	F[3]	Sig.[4]	F	Sig.	F	Sig.	Mean	Median	Mode	D_{10}	D_{25}	D_{50}	D_{75}	D_{90}	Value	CV%
Quiaios	F124A	59.71	1.21	3.07	5.67	10.58	3.73	23	58.1	0.00	32.2	0.00	101.1	0.00	610	607	860	374	477	607	765	854	6.07	28
	F124B	58.74	1.45	2.69	5.81	9.62	1.8								585	578	860	332	430	578	768	858	5.14	
	F124C	62.86	1.3	3.4	5.89	11.73	2.36								633	629	860	398	502	629	803	865	3.39	
	F125A	68.62	1.34	3.06	7.65	12.42	2.87	23	2.1	0.15	0.7	0.51	5.1	0.02	569	554	545	356	439	554	700	823	3.97	11.6
	F125B	68.7	0.86	3.03	7.57	12.57	2.62								535	517	546	343	418	517	639	778	4.79	
	F125C	69.27	0.76	2.92	10.29	11.96	4.21								527	505	498	335	406	505	630	781	4.96	
Figueira	F126A	64.68	0.7	3.04	6.79	11.58	3.48	23	3.7	0.04	1.8	0.19	4.2	0.03	562	561	785	312	412	561	719	822	4.77	20.1
	F126B	63.81	1.55	2.97	4.3	11.32	3.84								537	511	860	294	378	511	714	832	7.07	
	F126C	64.49	0.69	3.14	6.31	11.95	3.83								582	570	860	318	415	570	775	855	5.58	
Cabedelo	F127A	65.5	1.63	2.87	7	10.88	4.77	23	2.8	0.09	6.2	0.01	5.9	0.01	339	325	316	215	261	325	402	483	17.01	24.2
	F127B	64.2	2.05	2.61	2.35	10.17	4.89								438	407	379	251	313	407	536	683	11.39	
	F127C	64.29	1.78	2.61	6.9	10.09	4.04								398	378	379	246	300	378	475	581	11.5	
Espinho	F143A	50.93	1.09	2	6.28	7.38	3.9	20	2.3	0.13	2.0	0.16	0.4	0.67	547	559	860	263	425	559	681	846	35.32	13.9
	F143B	50.59	0.68	2.11	4.65	7.49	2.67								588	594	860	344	467	594	743	857	30.37	
	F143C	50.44	0.93	2.09	6.29	7.45	3.25								604	611	860	359	482	611	768	859	26.79	
	F144A	63.09	2.27	3.39	4.6	11.94	5.21	22	0.3	0.76	11.5	0.00	7.0	0.00	745	808	860	546	642	808	859	883	5.9	18.4
	F144B	63.7	5.26	3.76	4.8	13.25	5.72								746	809	860	548	644	809	859	883	4.84	
	F144C	62.9	1.63	3.46	4.47	12.52	5.68								758	830	860	566	652	830	865	886	4.09	
Afife	F129A	61.31	0.54	1.81	4.38	8.62	2.31	23	12.4	0.00	14.9	0.00	0.0	0.96	393	382	379	244	304	382	476	572	18.64	22.3
	F129B	61.65	1.15	1.6	3	8.59	2.96								365	356	379	231	285	356	440	526	26.81	
	F129C	60.41	0.52	1.66	6.36	8.6	3.2								394	383	379	246	305	383	475	571	29.28	
	F128A	58.11	1.75	1.14	4.6	6.32	2.71	17	8.5	0.00	8.3	0.00	13.0	0.00	523	503	498	317	398	503	633	793	22.85	31.3
	F128B	60.69	1.59	1.29	4.5	6.84	3.05								524	504	498	323	401	504	628	789	30.37	
	F128C	59.92	2.09	1.21	6.32	6.87	3.05								522	503	498	310	388	503	650	780	15.96	

Metasediments surrounding: Espinho, beach and dune samples. Granites surrounding: Afife, beach and dune samples. Dune samples are in normal text. Beach samples are in underlined text, while beach samples are in normal text.
[1] Coefficient of variation.
[2] Degrees of freedom.
[3] Value of the test.
[4] Significance.

variance (ANOVA) was performed to evaluate the null hypothesis that $L^*a^*b^*$ mean values measured at each site are statistically equal. Since particle size distribution in our study is an ordinal variable, a Kruskal–Wallis one-way analysis of variance was performed to evaluate the null hypothesis that the three sample medians collected at each site are drawn from the same population; that is, are statistically equal. For magnetic susceptibility measures, reproducibility was evaluated by the coefficient of variation.

A first hierarchical cluster analysis was performed in order to ascertain whether discrimination between samples using each analysis method alone could be achieved, and if so how significant it was. A second analysis was performed by combining the results of colour, particle size distribution and magnetic susceptibility analysis. The geochemical data from the ICP-MS analysis were not included owing to the lack of a sufficient amount of the <150 μm fraction in each of the three individual samples collected from every site. All the cluster analyses were performed on standardized values in order to reduce the effects of scale, and the number of clusters was determined using: (i) the squared Euclidean distance as a distance measure; and (ii) the Ward method as a linking method. In elemental analysis, elements that presented values below the lower detection limit in at least one sample were excluded from this study. The statistical analysis software that was used for all the analysis was SPSS (16.0) and the data interpretation followed the concerns of Morgan & Bull (2006).

Results and discussion

A total of 57 quantitative variables (colour analysis (three), particle size distribution (eight), elemental composition analysis (45) and magnetic susceptibility (1)) were obtained and considered in this study.

Colour analysis

Concerning the $L^*a^*b^*$ system colour sphere, the highest values were usually observed for dune and beach samples with carbonate rocks in their surrounding areas, while the lowest values were always observed for Afife beach (granites) and Espinho dune (metasediment) samples (Table 1). L^* values measured varied between 50.44 (measured on Espinho dune sample-metasediments) and 69.27 (measured on Quiaios beach sample carbonate rocks); a^* varied between 1.14 (measured on Afife beach sample granites) and 3.76 (measured on Espinho beach sample metasediments); and b^* varied between 6.32 (measured on Afife beach sample granites) and 13.25 (measured on Espinho beach sample metasediments).

The coefficient of variation of $L^*a^*b^*$ values obtained from the replicate samples were always lower than 10%, demonstrating measurement reproducibility. The lowest values were observed on L^*, while the parameter with the highest variation for all samples was a^* (Table 1).

The results of the ANOVA statistical analysis, which was performed to evaluate sampling reproducibility within the same sites, showed significant within-grid sampling variation between at least one colour parameter; although the exception was observed in the *Espinho* dune sample (F143). This possibly indicates that, in soil samples, three samples are not sufficient to represent the studied site in terms of colour properties and that more samples within the same site should be taken. Pye *et al.* (2006a) compared small-scale spatial colour variability of soil samples. They observed that a larger proportion of the total variation within one site was obtained with five samples, and suggested that a minimum of three samples should be performed to adequately represent within-site variation of an area of interest.

The results of the agglomeration hierarchical cluster analysis showed that samples from Espinho (metasediments: F143) and Afife (granites: F128 and F129) could be discriminated from samples from the Figueira–Quiaios–Cabedelo (carbonate rocks: F124, F125, F126 and F127) into two major distinct and separate groups (cluster distance close to 25) (Fig. 2). The exception was the beach sample from Espinho (F144 metasediments), which clustered closer to the Figueira–Quiaios–Cabedelo samples (carbonate rocks). This may be due to the fact that this beach is actually being artificially 'fed' with sand.

Some discrimination between dune and beach samples was observed. Furthermore, despite the differences observed in sampling reproducibility within the same site, samples F125 (Quiaios), F128 and F129, (Afife) and F143 (Espinho) (the first two beach and the other dune) could be distinctively discriminated between based on colour alone (Fig. 2).

Particle size distribution

For all of the samples, the particle size distribution patterns were mainly located in the coarser size classes (>300 μm) (Fig. 3). According to the results of the Kruskal–Wallis one-way ANOVA analysis, statistically significant sampling reproducibility was observed, considering particle size distribution by classes, in each set of three samples collected within the same site. Therefore, it is

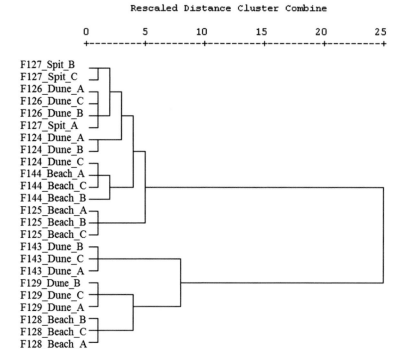

Fig. 2. Cluster analysis with $L*a*b*$ indices measured on samples from the different geological settings. Carbonates: Figueira da Foz, dune sample (F126); Quiaios, beach (F125) and dune (F124) samples; Cabedelo, spit sample (F127). Metasediments: Espinho, beach (F144) and dune (F143) samples. Granites: Afife, beach (F128) and dune samples (F129).

inferred that three samples were sufficient to represent the variability in terms of particle size distribution at our studied sites.

Particle size distribution by classes also reveal that Figueira, Afife and Espinho dune samples present a wider particle size distribution range; Quiaios beach and dune, and Cabedelo spit samples only contain sediments coarser than 100 μm, while the Espinho beach sample is characterized by material coarser than 180 μm.

Particle size descriptive parameters showed a mean distribution size between 339 (Cabedelo spit sample: F127) and 758 μm (Espinho beach sample; F144); a mode between 316 and 860 μm; and a median of 325 (Cabedelo spit and Afife dune samples) and 830 μm (Espinho beach sample). The mode for the Quiaios dune, the Figueira dune, and Espinho dune and beach samples are all similar. Sediment sorting varies between 268 (Cabedelo spit sample: F127) and 583 μm (Espinho dune sample: F143) (Table 1). Hierarchical cluster analysis, based on the descriptive parameters of particle size distribution, only confirms a clear discrimination between dune samples and samples collected from beach settings (Fig. 4).

Samples from the Afife (F129 granites) dune and the Cabedelo (F127 carbonate rocks) spit were clustered together and were separated from the other samples as they are composed of less coarse material. In contrast, the sample from Espinho beach (F144 metasediments) was discriminated in a single group, being the coarser sample. Afife and Quiaios beach samples (F128 and F125, respectively, with granites and carbonate rocks in their surroundings) were clustered together and have, as a common feature, very close mean, median and mode parameter values. Finally, Quiaios, Figueira and Espinho dune samples (F124, F126, carbonate rocks and F143, metasediments) present similar mode values, and were clustered in one group. Thus, the dune samples clustered together rather than being able to discriminate between each geological setting.

Chemical composition

The chemical composition of the samples was determined by ICP-MS analysis for total Fe, Ca, P, Mg, Ti, Al, Na, K, Mn, Cu, Pb, Zn, Ni, Co, As, U, Th,

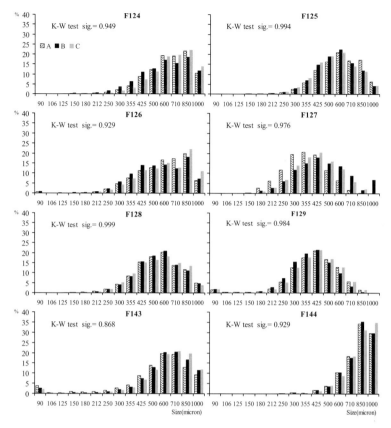

Fig. 3. Percentage of particle size distributions by classes and of sampling reproducibility evaluation by Kruskal–Wallis' one-way analysis of variance of the three samples collected within the same area of each of the eight sampling sites from different geological settings. Carbonates: Figueira da Foz, dune sample (F126); Quiaios, beach (F125) and dune (F124) samples; Cabedelo, spit sample (F127). Metasediments: Espinho, beach (F144) and dune (F143) samples. Granites: Afife, beach (F128) and dune samples (F129).

Sr, Cd, Sb, Bi, V, Cr, Ba, W, Zr, Sn, Be, Sc, Hf, Li, Rb, Ta, Nb, Cs, Ga, Y, La, Ce, Pr, Nd, Sm, Eu, Gd, Tb, Dy, Ho, Er, Tm, Yb and Lu concentrations (Table 2).

The composition in terms of major and trace elements showed that the highest concentrations of Mn, Fe, U, Th, Ti, W, Zr, Ta and Nb were obtained in samples F127 (carbonate setting), F128 and F129 (granitic settings). However, sample F127 also shows high concentrations of Ca, Mn, Mg, Cr and V. Sample F128 (beach granitic setting) gave the highest concentration of Pb, Ni, Fe, As, U, Th, W, Zr, Sn, Li, Ta and Ga in relation to all of the sample sets.

The highest concentration of Ca was obtained in samples F124, F125, F126 and F127 (carbonate setting). Sample F125 also shows the highest Cu, Zn and Sr concentrations, and sample F126 has the highest concentration of Ba and Rb of this set.

Sample F143 (metasedimentary setting) has higher values of Li and Rb than the other samples.

Based on the major and trace element composition, three sets of samples can be defined: a set with samples F128 and F129 (samples in a granitic setting and from a granitic origin); a set with samples F124, F125, F126 and F127 (samples in a limestone setting and, essentially, of a carbonate origin), with sample F127 showing similarities between both granitic and limestone settings; and, finally, a set with samples F143 and F144 (samples in a metamorphic settings), with no similarities with the other two sets.

In relation to the calculated REE concentrations, samples from granitic areas (F128 and F129) show the highest values of REEs, some of which being higher than the upper detection limit (La, Ce, Nd), and, although they can be compared directly for forensic purposes (Pye et al. 2007), a normalization

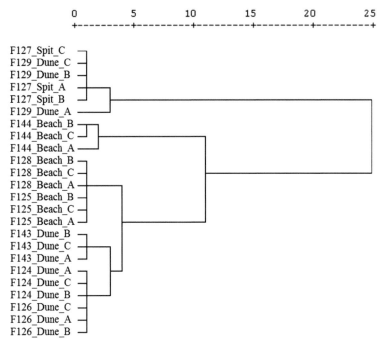

Fig. 4. Cluster analysis with particle size distribution indices (mean, median, mode, and D_{10}, D_{25}, D_{75}, D_{90} and $D_{90}-D_{10}$) measured on each of the three samples collected within the same site from different geological settings. Carbonates: Figueira da Foz, dune sample (F126); Quiaios, beach (F125) and dune (F124) samples; Cabedelo, spit sample (F127). Metasediments: Espinho, beach (F144) and dune (F143) samples. Granites: Afife, beach (F128) and dune samples (F129).

reported for chondritic meteorites (Taylor & Mclennan 1985) was conducted. The highest values observed can be explained by the presence of zircon and monazite as accessory heavy minerals (Fig. 5). Based on the REE concentration, and similar to the major and trace elements composition, the normalized REEs reveal the existence of three sets of samples with distinctive geochemical characteristics: one of high ΣREEs constituted by samples F127 (spit-limestone setting), and F128 and F129 (granitic setting); another with lower REE concentrations from samples F124, F125, F126 (carbonate setting) and F143 (metasedimentary setting); and, finally, one set showing the lowest REE concentration and composed of sample F144 (metasedimentary setting). Within the set containing F127 (limestone setting), F128 and F129 (granitic setting), the different samples are differentiated as they show different REE profiles, with samples F128 and F129 richer in ΣREE content, especially in light REE (LREE), than sample F127.

Comparing the REE profiles, sample F127 presents an intermediate profile position between samples F124 and F126 (limestone setting) and samples F128 and F129 (granite setting), thus suggesting a mixed origin. Within the set of samples F124, F125, F126 (carbonates) and F143 (metasediments), there is a similarity between the REE profiles. However, F143 has a low REE concentration. Sample F144 may also be discriminated on the basis of the REEs profile because, in contrast to the other samples, it shows a smoother REE pattern in the region Sm–Eu–Gd. This profile, together with the trace element composition, permitted discrimination of samples from a granite background from samples surrounded by limestones and metasediments.

The above-mentioned discrimination was confirmed by combining the trace and rare earth element data for cluster analysis (Fig. 6). The exception was the sample from *Cabedelo*, which clustered closer to the samples from a granitic background rather than to samples from a limestone bedrock. This may be explained as being due to the location of this sampling point next to the Mondego river mouth, which brings sediments from the central Portugal granite batholith with its W and U mineralization.

Table 2. Major, trace and rare earth element compositions of the <150 μm sediment fraction from eight composite samples collected in different geological settings

Surrounding lithology:	Limestones				Granites		Metasediments	
Location:	Quiaios		Figueira da Foz	Cabedelo	Afife		Espinho	
	Dune	Beach	Dune	Spit	Beach	Dune	Dune	Beach
Element/samples	F124	F125	F126	F127	F128	F129	F143	F144
Major (values in %)								
Fe	2.0	1.0	1.6	11.3	14.8	6.6	1.9	0.3
Ca	1.5	1.6	1.4	2.8	0.8	0.4	0.3	0.5
P	0.1	0.0	0.1	0.2	0.4	0.2	0.1	0.0
Mg	0.3	0.3	0.3	0.7	0.4	0.2	0.3	0.1
Ti	1.0	0.3	0.7	5.9	5.4	3.0	0.4	0.0
Al	2.2	2.8	2.7	3.8	4.1	3.0	3.1	1.8
Na	0.4	0.5	0.5	0.2	0.4	0.5	0.5	0.4
K	1.2	1.4	1.8	0.6	1.2	1.4	1.7	1.5
Mn	0.2	0	0.1	1	0.7	0.5	0.1	0
Trace (values in ppm)								
Cu	36.8	845.2	31.5	30.0	87.3	27.3	45.7	34.5
Pb	24.3	15.3	27.3	45.2	72.5	33.2	61.1	13.0
Zn	122.0	532.8	90.0	304.0	414.0	151.5	93.5	28.1
Ni	7.7	8.6	8.5	17.9	52.1	14.8	17.3	2.6
Co	4.4	3.3	4.1	17.3	25.9	9.1	6.7	0.8
As	4.9	6.6	7.3	13.6	91.9	23.1	11.8	3.5
U	4.4	1.5	2.5	31.0	80.5	43.2	4.4	0.9
Th	24.3	7.5	12.9	192.9	453.4	206.0	20.5	2.2
Sr	94.0	106.0	105.0	99.0	88.0	57.0	70.0	56.0
Sb	0.6	0.6	1.2	1.3	9.2	2.7	2.1	0.2
Bi	0.3	0.2	0.3	1.1	1.3	0.7	0.6	0.1
V	25.0	21.0	24.0	104.0	90.0	39.0	41.0	4.0
Cr	49.0	19.0	39.0	323.0	61.0	25.0	41.0	8.0
Ba	175.0	201.0	272.0	110.0	253.0	228.0	257.0	233.0
W	1.6	1.2	1.7	21.0	53.4	19.0	5.4	0.5
Zr	124.4	30.4	78.5	628.9	1212.5	1110.4	70.2	14.6
Sn	3.1	2.5	4.8	9.7	21.7	18.0	13.4	1.4
Sc	3.0	2.5	2.7	14.8	8.4	5.1	3.5	0.5
Hf	3.8	1.0	2.5	19.3	39.0	36.3	2.1	0.5
Li	19.9	19.9	29.1	19.4	49.8	29.1	41.7	15.8
Rb	65.6	66.9	102.0	30.3	73.4	79.7	108.2	78.2
Ta	2.2	0.7	1.9	19.1	42.1	27.0	2.7	0.1
Nb	24.1	9.7	20.0	168.8	159.5	86.1	16.9	1.2
Cs	2.8	2.7	4.8	1.4	4.2	3.7	5.7	2.7
Ga	5.6	6.2	7.4	10.0	14.7	8.6	10.0	3.7
Y	17.0	9.3	11.2	114.4	234.1	112.8	10.9	2.2
REE (values in ppm)								
La	51.2	20.9	33.6	463.7	**>2000**	1439.2	45.0	5.2
Ce	98.1	43.0	69.2	1069.8	**>2000**	**>2000**	86.8	9.6
Pr	13.3	5.8	9.2	120.4	1019.5	327.4	11.3	1.2
Nd	48.3	22.7	34.8	475.1	**>2000**	1395.4	41.2	4.3
Sm	7.9	4.5	5.8	76.7	560.8	203.4	6.8	0.6
Eu	0.8	0.7	0.8	6.3	86.4	29.1	0.7	0.2
Gd	5.6	3.1	4.3	49.3	272.4	112.2	5.1	0.6
Dy	3.8	2.1	2.7	29.2	93.1	39.7	2.9	0.6
Er	1.8	1.0	1.3	11.6	13.9	8.8	1.2	0.2
Yb	1.7	1.0	1.3	11.3	15.4	9.6	1.0	0.3

Bold: higher than the upper detection limit. Cd, Be, Tb, Ho, Tm and Lu presented values below the lower detection limits.

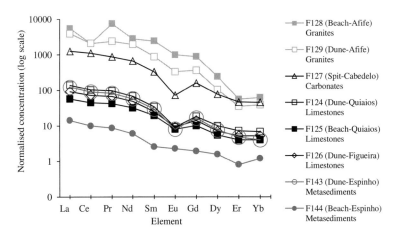

Fig. 5. Different REE profiles (chondrite normalized) in the analysed sediment samples.

The combination of these properties also clustered dune and beach samples apart. Dune samples from Quiaios, Figueira da Foz and Espinho, which have a different regional bedrock geology, combined to form one group and are separated from their respective beach samples. Afife's beach and dune sediments, in spite of being joined in one group, have a cluster distance greater than that separating the Quiaios, Figueira da Foz and Espinho dune sediments from the beach sediments.

Magnetic susceptibility

The magnetic susceptibility values range between 3.39×10^{-8} (Quiaios dune: F124C, limestone setting) and 35.32×10^{-8} m^3/kg (Espinho dune: F143A, metasedimentary setting), with the lowest values obtained on samples from dunes in limestone areas, and the highest in metasediments and granitic areas (Table 1). Variations in magnetic susceptibility values within the same area were observed, with CVs ranging between 12 and 31%, showing within-grid sampling variation that must be taken into account during a forensic investigation (Table 1). In previous cluster analysis, using only magnetic susceptibility, no discrimination between samples was observed. However, after combining the results from colour, particle size distribution and magnetic susceptibility in a cluster analysis, a clear discrimination between samples from different geological settings was observed (Fig. 7). As previously referred to in the Materials and Methods section, it was not possible to perform ICP-MS on the three individual samples collected from each site due to a lack of a sufficient amount of the <150 μm sediment fraction. In addition, since significant within-grid sampling variation was observed in colour and magnetic susceptibility results, the combination of data to obtain one average value of each parameter for each site is

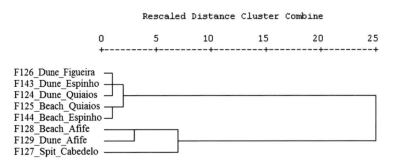

Fig. 6. Cluster analysis with major, trace and rare earth element values determined by ICP-MS analysis, present on each of the three sediment samples collected within the same site from different geological settings. Carbonates: Figueira da Foz, dune sample (F126); Quiaios, beach (F125) and dune (F124) samples; Cabedelo, spit sample (F127). Metasediments: Espinho, beach (F144) and dune (F143) samples. Granites: Afife, beach (F128) and dune samples (F129).

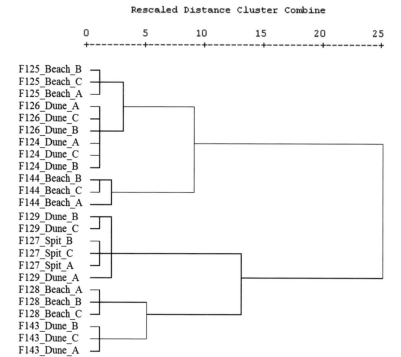

Fig. 7. Cluster analysis with $L^*a^*b^*$ indices, particle size distribution indices (mean, median, mode, and D_{10}, D_{25}, D_{75}, D_{90} and $D_{90}-D_{10}$) and magnetic susceptibility measured on each of the three sediment samples collected within the same site from different geological settings. Carbonates: Figueira da Foz, dune sample (F126); Quiaios, beach (F125) and dune (F124) samples; Cabedelo, spit sample (F127). Metasediments: Espinho, beach (F144) and dune (F143) samples. Granites: Afife, beach (F128) and dune samples (F129).

inadequate. Therefore, the incorporation of elemental composition results along with colour, particle size distribution and magnetic susceptibility into a cluster analysis was not possible in this study.

Conclusions

Our study reveals the importance of combining together different geological attributes for individual samples. Four geoforensic techniques were used in the discrimination of geological settings of coastal Portuguese sediments. Colour analysis allowed differentiation between different geographical areas; particle size distribution allowed separation between dune and beach sediments, but not between geological surroundings; chemical composition allowed both discrimination between different geological surroundings, and between dune and beach sediments; and, finally, the combination of magnetic susceptibility with colour and particle size distribution allowed for a clear discrimination between geological surroundings, and beach and sand dunes.

The results represent an important approach that may enable a specific site 'fingerprint' to be determined that could narrow future selections of coastal search areas in Portugal. The studied samples are part of a larger and growing collection of data, and a more extensive evaluation will continue to be made for a more definitive assessment of the use of different characteristics for discrimination. So far, the results give confidence that it is meaningful to use these characteristics, especially in combination. Finally, this study will contribute to the development of a forensic database of the characteristics of soils of the Portuguese mainland.

This work has been financially supported by the project PTDC/CTE-GEX/67442/2006 (FCT-Portugal).

References

DALPÉ, C., BLANCHARD, C., CHARTRAND, M., ST-JEAN, G. & WOJTYK, J. 2010. Trace element analysis of canadian surface soil and associated quartz by LA-ICP-MS: preliminary development of a geo-location database for forensic investigations. *In*: *Third International Soil*

Forensics Conference Abstracts, Long Beach, California, 31 October–4 November, 11–12.

DAWSON, L. A. & HILLIER, S. 2010. Measurement of soil characteristics for forensic applications. *Surface and Interface Analysis*, **42**, 363–77.

GUEDES, A., RIBEIRO, H., VALENTIM, B. & NORONHA, F. 2009. Quantitative colour analysis of beach and dune sediments for forensic applications: a Portuguese example. *Forensic Science International*, **190**, 42–51.

GUEDES, A., RIBEIRO, H., VALENTIM, B., RODRIGUES, A., SANT'OVAIA, H., ABREU, I. & NORONHA, F. 2011. Characterization of soils from the Algarve region (Portugal): a multidisciplinary approach for forensic applications. *Science and Justice*, **51**, 77–82.

MACAULAY LAND USE RESEARCH INSTITUTE. 2009. *SoilFUN – Soil Forensics University Network* (http://www.macaulay.ac.uk/forensics/soilfun/); *SoilFIT – Integration of Soil Fingerprinting Techniques for Forensic Applications* (http://www.macaulay.ac.uk/soilfit/); GIMI – Geoforensics and Information Management for crime Investigation (http://www.macaulay.ac.uk/geoforensic/).

MANRONG, C., LIZHONG, Y., XIANGFENG, N. & BIN, C. 2009. Application of environmental magnetism on crime detection in a highway traffic accident from Yangzhou to Guazhou, Jiangsu Province, China. *Forensic Science International*, **187**, 29–33.

MENCHACA, P., GRAHAM, R. & STAM, M. 2007. Update on the development of a searchable forensic soil database (SQUID) in California, USA. *In*: *2nd International Conference on Environmental and Criminal Soil Forensics Abstracts*, Edinburgh, 30 October–1 November, 24–25.

MORGAN, R. M. & BULL, P. A. 2006. Data interpretation in forensic sediment and soil geochemistry. *Environmental Forensics*, **7**, 325–334.

MORGAN, R. M. & BULL, P. A. 2007. The philosophy, nature and practice of forensic sediment analysis. *Progess in Physical Geography*, **31**, 43–58.

MURRAY, R. C. 2011. *Evidence from the Earth: Forensic Geology and Criminal Investigation*, 2nd edn, Mountain Press, Missoula, MT.

MURRAY, R. C. & TEDROW, J. 1992. *Forensic Geology*. Prentice-Hall, Englewood Cliffs, NJ.

PYE, K. 2007. *Geological and Soil Evidence: Forensic Applications*. CRC, Boca Raton, FL.

PYE, K. & BLOTT, S. J. 2004. Comparison of soils and sediments using major and trace element data. *In*: PYE, K. & CROFT, D. J. (eds) *Forensic Geoscience: Principles, Techniques and Applications*. Geological Society, London, Special Publications, **232**, 183–196.

PYE, K. & BLOTT, S. J. 2009. Development of a searchable major and trace element database for use in forensic soil comparisons. *Science and Justice*, **49**, 17–181.

PYE, K., BLOTT, S. J., CROFT, D. J. & CARTER, J. F. 2006a. Forensic comparison of soil samples: assessment of small-scale spatial variability in elemental composition, carbon and nitrogen isotope ratios, colour, and particle size distribution. *Forensic Science International*, **163**, 59–80.

PYE, K., BLOTT, S. J. & WRAY, D. S. 2006b. Elemental analysis of soil samples for forensic purposes by inductively coupled plasma spectrometry – precision considerations. *Forensic Science International*, **160**, 178–192.

PYE, K., BLOTT, S. J., CROFT, D. J. & WITTON, S. J. 2007. Discrimination between sediment and soil samples for forensic purposes using elemental data: an investigation of particle size effects. *Forensic Science International*, **167**, 30–42.

SAYE, S. E. & PYE, K. 2004. Development of a coastal dune sediment database for England and Wales: forensic applications. *In*: PYE, K. & CROFT, D. J. (eds) *Forensic Geoscience: Principles, Techniques and Applications*. Geological Society, London, Special Publications, **232**, 75–96.

TAYLOR, S. R. & MCLENNAN, S. M. 1985. *The Continental Crust: Its Composition and Evolution*. Blackwell, Oxford.

How useful are databases in environmental and criminal forensics?

JENNIFER McKINLEY

School of Geography, Archaeology and Palaeoecology, Queen's University Belfast, Elmwood Building, Belfast BT7 1 NN, UK (e-mail: j.mckinley@qub.ac.uk)

Abstract: Advances in computational and information technologies have facilitated the acquisition of geospatial information for regional and national soil and geology databases. These have been completed for a range of purposes from geological and soil baseline mapping to economic prospecting and land resource assessment, but have become increasingly used for forensic purposes. On the question of provenance of a questioned sample, the geologist or soil scientist will draw invariably on prior expert knowledge and available digital map and database sources in a 'pseudo Bayesian' approach. The context of this paper is the debate on whether existing (digital) geology and soil databases are indeed useful and suitable for forensic inferences. Published and new case studies are used to explore issues of completeness, consistency, compatibility and applicability in relation to the use of digital geology and soil databases in environmental and criminal forensics. One key theme that emerges is that, despite an acknowledgement that databases can be neither exhaustive nor precise enough to portray spatial variability at the scene of crime scale, coupled with expert knowledge, they play an invaluable role in providing background or reference material in a criminal investigation. Moreover databases can offer an independent control set of samples.

The generation of maps and databases to provide information at a regional and national scale has long been a focus for government mapping agencies to provide knowledge of a country's natural resources. Advances over the last 10 years in computational and information technologies have facilitated the acquisition of vast amounts of high-quality geospatial information, leading to the development of worldwide initiatives to create regional, continental and worldwide soil and geology databases (Lagacherie *et al.* 2007; Williams *et al.* 2009). Integral to this is synchronization with standard international classification (FAO-UNESCO-ISRIC 1990) and standard soil profiles and with compliance with recognized and emerging 3D standards for online geospatial information and compatibility with an ever expanding range of online data resources. The digital databases have been used for a wide range of purposes such as geological and soil mapping, baseline quality mapping, economic prospecting, land resource assessment, risk evaluation, environmental studies and prediction of soil provenance for forensic purposes. Regional ground-based and remotely sensed surveys (dependent on available funding) have become important methods of generating these spatial databases. Information collected by ground and remotely sensed surveys includes soil properties such as soil geochemistry and airborne geophysics. Examples of soil and geology databases at regional, national, continental and world scale are shown in Table 1. A number of comprehensive regional and national soil sampling programmes have been completed over the last decade across the UK to provide background and baseline quality mapping. These include the G-Base surveys conducted by the British Geological Survey (BGS, http://www.bgs.ac.uk/data/databases.html), the UK Soil and Herbage Pollutant Survey (Environment Agency 2007) covering England, Scotland, Wales and Northern Ireland as well as regional surveys such as the 1987–1997 Soil Atlas of Northern Ireland survey by the Department of Agriculture and Rural Development, DARD (Cruickshank 1997) and the 2004–2006 Tellus project, Geological Survey Northern Ireland (GSNI; funded by the Northern Ireland Department of Enterprise Trade and Investment and by the Rural Development Programme through the Northern Ireland Programme for Building Sustainable Prosperity). The Australian Soil Resource Information System (ASRIS, www.asris.csiro.au), a product of the Australian Collaborative Land Evaluation Progamme jointly funded by the Commonwealth Scientific and Industrial Research Organisation (CSIRO) land and water and the natural heritage trust with contributions from Australian state, territory and federal agencies, was developed to provide good quality online information on land and soil resources for economic, environmental and educational purposes (Table 1).

The rationale is that properties measured in the ground and remote sensed surveys are directly related to different aspects of the region such as underlying geology, superficial deposits and soil types. The resolution of ground-based sampling schemes is a balance between the scale of region

Table 1. *Examples of soil and geology databases at regional, national, multinational and global scale*

Scale / Country	Database	URL
Regional		
Northern Ireland	Tellus GSNI	http://www.bgs.ac.uk/gsni/tellus/
National		
UK	British Geological Survey (BGS)	http://www.bgs.ac.uk
		http://www.bgs.ac.uk/data/databases.html
	LandIS	http://www.landis.org.uk/data/natmap.cfm
	NATMAP	
	Macaulay Institute	
	The SoilFit Project	http://www.macaulay.ac.uk/soilfit/
	The SoilFun project	http://www.macaulay.ac.uk/forensics/soilfun/
Australia	ASRIS	www.asris.csiro.au
		http://www.asris.csiro.au/index_ie.html
USA	National Resources Conservation Service (NRCS) NASIS	http://soils.usda.gov/technical/nasis/
	USGS	http://ngmdb.usgs.gov/ngmdb/ngmdb_home.html
Canada	CanSIS	http://res.agr.ca/cansis/
	National Soil DataBase (NSDB)	http://res.agr.ca/cansis/nsdb/intro.html
Multinational/global		
Europe	European soil database (JRC)	http://eusoils.jrc.ec.europa.eu/esdb_archive/ESDB/Index.htm
	INSPIRE	http://inspire.jrc.ec.europa.eu/
	EuroGeoSurveys	http://www.eurogeosurveys.org/
	BGR and CGMW	http://www.bgr.de/karten/igme5000/igme5000.htm
World	One World Geology	http://www.onegeology.org/
	Digital Soil Map of the world (FAO)	http://www.fao.org/nr/land/soils/digital-soil-map-of-the-world/en/

covered and available funding. The G-BASE standard established by BGS comprises soil sampling at 20 and 50 cm depths on a regular grid at one site per 2 km^2. Ground-based soil geochemistry data generated as part of the Tellus project (GSNI) followed this sampling strategy. The ASRIS provides information at different scales: from general descriptions to more detailed information where mapping information is available to fully characterized soil profiles representative of key areas and environments. Lagacherie *et al.* (2007) propose that existing soil databases are neither exhaustive nor precise enough for promoting an extensive and credible use of the soil information within spatial data infrastructures that is being developed worldwide. They put this down to the fact that conventional soil survey methods are relatively slow and expensive. This arguably may be an equally valid assessment of geology mapping databases.

There is a long history of exploiting knowledge of differing geology and soil distributions in forensic work from the early pioneering work of Ehrenberg (1856, cited in Ruffell & McKinley 2008) and Heinrich (described by Block 1958) to more recent case studies (some examples include Croft & Pye 2004; Dawson *et al.* 2004; Stam 2004; Sugita & Marumo 2004; Pye 2007; Fitzpatrick 2008; Gradusova & Nesterina 2009; Ruffell & Dawson 2009; Murray 2011). Fitzpatrick *et al.* (2009) states that lack of relevant background or reference material can be a limitation in a criminal investigation. On being asked to establish or constrain the likely provenance of a seized sample from a forensic investigation, the well-informed geologist or soil scientist may draw on experiential knowledge, but for most, the first reference source will be a geological and/or soil map(s) or a digital map and database. The Bayesian statistical approach to weighing forensic evidence based on the likelihood ratio calculated on a ranking scheme of the strength of evidence (Aiken 2009) has been seen in a conservative light in courts (Sensabaugh 2009). However 'Bayesian' thinking is adopted by many if not all geoscientists working in forensics, in that intrinsic to the Bayesian approach is the integration of prior knowledge in weighing up the likelihood and ultimately deciding on some type of updated (posterior) decision or final ranking scheme. The choice of prior knowledge has been a source of contention in forensics owing to a lack of appropriate reference materials (Auchie 2009; Sensabaugh 2009). This paper suggests that, intuitively, prior knowledge for the geologist or soil scientist is experiential knowledge and reference to geology and soil maps along with, if available, digital maps and databases.

The use of reference material in forensic work has involved the use of both existing (government) mapping and database information (e.g. ASRIS, BGS) and specially generated databases. One example of an initiative designed to provide appropriately suitable soil forensic intelligence is the SoilFit project (Macaulay Institute 2008a), funded through the EPSRC and led by Macaulay Land Use Research Institute, now part of the James Hutton Institute. The project aims to do this through encompassing multi-disciplinary expertise in the form of the SoilFit team (Macaulay Institute 2008b) drawn from academic and law enforcement organizations across the UK and complemented by an extended network of expertise (the GIMI network – Geoforensics and Information Management for crime Investigation).

The question arises, are existing (digital) geology and soil databases useful and suitable for forensic inferences? Inherent in this question are issues of completeness, consistency, compatibility and applicability. This paper uses case study material (published and new) to discuss these themes.

Does the database provide complete spatial coverage and information on the property of interest to the forensic investigation?

Murray (2011) includes soil, mineralogical substances and glass as well as other natural and man-made substances incorporated into soil or other earth materials as examples of physical evidence used in forensic geology. This discussion primarily focuses on geology and soil databases but many other repositories of data information exist (e.g. on glass and mineralogy) and the same issue of completeness is pertinent for these other types of data. The concept of complete spatial coverage may be used to ask if sufficient knowledge, (digital) maps or databases exist to cover the whole of the area included in an investigation (local, regional to national), or indeed if sufficient detail is available at a local or small scale (that which more often describes the area extent of a scene of crime). Williams et al. (2009) discuss how available soil and geology databases relating to Wales were found to be insufficiently complete and not at an appropriate spatial resolution to allow a reliable assessment to be made for the selection of mass burial sites during and following the 2001 UK Foot and Mouth Disease outbreak (Anderson 2002).

If database reference material is to be used when evaluating any apparent similarity or difference between a questioned sample and that contained within a database, several issues need to be addressed. Regional and national surveys follow a standard procedure in the collection of samples. The G-BASE standard established by BGS comprises soil sampling at 20 and 50 cm depths on a regular grid at one site per 2 km^2. At each site the soil samples comprise a composite of five subsamples taken from the points and centre of a 20 m grid. Ground-based soil geochemistry data generated as part of the Tellus project, GSNI, followed this sampling strategy. As described by many authors (e.g. Rawlins et al. 2006; Lark & Rawlins 2008; Ruffell & McKinley 2008; Pye & Blott 2009) the sample support – the volume, shape and orientation – of forensic specimens may be unknown and uncontrolled. Likewise the quantity of soil material may be very different. Trace evidence may comprise <1 g compared with 12 g collected for laboratory-based analyses.

The relevance of maps and databases for a crime scene in an urban setting has been raised by several authors: natural urban topsoil is often removed only to be replaced with manufactured garden soil (Fitzpatrick et al. 2009), and very small volumes of soil and other urban materials are available (Gradusova & Nesterina 2009). Aimed at addressing the specific issue of a lack of information on urban soils, a 'daughter' project of SOILFIT, the SoilFUN network, is aimed at developing a coordinated network of universities to build a robust database of urban soil information from across the UK. More recently, where sufficient funding is available, regional (governmental) surveys are incorporating separate urban soil surveys. The 2004–2007 Tellus project included an 'urban' soil sampling programme (parallel to the rural survey) with a sample density of four samples per square kilometre across a number of selected urban areas across Northern Ireland. This more detailed dataset has an environmental forensic application in that it provides an opportunity to derive baseline values for urban centres to test the contribution of bedrock geology and historical land use to baseline contaminant concentrations (Nice 2010). The feasibility for urban crime scene investigations has yet to be tested.

Is the rationale for the generation of the existing database (economic, environmental, etc.) consistent with the purpose for which the forensic investigator is using it?

More often than not the answer *sensu stricto* is no. Geological and soil mapping, baseline quality mapping, economic prospecting, land resource assessment, risk evaluation and environmental studies are all cited as motivations for surveys. Although databases have been used for forensic provenance

(Lark & Rawlins 2008), this is rarely the original focus. One crucial component of the approach adopted by the SoilFit project was the development of a set of reference soils and databases to enable the estimation of the probability of obtaining accurate soil comparisons. Namely, in this project, the reference soils and related database are being developed with the forensic application in mind.

It could be argued, however, that the collection of 'control samples' in forensics represents baseline geological and soil characteristics and that these can be provided by regional and national databases. Murray (2011) states that known or control samples are of two types, those collected from a crime scene or alibi location and those from museum or professional collections. We could add geological and soil databases to this definition. In the identification of questioned samples and location of places with similar material, geological and soil maps and databases provide information to show the distribution of different kinds of soils and rocks to determine possible sources of the materials. Certainly, there are issues related to the non-contemporaneous timings of the criminal activity, the seizure of questioned samples and the collection of crime scene samples, all generally collected within two days of crime, compared with previously collected materials comprising the database. Sensabaugh (2009) identifies the need for an objective approach to assessing similarities and differences between samples. The use of previously collected samples in the form of a database may be looked at from two points of view: recognition that these have not been collected for the same purpose as the forensic investigator intends, but paradoxically, therefore, this provides an independent control set of samples to satisfy the requirement of investigator objectivity.

How can we ensure data and map compatibility?

Geographical information systems (GIS) are being used increasingly to integrate and interrogate data (Fig. 1; Ruffell & McKinley 2008; McKinley *et al.* 2009; Williams *et al.* 2009). Standards for 2D data operability and transfer exist (Open Geospatial Consortium Inc 2010) and 3D operability is currently under development and discussion (Gabriel *et al.* 2010); these are adopted by most regional and national surveys. Scanning and integration of historical maps and imagery may incorporate errors. These may be related to the conservation condition of historical paper maps – distortion of the paper is likely if the maps are not stored in climatically controlled conditions – and variable resolution quality of scanning technologies. The increasing availability of remote sensed data is to be welcomed but comes with the caution of varying resolution and interpretations of classification. If several different maps or databases are used in conjunction, the question needs to be addressed of whether they are compatible to ensure that, if differences are observed, the forensic investigator can be certain that these are real and not a function of differences in scale of measurement. The question of

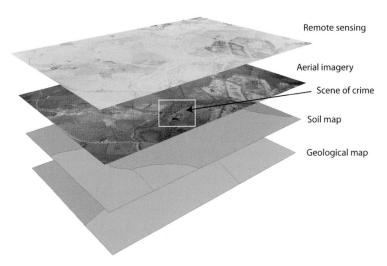

Fig. 1. Overlay in GIS of different types/layers of information (geology, soil, aerial imagery, remote sensed imagery) that may be used in the course of a forensic investigation. Aerial orthophotography © Crown copyright and database rights MOU203.

how valid for forensic investigations integration methods are (e.g. using spatial analysis, GIS, remote sensing techniques, geostatistics etc.), for combining different data generated at widely different spatial resolution scales of collection, is one that requires further research. Lagacherie & McBratney (2007) suggest that problems that occur in conventional soil survey, such as large areal extents and soil cover complexity, coarse resolutions and short-range variability representations, carry over to digital soil mapping. Their approach to maximize information in digital soil mapping is to populate spatial soil information systems (SSINFOS) using digital terrain models and a more exhaustive mapping of soil-forming characteristics from remote sensed imagery (Lagacherie et al. 2007). The integration of multiple data types such as incorporated in SSINFOS necessitates an assurance of adequate methods to resolve resolution and classification differences between datasets, as discussed above.

Finally, is the digital database applicable or suitable for the particular forensic investigation?

Can digital geology and soil databases based on surveys collected on a 2–10 km scale (more realistic for larger geographic areas) provide useful information for forensic investigations? In the example shown (Fig. 2), a sampling grid conducted on an actual crime scene is juxtaposed to a regional survey sampling scheme of one collated sample every 2 km². Although in this investigation the regional survey data were not used in the forensic case, the comparison of the data highlights the issue of suitability of use. In this example the national survey may offer a baseline control for the area.

As reflected by Murray (2011), the perpetrator rarely chooses the best sample to transfer to their trousers or shoes that is the most representative of the soil at the scene of crime. However this raises the question of how variable the soil at a scene of crime is, and whether soil and geological variability are adequately represented in a database. Lagacherie et al. (2007) highlight the issue that, when dealing with large areas, the chances increase of uncovering complex patterns of soil variation. Owing to limited sample spacing and coarse resolution surveying, digital mapping can only ever resolve part of total geological and related soil variability. This geological and soil variability can be seen across a region the size of Northern Ireland where the range of rocks represented forms a stratigraphic record commencing in the Mesoproterozoic (at least 600 myr ago) and includes

Fig. 2. A sampling grid undertaken for a crime scene compared with a regional survey sampling scheme of one collated sample every 2 km² (Tellus Geochemical Survey GSNI). Aerial orthophotography © Crown copyright and database rights MOU203.

examples of all geological systems up to and including the Palaeogene (Mitchell 2004). Moreover, the advance of ice sheets and meltwaters in the last 100 000 years has resulted in at least 80% of this bedrock being covered by superficial deposits, bringing further complexity to the pattern of soil variation. Soil surveyors have long recognized that landscapes are the result of many processes (Burrough 1993).

Soil genesis is assumed to be the same where the same combination of soil processes comes together (Kerry & Oliver 2011). The traditional approach in soil classification has been to describe spatial variation in soil with sharp boundaries between soil classes. Classification is a type of generalization to create order and allows us to organize data into meaningful structures; however, in doing so we lose some detail in an effort to gain clarity (Ahlqvist & Gahegan 2005). This generalization, recognized by Lagacherie et al. (2007) for traditional soil mapping, is carried over into digital databases. The use of classification boundaries implies that variation in soil properties is discontinuous. Gradusova & Nesterina (2009) highlight the difficulty in defining a soil boundary marker and how this limitation propagates to forensic work. A more imprecise but realistic description of soil mapping units is proposed using fuzzy logic and the creation of fuzzy membership sets (Fisher 1999). Whereas discrete boundaries are defined as mutually exclusive sets, soil transition zones can be handled by fuzzy membership where fuzzy boundary zones produce overlapping set boundaries and membership in more than one soil classification is possible (Fisher 2000; Ahlqvist & Gahegan 2005). How useful the fuzzy membership approach and the concept of more imprecise but realistic descriptions of soil units are in forensics has to be investigated.

Soil processes operate under different and independent scales (Cruickshank 1972). Complexity of interactions between soil-forming processes, incomplete knowledge of them and variation in time and space make it appear that soil variation is random (Webster 2000). Unmapped variability in soil surveys relates to variation in soil properties outside our available knowledge (this creates the random variability described by Webster 2000) but also to factors that operate over shorter distances than the resolution of the survey. As such digital (soil) mapping can be viewed as a combination of several soil mapping procedures operating at nested scales (Lagacherie et al. 2007). Geostatistical studies over large areas demonstrate this complexity (e.g. Kerry & Oliver 2011). Geostatistics assumes that variation of a property such as soil is continuous. The variogram, the central tool in geostatistics, can be used to indicate how the variance of a (soil) property changes on average with increasing distance between pairs of data. The parameters of the model fitted to the experimental variogram can be used to interpret the structure of variation and to predict soil properties at unsampled places (Kerry & Oliver 2011). Parameters of the variogram have been used to evaluate the importance of short-range variation in relation to field-scale parameters, for example, clay content and sand content. Local or short-range soil properties at a field scale were investigated (Fig. 3). Geo-corrected aerial orthophotography between 2003 and 2006 (OSNI®) in ESRI binary raster format (FLT) were used to extract RGB values for individual fields (completed by Mark Jackson, Brigham Young University, Utah, USA). The variogram was used to evaluate the spatial variation at different scales; parameters from the variogram were used to produce kriged maps (Fig. 3b, c). The modelled variogram showed periodical or cyclic behaviour that could be related to the field management practice (such as ploughing, Fig. 3b). Next, the residuals of the variogram modelling process were mapped (Fig. 3c). This produced a more variable pattern that describes short-scale variation, more likely to be related to clay and sand content. The significance of this for forensics links back to the issue of soil variability at a scene of crime and how realistic it is to expect a soil sample, collected by the forensic investigator, to be similar to a questioned sample.

To address this issue it would be useful to have knowledge of some measure of expected soil variability for different parent rock types. Geostatistics was used to map the soil geochemical data generated by the Tellus Survey analysed by XRF. The variogram ranges (correlation distances or average sizes of areas with similar properties; Kerry & Oliver 2011) were used to investigate the spatial variability in the distribution of geochemical soil properties for different lithological/pedological associations (Table 2). Modelled variogram ranges were found to be between 3 km (manganese) and 40–60 km (nickel, zinc) for basalt with a high degree of spatial variability shown for iron. Likewise, shale showed spatial variability but for different elements, nickel and copper showing a high degree of spatial variability contrasting with a range of almost 50 km for zinc (Table 2). This compares with work by Kerry & Oliver (2011), who found shorter variogram ranges for soil properties for sandstone and granite parent rocks (variogram range < 15 km) than for shale (range value = 30 km) and basalt (up to 50 km). So is the complexity of geological and soil properties an advantage or disadvantage for forensic investigations? The short-range unmapped variability in surveys is far from negative for forensic applications. The work of Rawlins et al. (2006) demonstrates this point.

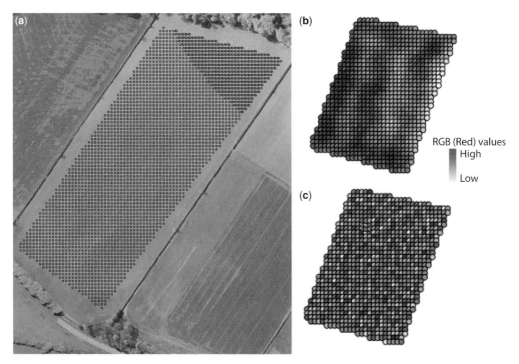

Fig. 3. (**a**) Geo-corrected aerial orthophotography (OSNI®) showing an individual field; (**b**) kriged interpolated map the field for RGB (red) values (periodical or cyclic behaviour can be be related to the field management practice, such as ploughing); (**c**) kriged interpolated map for RGB (red) residual values (more variable pattern describes short-scale variation, more likely to be related to clay and sand content). RGB values extracted from ESRI binary raster format (FLT) by Mark Jackson, Brigham Young University, UT, USA. Aerial orthophotography © Crown copyright and database rights MOU203.

The natural distribution of minerals in soil and material from plant communities was investigated to determine the characteristics of forensic sample locations. Four complimentary techniques were used (XRD, SEM, molecular organic matter signatures, palynology). The results determined the provenance of two out of the three sites with the highest degree of success at sites with distinctive mineralogical or vegetative properties. The importance of expert interpretation was also identified. This harmonizes with the idea that, as geoscientists, we intuitively use our prior expert knowledge and interpretation in a pseudo-'Bayesian' approach. The implications from the variogram results, demonstrated for the Tellus geochemical survey (Table 2), are that, for a crime scene located on a basalt parent rock, it would be more suitable to use iron oxide and the associated mineral suite as a distinctive mineralogy (as suggested by Rawlins *et al.* 2006) to determine provenance of a questioned sample. This will be examined further in the following case study.

An environmental forensic case published by Ruffell & Dawson (2009) was used to test the usefulness of a large-scale geological database and regional soil geochemistry survey in placing a perpetrator at an illegal waste dump and in doing so

Table 2. *Range parameter values from modelled variograms for geochemical elements from Tellus project soil ground survey (GSNI)*

Element	Rock types – variogram range or correlation distances (km)	
	Basalt	Shale
MgO%	40	30
P_2O%	10	30
K_2O%	40	15
CaO%	40	15
MnO%	3	25
FeO%	Variable	20
Cr (mg kg^{-1})	5	20
Ni (mg kg^{-1})	40	Variable
Cu (mg kg^{-1})	43	Variable
Zn (mg kg^{-1})	60	47

Variable indicates variability at finer resolution than measured by survey.

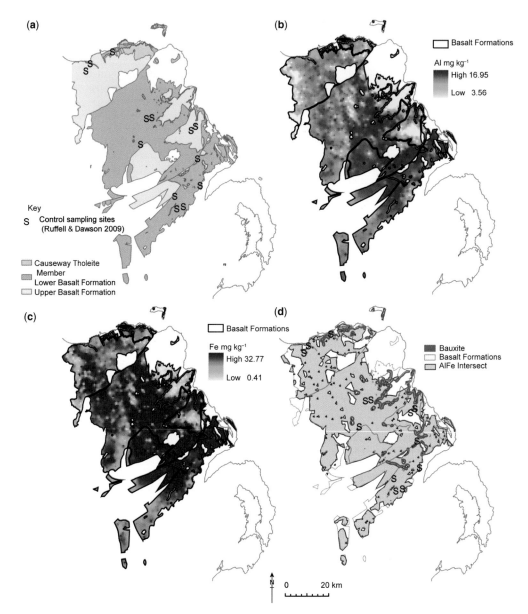

Fig. 4. (a) Geological map showing the Palaeogene Antrim Lava Group adapted from Mitchell (2004); (b) interpolated map of Al (mg kg^{-1}); (c) interpolated map of Fe$_2$O$_3$ (mg kg^{-1}) for the Palaeogene Antrim Lava Group; (d) map showing areas where elevated areas of Al and Fe intersect with distribution of bauxite. Sampling sites as indicated by Ruffell & Dawson (2009).

deny his alibi. XRD analysis of material retrieved from the suspect indicated the presence of gibbsite (aluminium hydroxide), hematite (iron oxide) and goethite (hydrated iron oxide). The spatial distribution of this mineral suite is limited to bauxites and laterites associated with the Causeway Basalts in Northern Ireland (Fig. 4a), and this knowledge allowed an illegal dump to be located and provided strong evidence to place the suspect at the same location. The success of this investigation was based on expert geological knowledge of the location of the bauxites and laterites and the distinctive mineralogy of the red (iron oxide) material found on the perpetrator's trousers and lorry. The test is whether, if expert knowledge was not available, the same conclusions would produce the same result, based solely on the Tellus ground geochemical survey and digital geology database (GSNI). Using the XRD results, the mineral suite suggested that the most useful geochemical elements to investigate were iron and aluminium. The indication from geostatistics discussed earlier was that short-range spatial variability observed for elemental iron for a basaltic parent rocks would be advantageous in this investigation. Kriged interpolated maps for iron and aluminium were produced from variogram analysis (Fig. 4b, c). GIS was used to select areas with elevated concentrations of iron (>3 mg kg^{-1}) and aluminium (>2 mg kg^{-1}) and then to intersect these areas with bauxite from the geology database. There is a strong correspondence between the final selected areas and those highlighted by Ruffell & Dawson (2009). The conclusion is that the Tellus dataset and digital geology map would have proved useful to provide locations for law enforcement to search to locate the illegal waste dump.

Conclusions

From the earliest pioneers (e.g. Ehrenberg, Heinrich) there have been many international case examples of exploiting knowledge of differing geology and soil distributions in forensic work. Indeed it is proposed that, on the question of provenance of a questioned sample, the geologist or soil scientist adopts a pseudo-Bayesian approach in drawing on prior expert knowledge and available digital map and database sources to achieve an updated (posterior) decision or final ranking scheme. It naturally follows, therefore, that with developing technologies, the acquisition of vast amounts of geospatial information for regional and national soil and geology databases is to be welcomed. Using published and new case study material, this paper has explored issues of completeness, consistency, compatibility and applicability in relation to the use of digital geology and soil databases in environmental and criminal forensics. Several themes emerge:

- Differences remain between questioned samples and database material in terms of spatial resolution, amount of material and condition of sample collection. Initiatives are underway (e.g. SoilFit Project) to build more robust databases (of rural and urban soil information) with a forensic application in mind. The use of regional urban soil databases has yet to be fully explored for criminal forensics.
- There is recognition that regional and national databases have not been collected for the same purpose as the forensic investigator intends, but conversely this may provide an independent or objective set of 'control' samples.
- Further research is needed on the implications for forensics of the use of integration methods for combining ground and remotely sensed data generated at widely different spatial resolution scales.
- Complexity and distinctive geological and soil properties can prove advantageous for forensic investigations, but some knowledge of expected soil or geological variability would be useful.

Despite an acknowledgement that databases can be neither exhaustive nor precise, on the scale of the scene of a crime, coupled with expert knowledge, they play an invaluable role in providing background or reference material in a criminal investigation.

Michael Young of the Geological Survey of Northern Ireland is thanked for arranging access to the Tellus data. The Tellus Project was funded by The Department for Enterprise, Trade and Investment (DETINI) and by the Building Sustainable Prosperity Scheme of the Rural Development Programme (Department of Agriculture and Rural Development of Northern Ireland). The Geocorrected aerial orthophotography, Ordnance Survey of Northern Ireland (OSNI®) was reproduced from Land and Property Services data with the permission of the Controller of Her Majesty's Stationery Office, Crown copyright and database rights MOU203. Ruth Kerry and Mark Jackson, Brigham Young University, UT, USA are thanked for providing the RGB values extracted from aerial orthophotography (OSNI®).

References

AHLQVIST, O. & GAHEGAN, M. 2005. Probing the relationship between classification error and class similarity. *Photogrammetric Engineering & Remote Sensing*, **71**, 1365–1373.

AIKEN, C. G. G. 2009. Some thoughts on the role of probabilistic reasoning in the evaluation of evidence. *In*: RITZ, K., DAWSON, L. & MILLER, D. (eds) *Criminal and Environmental Soil Forensics*. Springer, Berlin, 33–49.

ANDERSON, I. 2002. *Foot and Mouth Disease 2001. Lessons to be Learned Inquiry Report*. The Stationery Office, London.

AUCHIE, D. P. 2009. Expert scientific evidence in court: the legal considerations. *In*: RITZ, K., DAWSON, L. & MILLER, D. (eds) *Criminal and Environmental Soil Forensics*. Springer, Berlin, 13–33.

BLOCK, E. B. 1958. *The Wizard of Berkeley*. Coward-McCann, New York.

BURROUGH, P. A. 1993. *Principles of Geographical Information Systems for Land Resource Assessment*. Clarendon Press, Oxford.

CROFT, D. J. & PYE, K. 2004. Multi-technique comparison of source and primary transfer soil samples: an experimental investigation. *Science & Justice*, **44**, 21–28.

CRUICKSHANK, J. G. 1972. *Soil Geography*. David and Charles, Newton Abbot.

CRUICKSHANK, J. G. 1997. *Soil and Environment: Northern Ireland*. Queen's University Belfast, Belfast.

DAWSON, L. A., TOWERS, W., MAYES, R. W., CRAIG, J., VAISANEN, R. K. & WATERHOUSE, E. C. 2004. The use of plant hydrocarbon signatures in characterising soil organic matter. *In*: PYE, K. & CROFT, D. J. (eds) *Forensic Geoscience: Principles, Techniques and Applications*. Geological Society, London, Special Publications, **232**, 269–276.

ENVIRONMENT AGENCY 2007. *UK Soil and Herbage Pollutant Survey*, Summary SC000027. http://publications.environment-agency.gov.uk/PDF/SCHO0607 BMTE-E-E.pdf (accessed 21 November 2011).

FAO-UNESCO-ISRIC 1990. *Guidelines for Distinguishing soil Subunits in the FAO-UNESCO-ISRIC Revised Legend*. World Resource Report 60, http://library.wur.nl/isric/index2.html?url=http://library.wur.nl/WebQuery/isric/23807 (accessed 21 November 2011).

FISHER, P. F. 1999. Models of uncertainty in spatial data. *In*: LONGLEY, P., GOODCHILD, M., MAGUIRE, D. & RHIND, D. (eds) *Geographical Information Systems: Principles, Techniques, Management and Applications*. John Wiley and Sons, New York, **1**, 191–205.

FISHER, P. F. 2000. Fuzzy modelling. *In*: OPENSHAW, S., ABRAHART, R. & HARRIS, T. (eds) *Geocomputing*. Taylor & Francis, London, 161–186.

FITZPATRICK, R. W. 2008. Nature, distribution and origin of soil materials in the forensic comparison of soils. *In*: TIBBETT, M. & CARTER, D. O. (eds) *Soil Analysis in Forensic Taphonomy: Chemical and Biological Effects of Buried Human Remains*. CRC Press, Boca Raton, FL, 1–28.

FITZPATRICK, R. W., RAVON, M. D. & FORRESTER, S. T. 2009. A systematic approach to soil forensics: criminal case studies involving transference from crime scene to forensic evidence. *In*: RITZ, K., DAWSON, L. & MILLER, D. (eds) *Criminal and Environmental Soil Forensics*. Springer, Berlin, 105–129.

GABRIEL, P., GIETZEL, J., HA, L. H. & SCHAEBEN, H. 2010. A framework for a network-based datastore for spatial and spatio-temporal geoscience data. *Proceedings of the 30th GOCAD Meeting*, 2010, http://tu-freiberg.de/fakult3/IS4GEO/gOcad2011.pdf (accessed 25 October 2011).

GRADUSOVA, O. & NESTERINA, E. 2009. The current state of forensic soil examination in the Russian Federation. *In*: RITZ, K., DAWSON, L. & MILLER, D. (eds) *Criminal and Environmental Soil Forensics*. Springer, Berlin, 63–75.

KERRY, R. & OLIVER, M. 2011. Soil geomorphology: identifying relations between the scale of spatial variation and soil processes using the variogram. *In*: WARKE, P. & MCKINLEY, J. M. (eds) Special Issue on Scale. *Geomorphology*, **130**, 40–54.

LAGACHERIE, P. & MCBRATNEY, A. B. 2007. Spatial soil information systems and spatial soil inference systems: perspectives for digital soil mapping. *In*: LAGACHERIE, P., MCBRATNEY, A. B. & VOLTZ, M. (eds) *Digital Soil Mapping, an Introductory Perspective*. Developments in Soil Science. Elsevier, Amsterdam, **31**, 3–24.

LAGACHERIE, P., BRATNEY, A. B. & VOLTZ, M. (eds) 2007. *Digital Soil Mapping, an Introductory Perspective*. Developments in Soil Science. Elsevier, Amsterdam, **31**.

LARK, R. M. & RAWLINS, B. 2008. Can we predict the provenance of a soil sample for forensic purposes by reference to a spatial database? *European Journal of Soil Science*, **59**, 1000–1006.

MACAULAY INSTITUTE 2008a. *The SoilFit Project*. The James Hutton Institute, Aberdeen AB15 8QH, UK, http://www.macaulay.ac.uk/soilfit/ (accessed 12 November 2011).

MACAULAY INSTITUTE 2008b. *The SoilFun Project*. The James Hutton Institute, Aberdeen AB15 8QH, UK, http://www.macaulay.ac.uk/forensics/soilfun/ (accessed 12 November 2011).

MCKINLEY, J., RUFFELL, A., HARRISON, M., MEIER-AUGENSTEIN, W., KEMP, H., GRAHAM, C. & BARRY, L. 2009. Spatial thinking in search methodology: a case study of the 'No body murder enquiry', West of Ireland. *In*: RITZ, K., DAWSON, L. & MILLER, D. (eds) *Criminal and Environmental Soil Forensics*. Springer, Berlin, 285–303.

MITCHELL, W. I. 2004. *The Geology of Northern Ireland – Our Natural Foundation*. Geological Survey of Northern Ireland, Belfast.

MURRAY, R. C. 2011. *Evidence from the Earth: Forensic Geology and Criminal Investigation*, 2nd edn. Mountain Press, Missoula, MT.

NICE, S. E. 2010. *Inorganic Soil Geochemical Baseline Data for the Urban Area of the Belfast Metropolitan Area, Northern Ireland*. OR/08/021. British Geological Survey, Nottingham.

OPEN GEOSPATIAL CONSORTIUM INC 2010. *OpenGIS Geography Markup Language (GML) Encoding Standard*, Norm. 2010, http://www.opengeospatial.org/standards/gml

PYE, K. 2007 *Geological and Soil Evidence*. CRC Press, Abingdon.

PYE, K. & BLOTT, S. J. 2009. Development of a searchable major and trace element database for use in forensic soil comparisons. *Science & Justice*, **49**, 170–181.

RAWLINS, B. G., KEMP, S. J., HODGKINSON, E. H., RIDING, J. B., VANE, C. H., POULTON, C. & FREEBOROUGH, K. 2006. Potential and pitfalls in establishing the provenance of earth-related samples in forensic investigations. *Forensic Science International*, **51**, 832–845.

RUFFELL, A. & DAWSON, L. 2009. Forensic geology in environmental crime: illegal waste movement and burial in Northern Ireland. *Environmental Forensics*, **10**, 1–6.

RUFFELL, A. & MCKINLEY, J. 2008. *Geoforensics*, Wiley-Blackwell, Chichester.

SENSABAUGH, G. F. 2009. Microbial community profiling for the characterisation of soil evidence: forensic considerations. *In*: RITZ, K., DAWSON, L. & MILLER, D. (eds) *Criminal and Environmental Soil Forensics*. Springer, Berlin, 49–61.

STAM, M. 2004. Soil as significant evidence in a sexual assault/attempted homicide case. *In*: PYE, K. & CROFT, D. (eds) *Forensic Geoscience: Principles, Techniques and Applications*. Geological Society, London, Special Publications, **232**, 295–299.

SUGITA, R. & MARUMO, Y. 2004. 'Unique' particles in soil evidence. *In*: PYE, K. & CROFT, D. (eds) *Forensic*

Geoscience: Principles, Techniques and Applications. Geological Society, London, Special Publications, **232**, 97–102.

WEBSTER, R. 2000. Is soil variation random? *Geoderma*, **97**, 147–163.

WILLIAMS, A., TEMPLE, T., POLLARD, S. J., JONES, R. J. A. & RITZ, K. 2009. Environmental considerations for Common Burial Site Selection after Pandemic Events. *In*: RITZ, K., DAWSON, L. & MILLER, D. (eds) *Criminal and Environmental Soil Forensics*. Springer, Berlin, 87–105.

Instrumental neutron activation analysis (INAA) and forensic applications

G. N. EBY

Department of Environmental, Earth and Atmospheric Sciences, University of Massachusetts, Lowell, MA 01854, USA (e-mail: nelson_eby@uml.edu)

Abstract: Instrumental neutron activation analysis (INAA) is a non-destructive analytical technique that can be applied to a wide range of materials. The method requires no pretreatment of the sample, various geometries can be used and as many as 40 elements can be determined at the ppb–ppm level, depending on the characteristics of the specific sample. The method has been used in a number of forensic applications, such as identifying the source of ammunition, gunshot residue, arsenic in hair and the source of cigarettes. In this contribution, INAA is applied to characterizing ceramics, maple syrups and grasses, and based on the data it is possible to distinguish between various types of ceramics, sources of maple syrup and the geographical/geological locations of grass samples. As an example of an environmental forensic application, it is shown that a tree-ring record can be used to map the history of environmental pollution associated with a zinc smelter. Additional applications involve characterizing the products of nuclear detonations and the provenance of diamonds.

During forensic investigations, forensic scientists utilize a wide variety of scientific techniques ranging from simple comparisons of materials to DNA analysis to sophisticated elemental and isotopic analysis. Geoscientists have contributed their expertise to these investigations by identifying soil and sediment samples, mineral and rock identification, geophysical surveys, and detailed analytical studies. More recently, environmental forensics has become important in identifying the source and impact of contaminants. In many of these investigations the goal is to either include or exclude an individual or entity from a group. For example, soil evidence on the shoe of an alleged suspect can be compared with the soil at the crime scene. If the soil from the shoe differs to that of the crime scene then the footwear can be excluded. If the soil samples are similar then the individual remains in the group as a viable suspect.

Chemical data, both elemental and isotopic, can be used to characterize materials found at a crime scene with materials associated with a suspect. These techniques have proven to be very powerful, although often expensive and time-consuming, ways to relate various entities and a crime scene. Among these methods is instrumental neutron activation analysis (INAA), which is a non-destructive analytical method that in many applications requires only small amounts of material. The method has been used in forensics to characterize ammunition (Lukens & Guinn 1971; Desai & Parthasarathy 1983; Sedda & Rossi 2011), to identify gunshot residue (Gislason & Pate 1973) and copper wire (Chan 1972), in the determination of arsenic poisoning using hair samples (Kučera & Kofroňová 2011), to determine the illicit sale of tobacco (Giordani *et al.* 2005), in the forensic analysis comparison of brick stones (Scheid *et al.* 2009) and in many other applications. The following sections will describe the methodology of INAA and possible applications to forensic investigations (both criminal and environmental).

Instrumental neutron activation analysis (INAA)

INAA is a non-destructive analytical method that has the potential to determine the concentration of many elements in the ppb–ppm range in a wide variety of materials. Neutron activation analysis (NAA) was developed in the late 1950s (Guinn & Wagner 1960). In the early years of NAA, because of the low resolution of solid state detectors, radiochemical separation of radioactive isotopes was required. The development of GeLi and intrinsic Ge detectors negated the need for chemical separations, and led to the use of purely instrumental techniques (INAA). Detailed descriptions of the method are found in Gordon *et al.* (1968), Potts *et al.* (1985) and Koeberl (1993).

There are different variants of the INAA method (depending on the portion of the neutron flux spectrum that is used) but the most common makes use of thermal neutrons. In this scheme, a thermal neutron is captured by a target nucleus increasing the atomic mass by 1. The resulting isotope is radioactive and usually decays to a stable form of the

Table 1. *Typical detection limits for environmental and geological samples*

DL* (ppm)	Element
0.001	Au, Eu, Sm
0.01	As, Co, Cs, Hf, Ir, La, Lu, Sb, Sc, Se, Ta, Tb, Th, Yb, W
0.1	Ag, Br, Cd, Ce, Gd, Ho, Nd, Tm, U
1	Cr, Ni, Rb, Zn,
10	Ba, Na, Sr, Zr
100	Fe, K

*DL, detection limits for elements routinely determined in University of Massachusetts, Lowell (UML) experiments.

element by the emission of a beta particle and the accompanying emission of characteristic gamma rays. An example of such a reaction, referred to as an n, γ reaction, is shown below:

$$^{58}Fe + {}^{1}n \rightarrow {}^{59}Fe$$
$$\rightarrow {}^{59}Co + beta^- + gamma\ rays \quad (1)$$

In this case, the characteristic gamma rays that are used for INAA have energies of 142.4, 1099.2 and 1291.6 keV. About 40 elements in the periodic table can be determined by this method and these are listed, with approximate detection limits, in Table 1.

In a typical analytical procedure, samples are irradiated in a nuclear reactor, which serves as the neutron source. The length of the irradiation depends on the neutron flux and the amount of material. Typical irradiation times are between 1 and 4 h. Sample size is from 1 mg to several hundred milligrams. Samples are loaded in acid-cleaned polyethylene vials prior to irradiation. No pretreatment of the samples is required and any size sample that fits in the irradiation vial can be analysed. After a suitable decay period, the samples are counted using Ge detectors (Fig. 1). The characteristic gamma rays are used to determine the isotope and the intensities of the gamma rays, when referenced to suitable standards, are used to determine the absolute concentrations. A variety of corrections are applied for variations in the neutron flux, geometry, possible spectral interferences and decay. With a modern system, these corrections are all performed by computer and the final results are determined. In many cases, after a sufficiently long decay time, the radioactivity of the samples decreases to a safe level and the samples can be returned to the investigator.

Potential applications to forensic studies

The potential applications of INAA to environmental and criminal forensics are almost limitless.

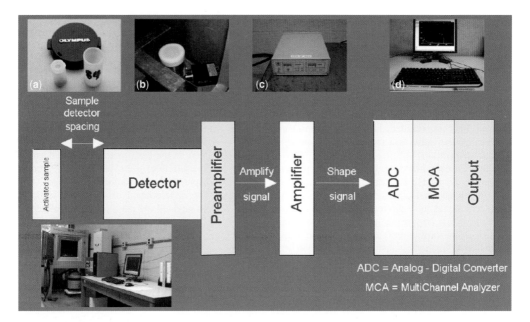

Fig. 1. Schematic showing the various components of an INAA spectroscopy system: (**a**) irradiation and counting vial; (**b**) BEGe detector with preamplifier; (**c**) spectroscopy amplifier and analogue–digital converter; (**d**) computer displaying a gamma ray spectrum; and (**e**) the INAA spectroscopy laboratory at the University of Massachusetts Lowell.

The following illustrate some of these potential applications.

Ceramics

Ceramics consist of mineral components and transition metals that are added for colour. The specific colours that are produced in part depend upon the temperature and oxidation conditions under which a ceramic is fired. Hence, the same metal(s) can produce different colours. For example, copper gives green in oxidizing conditions but red in reducing conditions. Among the most common colourant oxides (Table 2) are Cu, Co, Fe and Ti. Other additives include Cr, Ni and Zircopax (essentially zircon). The addition of Zircopax leads to a distinctive enrichment in the heavy rare earth elements (HREE). Numerous online references are available that discuss the fascinating topic of colourants in glazes. INAA has been used for decades in archaeology to determine the source of ceramic and other artefacts (Bishop & Blackman 2002; IAEA 2003). Hence, the methodology is well developed for the application of INAA to forensic investigations that involve ceramics.

In this study, seven samples of different types of pottery were acquired from a potter. For all of the samples essentially the same clay mixture was used but different components were used for the glazes. Small fragments of each piece of pottery were irradiated and their composition was determined by INAA. The analytical data, along with the colour of each glaze, is given in Table 3 and selected element data are plotted in Figure 2. Sample 7 (light tan in colour) gives the most distinctive result because Zircopax (zircon) was used in the glaze. Zircon is characterized by high Th relative to U and high HREE abundances, and these characteristics are reflected in the high Th/U (Fig. 2a) and HREE-enriched REE pattern (Fig. 2b) for Sample 7. Samples 3 and 4 (dark blue in colour) have high Co content (Table 3, Fig. 2d). Sample 6 (medium green in colour) is high in Zn (Table 3, Fig. 2d). Samples 1 and 5 (medium green and tan in colour, respectively) can be distinguished from the other samples in terms of their relatively higher Sb content (Fig. 2c) and lower Th/U ratio (Fig. 2a). While the sample set is small, the observed elemental variations indicate the usefulness of trace element data for characterizing ceramics.

Maple syrup

Maple syrup, a North American treat, is derived from the sap of the maple tree (*Acer saccharum*). The major maple syrup producing areas are the New England States (Maine, Massachusetts, New Hampshire and Vermont), New York, Michigan and Wisconsin in the United States, and the Canadian provinces of Ontario and Quebec. The sap is collected from the trees in early spring when daily temperatures fluctuate from freezing at night to thawing during the day. Given the required climatic conditions, the maple syrup season is relatively short. The sap is only slightly sweet and must be

Table 2. *Summary of materials used in glazes*

Material	Chemistry
Bentonite	$(Na,Ca)_{0.33}(Al,Mg)_2(Si_4O_{10})(OH)_2 \cdot nH_2O$
Cryolite	Na_3AlF_6
Dolomite	$CaMg(CO_3)_2$
Epsom salts	$MgSO_4 \cdot 7H_2O$
Fluorspar	CaF_2
Gerstley borate	$2CaO \cdot 3B_2O_3 \cdot 5H_2O$
Kaolin (kaolinite)	$Al_2Si_2O_5(OH)_4$
Lepidolite	$K(Li,Al)_3(Al,Si)_4O_{10}(F,OH)_2$
Lithium carbonate	$LiCO_3$
Nepheline syenite	Various Na–K–Al silicate minerals
Potash feldspar (K-spar)	$KAlSi_3O_8$
Silica (quartz)	SiO_2
Soda feldspar (albite)	$NaAlSi_3O_4$
Whiting (calcite)	$CaCO_3$
Wollastonite	$CaSiO_3$
Zircopax (zircon)	$ZrSiO_4$
Colourant oxides	
Cobalt	Co
Copper carbonate	$CuCO_3$
Hematite	Fe_2O_3
Rutile	TiO_2

Table 3. *INAA data for ceramic samples in ppm*

Sample	1	2	3	4	5	6	7
Glaze colour	Medium green	Dark brown	Dark blue	Dark blue	Tan	Medium green	Light tan
Fe	14 265	12 322	12 965	9925	8549	10 361	8075
Na	3300	1967	4564	2601	1785	4870	1415
K	17 758	16 420	16 896	18 415	10 879	18 212	12 929
Sc	17.8	19	18.6	19.1	23.8	20.3	24
Cr	119	131	133	131	171	141	110
Co	6.4	4.7	888	713	6	7.8	4.7
Ni	26.9	28.5	49.2	35.9	59.6	7.9	28.8
Zn	56	35.3	38	20.9	33.4	1627	61
Rb	104	101	110	138	48	106	86
Cs	bd	9.14	8.84	10.07	2.39	4.83	7.09
Sr	142	223	339	324	207	395	284
Ba	477	491	621	496	365	523	965
La	51.1	47.7	44.6	47.2	77.4	49.7	44.6
Ce	88.4	81.4	78.3	78.7	127.7	84.3	80.5
Nd	35.7	30.2	25.4	24.3	48.4	31.6	37
Sm	5.80	4.48	4.26	4.23	7.51	4.83	4.32
Eu	1.42	1.10	1.11	1.08	1.88	1.16	1.17
Gd	bd	6	6.3	6.2	8.3	6.3	7.0
Tb	1.13	0.91	1.12	1.11	1.38	1.02	1.72
Tm	0.51	0.49	0.53	0.52	0.64	0.57	2.69
Yb	3.80	3.77	3.88	3.96	4.73	4.26	21.3
Lu	0.63	0.62	0.63	0.63	0.73	0.71	3.55
Zr	313	333	408	357	454	382	39053
Hf	8.2	8.89	8.8	8.9	10.5	9.7	875
Ta	3.67	2.46	2.48	3.08	2.76	2.78	2.98
Th	18.5	22.5	22.9	23.1	23.8	24.6	32.6
U	5.12	5.16	5.16	5.67	6.58	5.79	27.9
As	3.82	9.7	9.6	10.4	3.99	10.4	14.3
Sb	1.55	1.31	1.30	1.32	1.66	1.47	1.15
W	3.25	3.47	3.69	3.66	4.28	4.10	5.01
Se	bd	645	469	729	1042	617	79368
Au	5.4	5.4	3.1	5	5	4.7	16.2

bd, below detection.

concentrated to yield maple syrup. This is accomplished by evaporating water until a Brix value of approximately 66.5° is achieved – 1° Brix is equivalent to 1 g of sucrose in 100 g of solution. The elemental composition of maple syrup is determined by the chemical composition of the sap, which reflects the underlying soil chemistry, elements introduced during the tapping of the tree and transport to the sugar house, and elements introduced during the boiling down of the sap to produce maple syrup (Fig. 3). Thus, the possibility exists that maple syrup from different sources can be chemically fingerprinted.

Stuckel & Low (1996) investigated the chemistry of maple syrup from most of the major producing areas listed above. This study was largely concerned with the organic compounds in maple syrup but the authors did report elemental data for Ca, Mg and K. In terms of these elements, there were no significant differences between the various producing regions. Bhandari & Amarasiriwardena (2000) determined the concentrations of V, Co, Cu, As, Cd, Pb, Zn and Mn in commercial maple syrup from Canada, Maine, Massachusetts, New York and Vermont. In this data set, V and Pb showed the most variability. Samples from one geographical area had an extremely high Cu content. The most complete study of the trace metal content of maple syrups was carried out by Greenough et al. (2010). These authors determined the concentrations of 39 elements (by inductively coupled plasma mass spectrometry (ICP-MS)) in samples from 16 producers in northern Nova Scotia (Canada). They found that it was possible to correlate the composition of the samples with the producers–woodlots. Hence, the elemental chemistry of maple syrup can be used to identify the source of the syrup.

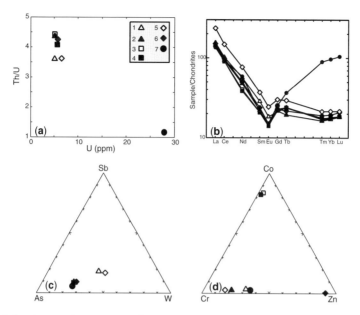

Fig. 2. Elemental plots for ceramics: (**a**) U v. Th/U; (**b**) REE chondrite-normalized plots (normalization values from Nakamura 1974); (**c**) As–Sb–W; and (**d**) Cr–Co–Zn.

In the study reported here, commercial maple syrup samples from Massachusetts, New York, Vermont and Quebec plus sap samples from an urban and a suburban maple tree were analysed for their trace element content. All of the commercial maple syrup samples were in plastic bottles. What was analysed was maple sugar and, in order to produce this sugar, the maple syrup (or sap) was further evaporated in Teflon beakers. The solid product was subsequently analysed by INAA. The results for selected elements are reported in Table 4. While the data set is small, a number of observations can be made.

- The sap samples from the maple trees (samples 7 and 8) can be distinguished from each other in terms of their Na, K, Sr and Ba content. Also of note is that there is no significant difference in the metal content of the sap from the suburban tree (Sample 7) and from the urban tree (Sample 8).
- The sample from Quebec, Sample 4, is differentiated from all other samples by the presence of measurable amounts of Se.
- Samples 6 and 7 are from the same sugar house and are distinguished from the other samples by their high Zn content.

Thus, even in this small data set, one can discern measurable differences that could be used to fingerprint the source of a commercial maple syrup or even the maple tree woodlot (in the case of the sap).

Grasses

Numerous books and papers have been published on the chemistry of grasses. Two comprehensive reference works are those of Adriamo (2001) and Kabata-Pendias (2001). For geologists/geochemists the relationship between the chemistry of plant material and the underlying soils is well established. Some of the earliest work on this relationship involved geochemical prospecting for ore deposits (Cannon 1960). The relationship between soils and plant materials was explored in the classic memoir edited by Cannon & Anderson (1971) on the relationship between environmental geochemistry and health and disease. Grass chemistry varies as a function of soil chemistry, which in turn is related to the chemistry of the underlying bedrock. Hence, in principle, it should be possible to distinguish between grasses of the same genus/species that are grown on soils overlying different bedrock.

Grass samples were collected from a number of locations in Serengeti National Park, Tanzania. The major grass types are *Digiteria*, *Sporobolus* and *Themeda*. Eleven sets of samples were analysed for a variety of trace elements by INAA. The locations of these sites, relative to the geology, are shown in Figure 4. Given that this area is well removed from any significant source of anthropogenic inputs, the grasses are expected to reflect the chemistry of the underlying soils, and the chemistry

Fig. 3. From sap to maple syrup: (**a**) tapping sugar maple using a galvanized bucket and tap; (**b**) holding tanks at sugar house; (**c**) initial evaporation; (**d**) boiler; (**e**) final evaporation; and (**f**) filtration.

of these soils is related to the chemistry of the bedrock. Numerous studies have been carried out on grass chemistry using solution-based methods, such as atomic absorption spectroscopy, ICP emission spectroscopy (ICP-ES) and ICP-MS. In most cases, these methods do not return the total metal content of the grass samples because of the presence in the grass of relatively insoluble silica particles. INAA returns the total content of metals in the soluble + insoluble phases. The results for selected trace elements are shown in Figure 5. The samples are grouped on the basis of the underlying bedrock, as shown on the Serengeti geological map. These elemental diagrams successfully differentiate between samples collected over the conglomerate, sandstone + quartzite and quartz reef. There is some overlap between marble and quartzite on all diagrams, although there is not complete overlap. Hence, it is possible to distinguish the location of most of the grass samples, in terms of geology, on

Table 4. *Data for maple sugar (1–6) and sap (7 and 8) in ppm*

Sample	1	2	3	4	5	6	7	8
Fe	45	70	60	73	15	65	32	31
Na	21	49	8	49	128	35	**501**	67
K	3161	3589	3014	4089	4439	4236	**7788**	2570
Sc	0.006	0.002	0.008	0.027	0.004	0.005	0.01	0.009
Cr	0.613	0.136	bd	**1.769**	0.878	0.993	0.632	0.67
Co	0.036	0.078	0.046	**0.119**	0.073	0.057	0.094	0.064
Ni	bd	bd	**1.240**	0.672	bd	bd	bd	bd
Zn	14.5	5.5	19.4	19.5	**50.8**	76.5	9.3	13.1
Rb	4.26	7.47	5.73	9.89	11.39	**17.90**	9.02	3.11
Cs	0.010	0.021	bd	0.022	0.048	0.024	0.012	0.009
Sr	12.71	10.54	bd	15.58	9.55	7.39	**25.48**	12.17
Ba	**40.88**	23.77	20.89	7.16	8.06	12.11	19.66	2.36
La	**0.580**	0.076	0.106	1.247	0.025	0.172	0.069	0.028
Ce	**1.079**	0.123	0.127	3.281	0.028	0.278	0.036	Bd
Th	0.009	0.004	bd	0.023	bd	bd	bd	Bd
U	bd	bd	bd	bd	0.124	bd	bd	Bd
As	bd	0.016	nd	0.02	0.026	0.028	0.034	0.017
Sb	0.010	0.002	0.004	0.004	0.035	0.01	0.018	0.011
Se	bd	bd	bd	**10.16**	bd	bd	bd	bd

bd, below detection. Numbers in bold are distinctive for the particular sample.

the basis of the elemental content of the grass samples. This approach holds the promise that in a forensic investigation grass samples of the same species, but from different locations, could be distinguished on the basis of their chemistry.

Tree rings as a record of environmental contamination (the Palmerton smelter)

Particularly in temperate climates, where there is a pronounced seasonality, tree-ring records (dendrochronology) can be used to look at past variations in climate and environmental pollution. Both isotopic and elemental variations have been investigated (dendrochemistry). Lepp (1974) was one of the first investigators to point out the potential use of tree rings to trace the temporal record of trace elements in the environment. According to Hagemeyer (2000) over 160 papers, as of 2000, have appeared on this topic. Hence, this is a fertile field of scientific investigation. However, Hagemeyer (2000) also cautions about the use of this record, without independent evidence, because of possible mobility of metals within and between the heartwood (dead central portion of the trunk) and sapwood (the living outer portion of the trunk). In a recent study, Liu et al. (2009) investigated the variation of Cd, Mn, P, Pb and Zn (emitted by a nearby steel plant) in a tree-ring record and concluded that element mobility was of limited significance. They also found a clear correlation between the emission history of the steel plant and the tree-ring record. Similarly, Baes & McLaughlin (1984) also found a relationship between the various trace metals in tree rings and emissions from both a coal-fired generating plant and a copper smelter. Most of the studies carried out on trace metals in tree rings have used analytical methods (atomic absorption and a variety of ICP emission methods) that require dissolution of the sample. This has two potentially undesirable results: (1) destruction of the sample so that it cannot be used for other studies; and (2) the possibility of incomplete dissolution of the sample. Both of these potentially undesirable results are obviated by INAA, which is non-destructive and requires no pretreatment of the sample.

Zinc mining began in northern New Jersey, USA, in the early 1700s. The various small mines were combined into a single entity, the New Jersey Zinc Company, in 1897. Two of these mines, Franklin Furnace (closed 1954) and Sterling Hill (closed 1986) are internationally known to mineralogists and mineral collectors because of their rich abundance of fluorescent minerals. A total of 345 minerals were identified in the two mines, 35 of which have no other known occurrence and 90 of these minerals are fluorescent, the largest number of fluorescent minerals found anywhere in the world. New Jersey Zinc built a smelter in Palmerton, Pennsylvania, USA, in 1898 to smelt the ores from these two mines. A second smelter was built at the same location in 1911. During the smelting process significant amounts of SO_2 and heavy metals (Zn, Cd, Pb and Cu, among

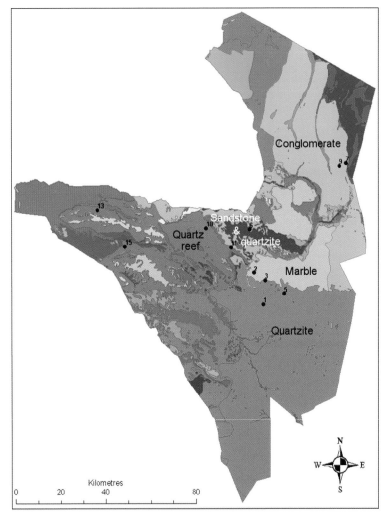

Fig. 4. Serengeti geological map. Sampled lithologies are labelled on the map. Source is anonymous.

others) were released to the environment. In 1953, an electrostatic precipitator was installed, and in 1980 the smelters were closed.

As part of a larger study (Li *et al.* 2010), tree-ring samples were collected from trees a few kilometres downwind of the Palmerton smelters. The details of the sampling method are found in Li *et al.* (2010), which deals largely with using tree-ring widths and $\delta^{13}C$ to measure the effects of industrial stress on temperate forests. A tree-ring core from a chestnut oak at the Palmerton site was divided into 10 increments and INAA was used to determine the metal content of individual increments. Concentration data for 27 elements, many in the ppb range, were obtained. The tree-ring data record the history of environmental contamination in this area. The data for some of these elements are plotted, as a function of age, in Figure 6.

The metals released during the smelter operation can be delivered to the trees either by direct deposition on the leaves (and stems) or through the soil–root interface. These interactions, particularly those involving the soil–root interface, are extremely complex and a discussion of these is beyond the scope of this paper. However, several preliminary conclusions can be drawn from the data.

- The onset of smelter operations led to an increase in metals in the tree rings. This increase is shown by all the metals plotted in Figure 6, and this is also the case for the metals that are not shown on the plots.

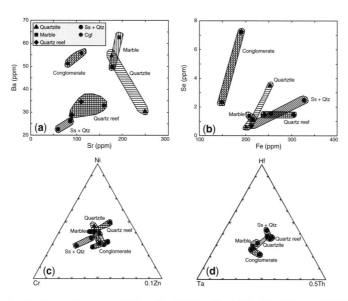

Fig. 5. Elemental plots for Serengeti grasses: (**a**) Sr v. Ba; (**b**) Fe v. Se; (**c**) Cr–Ni–0.1Zn; and (**d**) Ta–Hf–0.5Th.

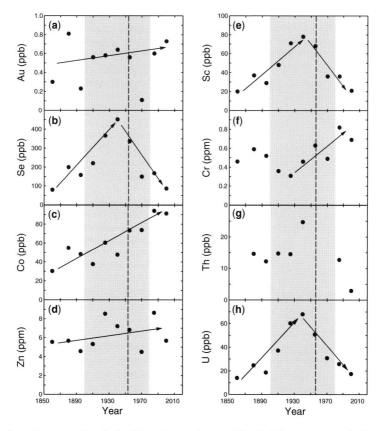

Fig. 6. Age v. element concentrations for the Palmerton tree-ring core. The shaded area represents the time the smelters were active. The vertical dashed line is the time the electrostatic precipitator was installed. (**a**) Au; (**b**) Se; (**c**) Co; (**d**) Zn; (**e**) Sc; (**f**) Cr; (**g**) Th; and (**h**) U.

- Temporally, there is a relationship between the concentrations of certain metals and the installation of the electrostatic precipitator. Se, Sc and U all decrease following the installation of the electrostatic precipitator, which suggests that these metals were mainly carried to the trees as airborne particles that were taken in through the leaves/stems or that these elements have a very short residence time in the soil.
- Th and U are decoupled in terms of their uptake by the trees. For the most recent tree ring (year 2000), the Th concentration is lower than the pre-smelter concentration.
- After cessation of smelting, Sc, Se, Zn and U return to pre-smelter values.
- After cessation of smelting, Au, Co and Cr concentrations are still significantly greater than pre-smelter values, an observation that suggests that these metals have a longer residence time in the biosphere.

From a forensic point of view what is notable is that the tree-ring record contains the history of environmental contamination from the smelters and reflects changes that were made to the smelters during their operation.

Other applications

The examples listed in the previous section are representative of the possible uses of INAA in criminal and environmental forensics. There are many other possible applications, a few of which are described below.

Recently, there has been a great deal of interest in nuclear forensics. The United States has initiated a nuclear detonation and detection programme with the aim of developing new techniques to detect nuclear materials and to characterize nuclear explosions. One of the products of a nuclear explosion is a glass generated by the melting of the immediately surrounding material. This glass contains a record of the nuclear explosion. Perhaps the most famous of these glasses is 'trinitite', the glass that was produced during the detonation of the first nuclear device at Alamogordo, New Mexico. The beads and dumbbell-shaped glasses produced by this event resemble tektites, glasses produced by a meteorite impact. Several recent papers (Eby et al. 2010; Fahey et al. 2010) have dealt with the application of various analytical techniques, among these gamma ray spectroscopy and INAA, to the characterization of this glass. The INAA measurements of the bulk material were used to determine the natural v. bomb-produced fractions of elements that can be formed as fission products and the addition of other elements during the nuclear detonation. It is also possible, via a reactor–INAA experiment, to determine the neutron dose delivered to the glass during the nuclear detonation. Hence, INAA is one of the analytical tools that can be used in nuclear forensic investigations.

Another potential application is identifying the source of diamonds (a very useful application for forensic investigations of diamond thefts). Using INAA, Damarupurshad et al. (1997) were able to analyse small diamonds ranging in size from 0.01 to 0.5 carat. They were able to detect as many as 40 elements at the ppb–ppt level. The data were used to divide the diamonds into two paragenetic groups, eclogitic and peridotitic, and it was also possible to chemically group diamonds from different mines. Hence, 'fingerprinting' of diamonds is a viable proposition.

Conclusions

INAA is a sensitive and non-destructive analytical method that can be used to determine as many as 40 elements in the ppm–ppb range in a variety of materials. The advantages of the method are that no pretreatment of the sample is required (the method is non-destructive), small samples (1 mg up to several hundred milligrams) can be analysed, a specific geometry is not required (if the material can be accommodated in the irradiation vial it can be analysed) and detection limits are low for many elements in a variety of matrices. As illustrated earlier, INAA can be used to record the history of environmental pollution, determine the characteristics of a nuclear explosion, and in comparison studies of ceramics, grasses, natural food products (in this case, maple syrup) and diamonds.

The nuclear irradiations and gamma ray spectroscopy were carried out at the University of Massachusetts Lowell radiation laboratory. L. Bobek provided support for the irradiations. L. James provided the ceramic samples. S. Eby (Yale University) provided the Serengeti grass samples. The Palmerton tree-ring core was collected by Z. Yu (Lehigh University). D. Pirrie and an anonymous reviewer are thanked for their helpful comments.

References

Adriamo, D. C. 2001. *Trace Metals in Terrestrial Environments: Biogeochemistry, Bioavailability, and Risks of Metals*, 2nd edn. Springer, New York.

Baes, C. F., III. & McLaughlin, S. B. 1984. Trace elements in tree rings: evidence of recent and historical air pollution. *Science*, **224**, 494–497.

Bhandari, S. A. & Amarasiriwardena, D. 2000. Closed-vessel microwave acid digestion of commercial maple syrup for the determination of lead and seven other trace elements by inductively coupled

plasma-mass spectrometry. *Microchemical Journal*, **64**, 73–84.
BISHOP, R. L. & BLACKMAN, M. J. 2002. Instrumental neutron activation analysis of archaeological ceramics: scale and interpretation. *Accounts of Chemical Research*, **35**, 603–610.
CANNON, H. 1960. *The Development of Botanical Methods of Prospecting for Uranium on the Colorado Plateau*. United States Geological Survey, Bulletin, **1085-A**.
CANNON, H. L. & ANDERSON, B. M. 1971. *Environmental Geochemistry in Health and Disease*. American Association for the Advancement of Science, Washington, DC.
CHAN, R. K. H. 1972. Identification of single-stranded copper wires by nondestructive neutron activation analysis. *Journal of Forensic Sciences*, **17**, 93–96.
DAMARUPURSHAD, A., HART, R. J., SELLSCHOP, J. P. F. & MEYER, H. O. 1997. The application of INAA to the geochemical analysis of single diamonds. *Journal of Radioanalytical and Nuclear Chemistry*, **219**, 33–39.
DESAI, H. B. & PARTHASARATHY, R. 1983. A radiochemical neutron activation analysis method for the determination of tin, arsenic, copper and antimony for the forensic comparison of bullet lead specimens. *Journal of Radioanalytical and Nuclear Chemistry*, **77**, 235–240.
EBY, N., HERMES, R., CHARNLEY, N. & SMOLIGA, J. A. 2010. Trinitite – the atomic rock. *Geology Today*, **26**, 181–186.
FAHEY, A. J., ZEISSLER, C. J., NEWBURY, D. E., DAVIS, J. & LINDSTROM, R. M. 2010. Postdetonation nuclear debris for attribution. *Proceedings of the National Academy of Sciences*, **107**, 20207–20212.
GIORDANI, L., RIZZIO, E. & BRANDONE, A. 2005. Neutron activation analysis in forensic investigations: trace elements characterization of cigarettes. *Journal of Radioanalytical and Nuclear Chemistry*, **263**, 739–744.
GISLASON, J. & PATE, B. D. 1973. Studies of gunshot residue. *Journal of Radioanalytical and Nuclear Chemistry*, **15**, 103–113.
GORDON, G. E., RANDLE, K., GOLES, G. G., CORLIS, J. B., BEESON, M. H. & OXLEY, S. S. 1968. Instrumental activation analysis of standard rocks with high-resolution γ-ray detectors. *Geochimica et Cosmochimica Acta*, **32**, 369–396.
GREENOUGH, J. D., FRYER, B. J. & MALLORY-GREENOUGH, L. 2010. Trace element geochemistry of Nova Scotia (Canada) maple syrup. *Canadian Journal of Earth Sciences*, **47**, 1093–1110.
GUINN, V. P. & WAGNER, C. D. 1960. Instrumental neutron activation analysis. *Analytical Chemistry*, **32**, 317–323.
HAGEMEYER, J. 2000. Trace metals in tree rings: what do they tell us? *In*: MARKERT, B. & FRIESE, K. (eds) *Trace Elements – Their Distribution and Effects in the Environment*. Elsevier Science, Amsterdam, 375–385.
IAEA 2003. *Nuclear Analytical Techniques in Archaeological Investigations*. International Atomic Energy Agency, Technical Report Series, **416**.
KABATA-PENDIAS, A. 2001. *Soils and Plants*. CRC Press, Boca Raton, FL.
KOEBERL, C. 1993. Instrumental neutron activation analysis of geochemical and cosmochemical samples: a fast and reliable method for small sample analysis. *Journal of Radioanalytical and Nuclear Chemistry*, **168**, 47–60.
KUČERA, J. & KOFROŇOVÁ, K. 2011. Determination of As by instrumental neutron activation analysis in sectioned hair samples for forensic purposes: chronic or acute poisoning? *Journal of Radioanalytical and Nuclear Chemistry*, **287**, 769–772.
LEPP, N. W. 1974. The potential use of tree-ring analysis for monitoring heavy metal pollution patterns. *Environmental Pollution*, **9**, 49–61.
LI, L., ZICHENG, Y., BEBOUT, G. E., STRETTON, T., ALLEN, A. & PASSARIS, P. 2010. Tree-ring width and $\delta^{13}C$ records of industrial stress and recovery in Pennsylvania and New Jersey forests: Implications for CO_2 uptake by temperate forests. *Chemical Geology*, **273**, 250–257.
LIU, Y., TA, W-Y. ET AL. 2009. Trace elements in tree rings and their environmental effects: a case study in Zi'an City. *Science in China Series D: Earth Sciences*, **52**, 504–510.
LUKENS, H. R. & GUINN, V. P. 1971. Comparison of bullet lead specimens by nondestructive neutron activation analysis. *Journal of Forensic Sciences*, **16**, 301–308.
NAKAMURA, N. 1974. Determination of REE, Ba, Fe, Mg, Na and K in carbonaceous and ordinary chondrites. *Geochimica et Cosmochimica Acta*, **38**, 757–775.
POTTS, P. J., WILLIAMS THORPE, O., ISAACS, M. C. & WRIGHT, D. W. 1985. High-precision instrumental neutron-activation analysis of geological samples employing simultaneous counting with both planar and coaxial detectors. *Chemical Geology*, **48**, 145–155.
SCHEID, N., BECKER, S. ET AL. 2009. Forensic investigation of brick stones using instrumental neutron activation analysis (INAA), laser ablation-inductively coupled plasma-mass spectrometry (LA-ICP-MS) and X-ray fluorescence analysis (XRF). *Applied Radiation and Isotopes*, **67**, 2128–2132.
SEDDA, A. F. & ROSSI, G. 2011. Bullets fragments identification by comparison of their chemical composition obtained using instrumental neutron activation analysis. *Forensic Science International*, **206**, e5–e7.
STUCKEL, J. G. & LOW, N. H. 1996. The chemical composition of 80 pure maple syrup samples produced in North America. *Food Research International*, **29**, 373–379.

Lichen monitoring as a potential tool in environmental forensics: case study of the Cu smelter and former mining town of Karabash, Russia

O. W. PURVIS[1]*, B. J. WILLIAMSON[1], B. SPIRO[2], V. UDACHIN[3],
I. N. MIKHAILOVA[4] & A. DOLGOPOLOVA[2]

[1]*College of Engineering, Mathematics and Physical Sciences, University of Exeter – Cornwall Campus, Camborne School of Mines, Penryn TR10 9EZ, UK*

[2]*The Natural History Museum, Cromwell Road, London SW7 5BD, UK*

[3]*Institute of Mineralogy, Russian Academy of Sciences, Miass 456317, Russia*

[4]*Institute of Plant and Animal Ecology, 8 Marta Street 202, Ekaterinburg 620144, Russia*

Corresponding author (e-mail: owpurvis@gmail.com)

Abstract: The aim of this paper is to showcase the use of lichens in environmental forensics from an assessment of atmospheric deposition in and around the Cu smelter and former mining town of Karabash, Ural Mountains of Russia. *Hypogymnia physodes* was collected on its bark substrate in July 2001 from a 'reference' site (*c*. 25 km SW of Karabash) and transplanted to 10 stations along an approximately 60 km SSW–NNE transect centred on Karabash. Transplants were collected after 2 and 3 month exposure periods. The elemental compositions of *Hypogymnia* and potential sources of particulates in the study area (smelter blast furnace and converter dusts, wastes, tailings, road dusts, metallurgical slags and top soils) were determined by inductively coupled plasma optical emission spectrometry (ICP-OES) and quadrupole ICP mass spectrometry (ICP-MS), and the Pb isotope compositions of the lichens and smelter dusts by multicollector ICP-MS. Particulates on lichen surfaces were analysed by scanning electron microscopy with energy-dispersive X-ray analysis (SEM-EDX).

The method of lichen transplantation, combined with multi-element and surface particle elemental analysis, high-precision Pb isotope ratio determinations and modelling, was shown to be useful for the tracing of the smelter signal, and components from different smelter processes, for more than 25 km from Karabash town. The lichen monitoring methodology is discrete and comparatively low cost, enabling atmospheric deposition from natural and anthropogenic sources to be determined over short (<3 month) periods, and is therefore a valuable qualitative tool for environmental forensics.

Key aspects of the nature and biology of lichens

Lichens are mutualistic associations involving at least one fungus (mycobiont) and an alga or cyanobacterium (photobiont). Overall, this relationship is very successful as lichens are found in almost all terrestrial habitats from the tropics to polar regions. Lichens are classified as fungi and vary from less than 1 mm^2 to over 2 m long. They are poikilohydric organisms whose water status varies passively with the surrounding environmental conditions. Most species live for decades or hundreds of years and, as perennials, are subject to the cumulative effects of pollutants. Unlike many vascular plants, lichens have no deciduous parts and, hence, cannot shed these to expel potential toxins. Furthermore, the lack of stomata and a protective cuticle in lichens means that atmospheric deposits may be absorbed over the entire thallus surface (Fig. 1). During dry periods, solutions may be concentrated to the point that levels of toxic substances are sufficient to alter the balance between photobiont and mycobiont. This may lead to a breakdown in the association and death of the lichen. Lichen species distribution patterns may therefore reflect varying levels of air pollutants, as well as a variety of unrelated biotic and abiotic factors. Lichens are efficient accumulators of metals and persistent organic pollutants, and are frequently used to monitor the deposition of such substances (e.g. Farmer 1993; Nimis *et al.* 2002; Nash 2008), and have even been used in minerals prospecting (Chettri *et al.* 1997). Some species are highly tolerant of pollution, including from smelters (e.g. Purvis *et al.* 2000).

Historical background

Pioneering biomonitoring using lichens investigated fallout from nuclear testing in the 1950s (James 1973). Studies in the late 1960s and the early 1970s identified a correlation between atmospheric

MECHANISMS OF METAL FIXATION

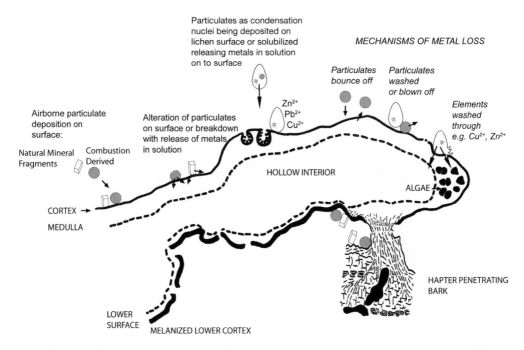

Fig. 1. Diagrammatic representation of section of *Hypogymnia physodes* showing the main mechanisms by which metals may be deposited on or lost from surfaces (adapted from Beltman 1978; Williamson *et al.* 2004a), not to scale.

SO_2 concentrations and sulphur content in samples of the lichens *Parmelia saxatilis* (Gilbert 1965, 1968, 1969) and *Hypogymnia physodes* (Hawksworth *et al.* 1973; James 1973), and a correlation between the atmospheric deposition of heavy metals and metal content in lichens (Jenkins & Davies 1966). Lichens have since been increasingly used to monitor spatial and temporal patterns of heavy metal dispersion from industrial sources and roads, for example by Seaward (1973), Nieboer *et al.* (1975), Richardson (1995), Garty (2001) and many others; see the list of studies abstracted in the Recent Literature on Air Pollution Lichen series published in *The Lichenologist* over the period 1974 to 2000 (Hawksworth 1974; Henderson 2000). In relation to a smelter, Nieboer *et al.* (1975) were amongst the first to show a curvilinear relationship for Cu, Ni, Zn and Fe content in metal-tolerant lichen species, with distance confirming this as their point source. There is a general consensus that the trapping of relatively large particles in large medullary intercellular spaces accounts for the bulk of accumulated metals in thalli near industrial sites (Seaward 1973; Garty *et al.* 1979).

The widely distributed epiphytic foliose (leaf-like) macrolichen *Hypogymnia physodes* has been extensively used in the source apportionment of anthropogenic and natural substances deposited in the environment. These substances may originate in the lichens either by dry and/or wet atmospheric deposition or stem flow and/or canopy through-fall (Bargagli 1998). In spatially extensive surveys of remote regions where it is impractical or too costly to employ large numbers of recording/sampling devices, lichen biomonitoring may be one of the only ways to determine patterns of atmospheric deposition (Nimis *et al.* 2002; Nash 2008). Various methods may be applied including monitoring species abundance, diversity and health, which is rapid and relatively low cost. Where lichens are not naturally present (e.g. the environment being too contaminated), transplants may be used. These are sampled from a 'background' or 'reference' site and transplanted to monitoring stations elsewhere for a fixed exposure period. The lichen transplants can then be analysed, and their compositions compared with samples from the reference site to determine and/or map relative levels and sources of metals and other species. It should be noted, however, that *Hypogymnia* and other lichens should not be regarded as 'inert depositimiters'; that is, they are not 'passive filters' (Johnsen 1981; Glenn *et al.* 1991; Rossbach *et al.* 1999; Nimis *et al.* 2002). Complex, often poorly understood, biogeochemical processes may contribute to the fixing and localization of metals in lichens (e.g. Garty 2001; Haas & Purvis 2006; Purvis & Pawlik-Skowrońska 2008), and there are a number of ways metals can be lost and/or gained during the transplant exposure period (Fig. 1). The method of lichen transplant monitoring therefore should be regarded as qualitative, although it may be possible to quantify the relative signature of different source inputs (see the subsection on 'The modelling approach' later). The earliest transplant studies using *H. physodes* were by Schönbeck (1969) and others in the early 1970s (Mikhailova 2002; Rusu *et al.* 2006a, b). Subsequent developments in analytical techniques have lead to further innovation.

Sampling and analytical methods used to assess metal contents in lichens

The analysis of whole-lichen samples is practical, sensitive and is relatively cost effective. It generally requires sample homogenization and dissolution followed by wet chemical analysis, most commonly by ICP-OES and ICP-MS (Rusu 2002). During the sampling and analysis of lichens, rigorous quality assurance/control procedures should be followed at all times. Lichens must be sampled from equivalent sites, including distance from roads, vegetation/ woodland type and similar substrates (Bargagli 1998; Garty 2001; Bargagli & Nimis 2002). Adequate numbers of replicate samples must be collected in order to establish the precision of the methodology. Certified reference materials should be analysed at the same time as the collected samples to establish the accuracy of the analytical procedures being used. Data obtained from whole-lichen analyses can be visualized in a number of different graphical plots (see example figures in this paper) or subject to multivariate statistical analyses to help identify the sources of contained elements (Sloof 1995; Faus-Kessler *et al.* 2001; Yenisoy-Karakas & Tuncel 2004, 2008).

Data from whole-sample wet chemical analysis, however, cannot generally be used to identify the chemical and mineralogical form of contained elements, which may provide important clues as to their source and mode of deposition. The presence and nature of metal-rich particles on thallus surfaces and within their interiors has been directly determined using SEM, often combined with X-ray elemental analysis (Garty *et al.* 1979; Olmez *et al.* 1985; Bargagli & Mikhailova 2002; Williamson *et al.* 2004a).

Increasingly, isotopic analysis including that for sulphur (Krouse 1977; Spiro *et al.* 2002), lead and zinc isotopes (Monna *et al.* 1999; Spiro *et al.* 2004, 2012; Weiss *et al.* 2004; Dolgopolova *et al.* 2006b; Sondergaard *et al.* 2010), and rare earth element signatures have been utilized in source apportionment (Chiarenzelli *et al.* 2001; Zschau *et al.* 2003; Dolgopolova *et al.* 2006a; Purvis *et al.* 2010).

The aim of this paper is to showcase the use of lichens in environmental forensics, which follows on from a recent study on the use of fungi for such studies (Hawksworth & Wiltshire 2011). We present a case study illustrating the assessment of atmospheric deposition from a copper smelter and mining-contaminated areas in the town of Karabash, Ural Mountains of Russia (Fig. 2). Such point sources can provide natural laboratories to investigate the effects of pollutants and geology on vegetation (Bell & Treshow 2002). Methods used in our study included multi-element, isotopic and microscopical analysis of native lichens (*Hypogymnia physodes*), and transplants of this species on birch (*Betula*), with the transplants exposed at the *transplant* stations for 2 and 3 month periods. Previous publications from the Karabash study include general pollution aspects (Udachin *et al.* 2003), characterization of particulates on lichen transplant surfaces (Williamson *et al.* 2004a), biogeochemical signatures and source apportionment (Purvis *et al.* 2004, 2006; Williamson *et al.* 2008), and the application of Pb isotopes to determine the sources of Pb in lichen transplants (Spiro *et al.* 2004, 2012).

Study area

The smelter and former mining town of Karabash lies in the Chelyabinsk region of the South Urals of Russia. It sits within a NE–SW-trending flat-bottomed valley with altitudes ranging from 250 to 650 m. The climate is moderately continental with cold winters and warm summers (mean January and July temperatures are -16 and $+18\ °C$, respectively). Precipitation in winter and summer is approximately 130 and 500 mm, respectively, with a snow depth in winter of about 40 cm. Prevailing winds are northwesterly (Stepanov *et al.* 1992). Phytogeographically, the region lies on the border between taiga and forest-steppe zones, consisting of a mosaic of mesic pine woods (*Pinus sylvestris* mixed with *Betula pendula*, *B. pubescens* and *Larix sibirica*) and secondary mesic birch woods that have replaced pine following logging or forest fires.

The Karabash smelter, which lies close to the centre of the town, mainly produces blister copper, around 41 700 tonnes in 2001 (Voskresensky 2002). During the period of the lichen transplant experiment, the smelter was operating blast furnaces and a converter, each with separate stacks. In the

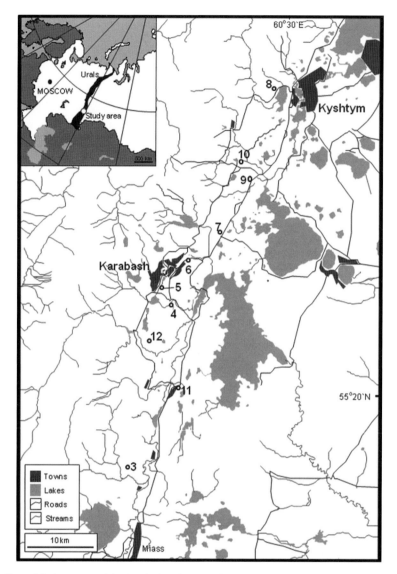

Fig. 2. Locality and drainage map of the Karabash and surrounding areas showing the location of lichen transplant stations.

blast (reverberatory) furnace copper concentrates are melted and silicate slag drawn off to produce a molten sulphide matte. In the converter, however, remaining elements such as iron and sulphur are removed to produce blister copper (c. 99% Cu: Williamson et al. 2008). From studies of airborne total suspended particulate (TSP) collected in Karabash town in 2001, using air pump apparatus (Williamson et al. 2004b), particulates downwind from the smelter was mainly composed of anglesite ($PbSO_4$), zincite (ZnO), gunningite ($(Zn,Mn)SO_4 \cdot H_2O$), a phase with the composition Zn_2SnO_4, and poorly ordered Zn sulphates, with lesser amounts of pyrite, sphalerite, chalcopyrite and galena. This material had a maximum particle size (equivalent spherical diameter) of around 2 μm (average of 0.5 μm, SD = 0.2). From their composition and size, these particles were thought to mostly originate from the smelter converter (Williamson et al. 2004b). More Fe–Cu-rich particles on lichen surfaces, with a mean equivalent spherical diameter of 2.2 μm (SD = 2.4) and not present in the air filters, are thought to have been mainly derived from the blast furnace (Williamson et al. 2004b).

Other possible sources of metal-rich airborne particulates in Karabash are waste dumps and tailings, which lie to the NE, south and SE of the town. Three tailings dumps cover a total area of around 2.1 km^2. In two of the tailings ponds, around 65% of the material has a grain size <0.074 mm. In the third, the 'Sak-Elga' tailing, around 11% is <0.074 mm. The mineralogy of the tailings is dominated by pyrite, chalcopyrite, sphalerite, tennantite and magnetite, with quartz, feldspar, sericite, chlorite and barite, and a variety of alteration phases. Metallurgical waste dumps cover an area of around 170 000 m^2 and contain granulated (<0.5 mm) Ca-, Fe-rich silicate glass (slags) and minor amounts of sulphides (Udachin et al. 1998). Dumps of pyrite-containing mine waste material cover an area of around 150 000 m^2. Other major sources of dusts include local roads (commonly gritted using metallurgical slags) and natural wind-blown soil particles from de-vegetated hillsides around the town (Williamson et al. 2004b), such as Karabash 'Mountain' (Fig. 3a).

Materials and methods

The lichen transplant methodology

Healthy *Hypogymnia physodes* thalli, about 3–4 cm in diameter, were cut with underlying bark from birch trunks sampled in June 2001 in a birch (mixed *Betula pendula*/*Betula pubescens*) stand from Site 3 ('reference site'), approximately 25 km from Karabash copper smelter (Fig. 2). During sampling, the general guidance in Bargagli & Nimis (2002) was followed, such that: (i) each lichen thallus should have the same probability of being selected from a total population of the same species; and (ii) collected samples suffered no alteration in chemical composition from the original population. Lichens were always removed from the tree bark using a stainless steel knife and plastic gloves to avoid contamination. The transplant stations were established in birch stands along an approximately 60 km transect, with Karabash in the centre (Fig. 2). The predominant wind direction is from the NW; however, due to practical considerations, it was necessary to sample across a NE–SW transect, coincident with the main road in the area. Six birch trees were selected at each transplant site and 10 thalli attached to each tree by carefully gluing (with the all purpose glue 'Moment') their bark substrate (to avoid contaminating the lichen) to within 1 m of the tree base. The transplants were placed in two rows of five thalli facing towards the smelter (see example in Fig. 3b). The transplants were removed after 2 and 3 month exposure periods (September and October 2001). The material was transported to the laboratory in paper bags, allowed to air-dry, and then sealed in plastic bags before hand-cleaning under a microscope to remove bark flakes and other debris prior to elemental and isotopic analysis (Purvis et al. 2004).

Imaging and elemental analysis of particulates on transplant surfaces

Material collected at sites 5 (Karabash) and 3 (transplant/reference site at Severnye Peche) in September 2001 was subsampled for examination. Four thalli collected from different trees were selected from each site and tips of upwards-directed lobes 2–3 mm in length were taken (one lobe from each thallus). Samples were mounted on their lower surfaces onto aluminium 12.5 mm-diameter SEM stubs (Agar, UK), using a thin layer of Araldite, and then each stub was carbon coated. The stubs were loaded into a JEOL 5900LV SEM. The surfaces were imaged in backscattered electron mode in order to distinguish non-organic material from the sometimes complex lichen surface and accumulated organic airborne debris (Fig. 4; Purvis et al. 2002). An accelerating voltage of 20 kV and a beam current of approximately 1.4 nA produced a reasonable level of contrast between the thalli (organic) and inorganic surface materials, and a high enough resolution to image particles of <1 μm. Images (×1000 magnification) were obtained for three randomly chosen areas on thalli from each of the four lichen stations from sites 5 (Karabash) and 3 (reference site) (i.e. 12 imaged areas from each site). Elemental analysis was undertaken using an Oxford Instruments INCA energy-dispersive X-ray analysis system. The elemental data obtained are far from quantitative as, amongst other factors, the sample surface is not flat and the particles are irregularly shaped. The data can therefore only be used to infer the mineralogy of individual particles from inter-element relationships (Williamson et al. 2004a).

Whole-lichen multi-element analysis

For all transplant stations (except where replicate samples were taken), five thalli from each tree were bulked for chemical analysis for each collection period; that is, 30 thalli per site per sampling period. To determine natural variation at each station, samples of five thalli were bulked (September collection) from each tree from the 'reference' site (Site 3), an 'intermediate' site (Site 12) and the 'impacted' site (Site 5). Native thalli were bulked from each of sites 1, 2, 3 and 8 (July 2001), and both transplants and native thalli from sites 3, 8, 9, 10 and 11 (September and October

Fig. 3. (**a**) Photograph looking north towards the Karabash smelter. Note the eroded and de-vegetated slopes of Karabash 'Mountain' to the east; and (**b**) *Betula* trunk at Site 12 with 10 *Hypogymnia physodes* transplants adhered to the bark with glue. The trunk is colonized by *Cladonia coniocraea* at the base. The understorey was dominated by vascular plants characteristic of acidic habitats (*Vaccinium myrtillus* and *Deschampsia caespitosa*).

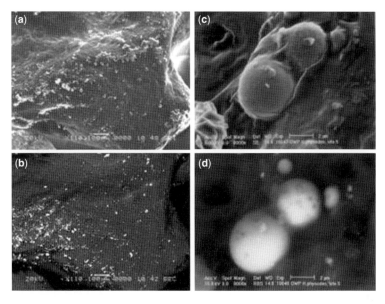

Fig. 4. SEM images of surface of *Hypogymnia physodes* thalli transplanted near the Karabash smelter: (**a**) and (**b**) general view from above; and (**c**) and (**d**) showing particulates (PM10s), becoming incorporated into the melanized lower surface. (a) and (c) are secondary electron images; and (b) and (d) backscattered (bright areas indicating high atomic number) electron images.

2001). Around 250 mg of *Hypogymnia* thalli were digested in HNO_3–H_2O_2 in open vessels under reflux according to method B (Bettinelli *et al.* 1996). Sample solutions were filtered, and the elements Al, Ba, Ca, Cd, Cr, Cu, Fe, K, Mg, Mn, Na, Ni, P, Pb, S, Sn, Sr, Ti, V and Zn determined by ICP-OES, and As, Bi, Co, Cs, Ga, Ge, Li, Mo, Rb, Sb, Se, Te, Th, Tl and U by mass spectrometry (ICP-MS). Accuracy was assessed by analysing the lichen reference material CRM482 (Rusu 2002) and was found to be within 10%.

Pb elemental and isotopic determinations

Sample preparation followed the methods of Rusu (2002) and Dolgopolova *et al.* (2004). Samples of lichens were digested using a high-pressure and high-temperature microwave MarsX system with a H_2O_2–HF–HNO_3 acid mixture. The Pb was separated from the sample matrix using standard anion-exchange column chemistry with an EiChrom Sr-resin and varying HCl acid strengths. Samples were then measured using a Micromass IsoProbe multi-collector ICP-MS. The mass bias was corrected using thallium (Tl) as external dopant. The long-term 2σ reproducibility of the Pb isotope ratio measurements is below 300 µg/g for all measured isotopes ($^{206}Pb/^{204}Pb$, $^{207}Pb/^{204}Pb$, $^{208}Pb/^{204}Pb$) and the accuracy after daily adjustment of the Tl ratio was more than 100 µg/g for all ratios (Spiro *et al.* 2004). Full details of the procedures for mass spectrometry (e.g. regarding mass bias correction, blank contribution and error propagation) are given elsewhere (Kylander *et al.* 2004; Weiss *et al.* 2004).

Statistical analyses and modelling

Multivariate analysis was carried out using PRIMER 5, and statistical tools within MS Excel and spatial relationships of element concentrations explored. Enrichment factors (EFs) were calculated for elements in relation to average element concentrations in the Earth's upper continental crust (UCC) (Taylor & McLennan 1985) according to the formula: EF_{UCC} [Element/Ti]$_{sample}$/[Element/Ti]$_{UCC}$. Titanium was selected as the reference element because Sc, frequently used as an indicator of soil contamination and to calculate enrichment factors, was below detection limits of ICP-MS (<0.4 mg/kg) in all samples (Purvis *et al.* 2006).

To determine which process in the smelter was emitting the highest relative levels of potentially toxic metals (either the blast furnaces or converter), and whether particulates from each showed a different spatial dispersion pattern, a multiple linear least-squares regression mixing model was used (Le Maitre 1979; Williamson *et al.* 2008). The objective of this was to provide data to aid in any future prioritization of measures to reduce emissions. The assessment involved modelling ('fitting') the elemental compositions (Al, As, Cd, Co, Cr, Cu, Fe, Ni, Pb, Sn, Ti and Zn) of lichen transplants from different distances from the smelter by mixing the pattern of the 'reference' transplant lichens with varying proportions of the element compositions of the blast furnace and converter dusts. The model yielded the proportions of: (a) lichen 'reference' composition; (b) smelter blast furnace; and (c) smelter converter compositions to best fit the pattern for the lichen transplants. This type of analysis was not found to be appropriate for determining the relative contributions from minor sources (e.g. waste dumps and tailings). Owing to natural variability between replicate samples from sites 3, 5 and 12, weak source signatures will have been lost amongst the 'noise' (Williamson et al. 2008).

Results and discussion

Whole-lichen multi-element analysis

The distribution of 25 elements along the transect showed a characteristic curvilinear decrease with distance from the Karabash smelter suggesting influence from the point source (Purvis et al. 2004). In the case of Pb (Fig. 5), over the 3 month exposure period transplants showed more than a five times increase in concentration near the smelter (Site 5, 1200 µg/g Pb) compared with those from the 'reference' site 3 (average of 213 µg/g Pb, $n = 6$) approximately 25 km to the SW. The average concentration of 15 elements with the highest enrichment factors (see the subsection on 'Statistical analyses and modelling') in replicate transplants from sites 5 and 3 were compared. The most enriched elements (in order of decreasing enrichment) were Te, Cd, Cu, Pb, Se, Zn and Sb. All elements showed curvilinear decreases from Karabash, even those at concentrations below 1 µg/g.

Interestingly, 14 of the top 15 elements (Te, Cd, Cu, Pb, Se, Zn, Sb, Sn, As, Bi, S, P, Mo and Ca) with the highest enrichment factors in replicate transplants established at Karabash site 5 were also the most enriched in Hypogymnia transplants from along a transect centred on the Cu smelter in Zlatna town, Romania (Rusu et al. 2006a). This confirms the association of these particular elements with Cu ore smelting. In contrast, little or no enrichment was observed for Na, Th, Ti, Al, Li, U, Cs and V (Purvis et al. 2006). By analogy with moss biomonitoring studies carried out elsewhere (Faus-Kessler et al. 2001), the data strongly suggested that the elements U, Th, Cs, Na, Ti and Al were at least partly derived from fossil fuel combustion (Purvis et al. 2006). Negative correlations between concentrations of K with S and potentially toxic metals (e.g. Cu, Pb, As) was interpreted to reflect element leakage from lichen tissues as a result of stress (Purvis et al. 2004; Williamson et al. 2004a).

Element variability amongst Hypogymnia transplants from the same site was highlighted through analysing replicate samples from sites 3, 12 and

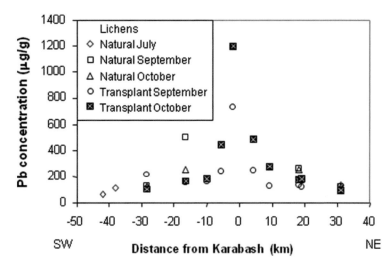

Fig. 5. Concentration of Pb (µg/g) in *Hypogymnia* collected from the SW to NE 60 km-long transect centred at Karabash. Native (natural) lichen were collected in July, September and October in 2001. Transplants were deployed for 2 months (September) and for 3 months (October). The transplants deployed for 3 months and the native samples from reference site 3 collected in July were used for Pb isotope analysis.

5. At the 'reference' site 3, Pb content ranged from 68 to 760 μg/g; that is, a more than 11 times variation in Pb content amongst the five samples analysed. Nevertheless, the similar trends in enrichment factors at sites 3 and 5 indicate the robustness of the averaging approach.

Imaging and analysis of particles on lichen transplant surfaces

Particulates in the Karabash samples were mainly in the form of 5–200 μm sized, commonly spherical sulphide and silicate metallurgical slag particles. Approximately 20–30% of particulates consisted of aggregates of materials of variable composition, occasionally up to 40 μm in diameter, which appeared to represent surface deposits formed *in situ* on the lichen surface. This may be either through the breakdown of primary particulate material or precipitation from solution. The Karabash samples contained seven times the number of chalcopyrite particles, over nine times the proportion of the Zn–Sn-rich phase, and around half and two-thirds the number of quartz and albite particulates, respectively, compared with the 'reference site'. The discovery of particles of the Zn–Sn-rich phase (thought to be exclusively of smelter origin) in the 'reference' sample (25 km south of Karabash), together with chalcopyrite, indicates wide spatial dispersion towards the SW of particles from the Karabash smelter.

Particulates on the surfaces of the transplants from Karabash were found to be relatively coarse grained and to have compositions similar to dusts from the blast furnace. This observation is at odds with the size (<2 μm) and composition (Pb- and Zn-rich) of particles collected on air filters in the town, which are thought to have been derived from the converter process (Williamson *et al.* 2004*b*). The most likely explanation for this is that the relatively small particles from the converter have a longer atmospheric residence time compared with the generally larger Cu–Fe-rich particles from the blast furnace. Spatial fractionation of airborne particulates according to particle size is considered a major factor affecting elemental inventories in lichens (Zschau *et al.* 2003). This hypothesis was tested by carrying out source apportionment modelling for metals in lichen transplants from different distances from the smelter (see the subsection on 'The modelling approach').

Interestingly, thallus surfaces from the Karabash and 'reference' sites showed similar particle numbers per unit area (Fig. 6). This was interpreted as indicating that over the 2 month exposure period, the 'input' of particles to the Karabash site samples was apparently balanced by removal of 'background' particles, possibly being washed or blown off, or some types of materials being solubilized (Williamson *et al.* 2004*a*). A sketch diagram outlining the main mechanisms by which metals may be deposited on the surfaces of *Hypogymnia physodes* is presented in Figure 1.

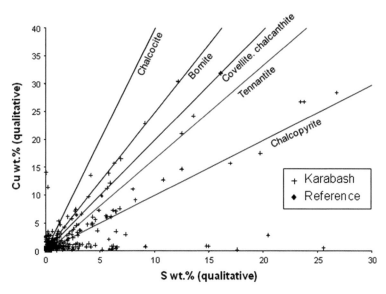

Fig. 6. Diagram of S v. Cu wt% (qualitative) for particulates on transplant thallus surfaces from the Karabash and the 'reference' site. Lines are shown for the stoichiometric ratios of different common Cu-bearing sulphides (modified after Williamson *et al.* 2004*a*).

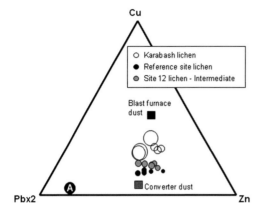

Fig. 7. Ternary diagram for Cu v. Pb × 2 v. Zn for lichen transplants, TSP filter collections and samples from the blast furnace and converter stacks. The sizes of symbols for lichen transplants and TSP filter collections are proportional to Pb content (relative to other samples from each media type). Point A is an anomalous value for lichen transplants from the 'reference' site (adapted from Williamson et al. 2008).

The modelling approach

On the triangular diagram showing relative concentrations of Cu, Zn and Pb in the lichen samples from wet chemical analysis (Fig. 7), the replicate transplants from the Karabash site lie further towards the Cu apex compared with the 'reference' site samples, with the intermediate site transplants in between. Samples from the Karabash site therefore have a relatively high ratio of Cu to Zn and Pb (Williamson et al. 2008). There is a trend in the compositions of the lichen transplants from the 'reference' site to the Karabash samples (Site 5) towards the composition of the blast furnace dust. This was interpreted as indicating that the dominant metal from the smelter is Cu, and that the dominant source of this metal is the smelter blast furnace. Other potential sources (tailings, metallurgical slags, road dusts and soils) were also plotted on this diagram (Williamson et al. 2008) but owing to a lack of detailed trends within the replicate datasets, and because some of the compositions for minor sources overlap with those of the smelter dusts, no non-smelter source signatures could be identified with any confidence.

The results of the least-squares modelling are graphically presented in Figure 8, which shows the modelled percentage of blast furnace and converter signature for lichen transplants collected at different distances from the smelter. From this figure, it is clear that close to the smelter the signature of the blast furnace is dominant over that of the converter in the transplants. Further from the smelter, the converter signature dominates (Williamson et al. 2008). The same pattern was observed in the September and October sample datasets.

Sources from Pb isotopes data

The blast furnace and converter dusts (smelter dust in Fig. 9) from the Karabash smelter show a $^{208}Pb/^{204}Pb$ isotope ratio characteristic of Ural-type

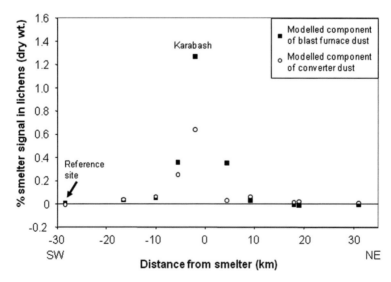

Fig. 8. Results of multi-element (Al, As, Cd, Co, Cr, Cu, Fe, Ni, Pb, Sn, Ti and Zn) least-squares modelling for October collections (adapted from Williamson et al. 2008).

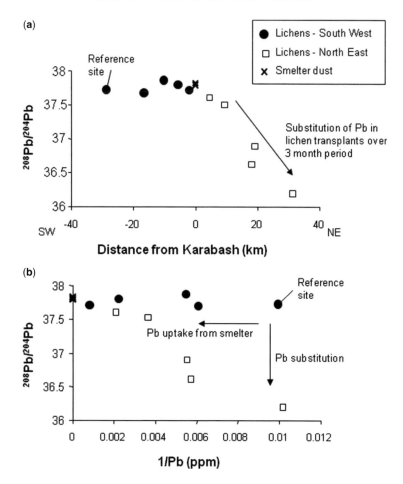

Fig. 9. (**a**) Pb isotope composition of *Hypogymnia* transplants exposed for 3 months (2001). Each lichen sample is bulked from five individual thalli from each of six trees; $^{208}Pb/^{204}Pb$ isotope ratios are relatively constant in the SW and change NE of the smelter (Spiro *et al.* 2004). (**b**) Plot of $^{208}Pb/^{204}Pb$ v. 1/Pb in lichen transplants and smelter dust showing that the isotopic composition in lichen transplants are independent of the Pb concentration in samples collected from the SW part of the transect (black circles) but not in the NE part (open quadrangles). This suggests the presence of an additional dispersed source to the NE, most probably soil.

volcanogenic massive sulphide deposits (Tessalina *et al.* 2001). To the SW of the transect, isotope ratios in transplanted and native lichens show this same signature, indicating that the smelter was the main source of Pb in these samples. However, to the NE, the Pb isotope ratios of the lichens changes markedly indicating a contribution of Pb from alternative sources. This is most clearly discernible on the plot of $^{208}Pb/^{204}Pb$ v. 1/Pb, which presents two arrays (Fig. 9). The array of transplants to the NE shows a decreasing isotope ratio with decreasing 1/Pb, whereby at low Pb concentrations the $^{208}Pb/^{204}Pb$ ratio decreases to the value of 36.199, and at high Pb concentrations it reaches that of the smelter dust samples (37.821). This suggests a mixing of two end members; one end member being the smelter dust, while the other has a low concentration of Pb that may be associated with soil. This indicates that the Pb content from the background site was largely replaced, and that the transplants therefore provide a transient record reflecting a continuous fixation and loss of environmental Pb (Spiro *et al.* 2004). By using tri isotope plots $^{208}Pb/^{204}Pb$, $^{207}Pb/^{204}Pb$ and $^{206}Pb/^{204}Pb$, distinct sources of lead in the native and transplanted lichen were identified (for details see Spiro *et al.* 2004). A study of the Orlovka–Spokoinoe mining site, Transbaikalia, Russia identified atmospheric transfer of lead from distant sources to the lichen, in addition to local components mobilized by mining and ore-processing activities (Dolgopolova *et al.* 2006*a, b*).

Conclusions

This study of transplants and native thalli of the lichen *Hypogymnia physodes* as a potential forensic tool focused on the area in and around the Cu smelter and former mining town of Karabash, South Urals of Russia. The main source of heavy metals was known (the smelter), and therefore the study focused on the indicative, largely spatial, characteristics of atmospheric dispersion, the relative contribution of metals in the lichens (with distance) from each of the main processes in the smelter (blast furnace or convertor) and what other sources contributed to the environmental load of heavy metals in the area. A smelter signal, and relative contributions from the blast furnace and converter, could be identified for up to 25 km from Karabash town. The most prevalent additional source signature was from the burning of coal (Spiro *et al.* 2004).

In conclusion, *Hypogymnia* transplants are dynamic systems that trap and lose particles deposited on their surfaces. They are therefore useful as monitors of deposition for limited time periods when transplanted to environments where they would not naturally grow. The present study suggests that a 2–3 month period is useful for the heavily polluted area in the vicinity of the Karabash smelter. However, other periods and other species may be useful for the biomonitoring and sampling of atmospheric deposition in other settings. This case study demonstrates the efficacy of the approach in general, and that it could be used for different compounds and with different lichens. Provided there is sufficient background research, and rigorous sampling and analytical protocols are followed, we suggest that the information derived from lichen studies could be used in a forensic situation and challenged in court.

We thank G. Jones and T. Jeffries for ICP-AES and ICP-MS analyses. The transplant studies were carried as part of a 3 year (2000–2002), EU FP5 INCO-Copernicus 2 contract 'Mineral resources of the Urals: Origin, Development and Environmental Impacts' (MinUrals, contract: ICA2-CT-2000-10011) awarded to a consortium, including the Natural History Museum where this work was originally undertaken, with additional funding provided by a Royal Society Joint Project Collaborative Grant with Russia (gt/fSU/JP). Follow-up studies (including the drafting of this article) were funded by a 3 year (2010–2012) EU FP7 contract 'Impact Monitoring of Mineral Resources Exploitation' (ImpactMin, contract: 244166) awarded to a consortium including the University of Exeter. This article represents the views of the authors alone and not those of the European Community or Royal Society. The European Community and Royal Society accept no responsibility for the use of any data contained in this article.

References

BARGAGLI, R. 1998. *Trace Elements in Terrestrial Plants: An Ecophysiological Approach to Biomonitoring and Biorecovery*. Springer, Berlin.

BARGAGLI, R. & MIKHAILOVA, I. 2002. Accumulation of inorganic contaminants. *In*: NIMIS, P. L., SCHEIDEGGER, C. & WOLSELEY, P. A. (eds) *Monitoring with Lichens – Monitoring Lichens*. NATO Science Series, IV. Earth and Environmental Science, Kluwer, Dordrecht, **7**, 65–84.

BARGAGLI, R. & NIMIS, P. L. 2002. Guidelines of the use of epiphytic lichens as biomonitors of atmospheric deposition of trace elements. *In*: NIMIS, P. L., SCHEIDEGGER, C. & WOLSELEY, P. A. (eds) *Monitoring with Lichens – Monitoring Lichens*. NATO Science Series, IV. Earth and Environmental Science, Kluwer, Dordrecht, **7**, 295–299.

BELL, J. N. B. & TRESHOW, M. 2002. *Air Pollution and Plant Life*. Wiley, Chichester.

BELTMAN, H. A. 1978. *Vegetative Strukturen der Parmeliaceae und ihre Entwicklung*. J. Cramer, Vaduz.

BETTINELLI, M., SPEZIA, S. & BIZZARRI, G. 1996. Trace element determination in lichens by ICP-MS. *Atomic Spectroscopy*, **17**, 133–141.

CHETTRI, M. K., SAWIDIS, T. & KARATAGLIS, S. 1997. Lichens as a tool for biogeochemical prospecting. *Ecotoxicology and Environmental Safety*, **38**, 322–335.

CHIARENZELLI, J., ASPLER, L., DUNN, C., COUSENS, B., OZARKO, D. & POWIS, K. 2001. Multi-element and rare earth element composition of lichens, mosses, and vascular plants from the Central Barrenlands, Nunavut, Canada. *Applied Geochemistry*, **16**, 245–270.

DOLGOPOLOVA, A., WEISS, D. J., SELTMANN, R., STANLEY, C. J., COLES, B. & CHERBURKIN, A. K. 2004. Closed-vessel microwave digestion technique for lichens and leaves prior to determination of trace elements (Pb, Zn, Cu) and stable Pb isotope ratios. *International Journal of Environmental and Analytical Chemistry*, **84**, 889–899.

DOLGOPOLOVA, A., WEISS, D. J., SELTMANN, R. & DULSKI, P. 2006*a*. Dust dispersal and Pb enrichment at the rare-metal Orlovka-Spokoinoe mining and ore processing site: insights from REE patterns and elemental ratios. *Journal of Hazardous Materials*, **132**, 90–97.

DOLGOPOLOVA, A., WEISS, D. J., SELTMANN, R., KOBER, B., MASON, T. F. D., COLES, B. & STANLEY, C. J. 2006*b*. Use of isotope ratios to assess sources of Pb and Zn dispersed in the environment during mining and ore processing within the Orlovka-Spokoinoe mining site (Russia). *Applied Geochemistry*, **21**, 563–579.

FARMER, A. M. 1993. The effects of dust on vegetation – a review. *Environmental Pollution*, **79**, 63–75.

FAUS-KESSLER, T., DIETL, C., TRITSCHLER, J. & PEICHL, L. 2001. Correlation patterns of metals in the epiphytic moss *Hypnum cupressiforme* in Bavaria. *Atmospheric Environment*, **35**, 427–439.

GARTY, J. 2001. Biomonitoring atmospheric heavy metals with lichens: theory and application. *Critical Reviews in Plant Sciences*, **20**, 309–371.

GARTY, J., GALUN, M. & KESSEL, M. 1979. Localization of heavy metals and other elements accumulated in the lichen thalli. *New Phytologist*, **82**, 159–168.

GILBERT, O. L. 1965. Lichens as indicators of air pollution in the Tyne Valley. *In*: GOODMAN, G. T., EDWARDS, R. W. & LAMBERT, J. M. (eds) *Ecology and the Industrial Society*. Blackwell Scientific, Oxford, 35–47.

GILBERT, O. L. 1968. Bryophytes as indicators of air pollution in the Tyne Valley. *New Phytologist*, **67**, 15–30.

GILBERT, O. L. 1969. The effect of SO_2 on lichens and bryophytes around Newcastle upon Tyne. In: *Air Pollution, Proceedings of the First European Congress on the Influence of Air Pollution on Plants and Animals*, April 22–27, 1968. Centre for Agricultural Publishing and Documentation, Wageningen, 223–235.

GLENN, M. G., ORSI, E. V. & HEMSLEY, M. E. 1991. Lichen metal contents as correlates of air filter measurements. *Grana*, **30**, 44–47.

HAAS, J. R. & PURVIS, O. W. 2006. Chapter 15. Lichen biogeochemistry. *In*: GADD, G. M. (ed.) *Fungi in Biogeochemical Cycles*. Cambridge University, Cambridge, 344–376.

HAWKSWORTH, D. L. 1974. Literature on Lichens. 1. *Lichenologist*, **6**, 122–125.

HAWKSWORTH, D. L. & WILTSHIRE, P. E. J. 2011. Forensic mycology: the use of fungi in criminal investigations. *Forensic Science International*, **206**, 1–11.

HAWKSWORTH, D. L., ROSE, F. & COPPINS, B. J. 1973. Chapter 16. Changes in the lichen flora of England and Wales attributable to pollution of the air by sulphur dioxide. *In*: FERRY, B. W., BADDELEY, M. S. & HAWKSWORTH, D. L. (eds) *Air Pollution and Lichens*. Athlone Press of the University of London, London, 330–367.

HENDERSON, A. 2000. Literature on Lichens. XXVII. *Lichenologist*, **20**, 191–196.

JAMES, P. W. 1973. The effect of air pollutants other than hydrogen fluoride and sulphur dioxide on lichens. *In*: FERRY, B. W., BADDELEY, M. S. & HAWKSWORTH, D. L. (eds) *Air Pollution and Lichens*. Athlone Press of the University of London, London, 143–175.

JENKINS, D. & DAVIES, R. 1966. Trace element concentrations of organic accumulations. *Nature*, **214**, 1296–1297.

JOHNSEN, I. B. 1981. Heavy metal deposition on plants in relation to immission and bulk precipitation. *Silva Fennica*, **15**, 444–445.

KROUSE, H. R. 1977. Sulphur isotope abundance elucidate uptake of atmospheric sulphur emissions by vegetation. *Nature*, **265**, 45–46.

KYLANDER, M., WEISS, D. J., JEFFRIES, T. & COLES, B. 2004. Sample preparation methods for Pb concentration and isotopic analysis in peat by MC-ICP-MS. *Journal of Analytical Atomic Spectrometry*, **2004**, 1275–1277.

LE MAITRE, R. W. 1979. A new generalised petrological mixing model. *Contributions to Mineralogy and Petrology*, **71**, 133–137.

MIKHAILOVA, I. 2002. Transplanted Lichens for Bioaccumulation Studies. *In*: NIMIS, P. L., SCHEIDEGGER, C. & WOLSELEY, P. A. (eds) *Monitoring with Lichens – Monitoring Lichens*. NATO Science Series, IV. Earth and Environmental Science, Kluwer, Dordrecht, **7**, 301–309.

MONNA, F., AIUPPA, A., VARRICA, D. & DONGARRA, G. 1999. Pb isotope composition in lichens and aerosols from eastern Sicily: insights into the regional impact of volcanoes on the environment. *Environmental Science & Technology*, **33**, 2517–2523.

NASH, T. H. 2008. *Lichen Biology*. 2nd edn. Cambridge University Press, Cambridge.

NIEBOER, E., AHMED, H. M., PUCKETT, K. J. & RICHARDSON, D. H. S. 1975. Heavy metal content in lichens in relation to distance from a nickel smelter in Sudbury, Ontario. *The Lichenologist*, **5**, 292–304.

NIMIS, P. L., SCHEIDEGGER, C. & WOLSELEY, P. A. (eds) 2002. *Monitoring with Lichens – Monitoring Lichens*. NATO Science Series, IV. Earth and Environmental Science, **7**. Kluwer, Dordrecht.

OLMEZ, I., GULOVALI, M. C. & GORDON, G. E. 1985. Trace-element concentrations in lichens near a coal-fired power-plant. *Atmospheric Environment*, **19**, 1663–1669.

PURVIS, O. W. & PAWLIK-SKOWROŃSKA, B. 2008. Chapter 12. Lichens and metals. *In*: AVERY, S., STRATFORD, M. & VAN WEST, P. (eds) *Stress in Yeasts and Filamentous Fungi*. British Mycological Society, Symposium Series, 175–200.

PURVIS, O. W., WILLIAMSON, B. J., BARTOK, K. & ZOLTANI, N. 2000. Bioaccumulation of lead by the lichen *Acarospora smaragdula* from smelter emissions. *New Phytologist*, **147**, 591–599.

PURVIS, O. W., MIKHAILOVA, I., WILLIAMSON, B. J., UDACHIN, V. & CHIMONIDES, P. J. 2002. *Accumulation of Particles in Lichens from Smelter Emissions in Urals*. First Year Report. Royal Society Joint Collaborative Grant with Russia (ref: gt/fSU/JP, awarded February 2001. 19th March 2002: unpublished, authors not listed). Natural History Museum, London and Insititute of Plant and Animal Ecology, Ekaterinburg.

PURVIS, O. W., CHIMONIDES, P. J., JONES, G. C., MIKHAILOVA, I. N., SPIRO, B., WEISS, D. J. & WILLIAMSON, B. J. 2004. Lichen biomonitoring near Karabash Smelter Town, Ural Mountains, Russia, one of the most polluted areas in the world. *Proceedings of the Royal Society of London Series B, Biological Sciences*, **271**, 221–226.

PURVIS, O. W., LONGDEN, J. ET AL. 2006. Biogeochemical signatures in the lichen *Hypogymnia physodes* in the mid Urals. *Journal of Environmental Radioactivity*, **90**, 151–162.

PURVIS, O. W., GONZÁLEZ-MIQUEO, L., DOLGOPOLOVA, A., DUBBIN, W. & UNSWORTH, C. 2010. Use of rare earth element signatures in lichen, bark and adjacent soils as indicators of sources of geological materials. *In*: NASH, T. H., III, GEISER, L., MCCUNE, B., TRIEBEL, D., TOMESCU, A. M. F. & SANDERS, W. B. (eds) *Biology of Lichens, Symbiosis, Ecology, Environmental Monitoring, Systematics, Cyber Applications*. Bibliotheca Lichenologica, **105**. J. Cramer in der

Gebrüder Borntraeger Verlagsbuchhandlung, Stuttgart, 65–73.

RICHARDSON, D. H. S. 1995. Metal uptake in Lichens. *Symbiosis*, **18**, 119–127.

ROSSBACH, M., JAYASEKERA, R., KNIEWALD, G. & THANG, N. H. 1999. Large scale air monitoring: lichen vs. air particulate matter analysis. *Science of the Total Environment*, **232**, 59–66.

RUSU, A. M. 2002. Sample preparation of lichens for elemental analysis. *In*: NIMIS, P. L., SCHEIDEGGER, C. & WOLSELEY, P. A. (eds) *Monitoring with Lichens – Monitoring Lichens*. NATO Science Series, IV. Earth and Environmental Science, Kluwer, Dordrecht, **7**, 305–309.

RUSU, A. M., CHIMONIDES, P. D. J., JONES, G. C., GARCIA-SANCHEZ, R. & PURVIS, O. W. 2006a. Multi-element including rare earth content of lichens, bark, soils, and waste following industrial closure. *Environmental Science & Technology*, **40**, 4599–4604.

RUSU, A. M., JONES, G. C., CHIMONIDES, P. D. J. & PURVIS, O. W. 2006b. Biomonitoring using the lichen *Hypogymnia physodes* and bark samples near Zlatna, Romania immediately following closure of a copper ore-processing plant. *Environmental Pollution*, **143**, 81–88.

SCHÖNBECK, H. 1969. Eine Methode zur Erfassung der biologischen Wirkung von Luftverunreinigungen durch transplantierte Flechten. *Staub-Reinhaltung der Luft*, **29**, 14–18.

SEAWARD, M. R. D. 1973. Lichen ecology of the Scunthorpe Heathlands. I. Mineral accumulation. *The Lichenologist*, **5**, 423–433.

SLOOF, J. E. 1995. Pattern-recognition in lichens for source apportionment. *Atmospheric Environment*, **29**, 333–343.

SONDERGAARD, J., ASMUND, G., JOHANSEN, P. & ELBERLING, B. 2010. Pb isotopes as tracers of mining-related Pb in lichens, seaweed and mussels near a former Pb-Zn mine in West Greenland. *Environmental Pollution*, **158**, 1319–1326.

SPIRO, B., MORRISSON, J. & PURVIS, O. W. 2002. Sulphur isotopes in lichens as indicators of sources. *In*: NIMIS, P. L., SCHEIDEGGER, C. & WOLSELEY, P. (eds) *Monitoring with Lichens – Monitoring Lichens*. NATO Science Series, IV. Earth and Environmental Science, Kluwer, Dordrecht, **7**, 311–315.

SPIRO, B., WEISS, D. J., PURVIS, O. W., MIKHAILOVA, I., WILLIAMSON, B. J., COLES, B. J. & UDACHIN, V. 2004. Lead isotopes in lichen transplants around a Cu smelter in Russia determined by MC-ICP-MS reveal transient records of multiple sources. *Environmental Science & Technology*, **38**, 6522–6528.

SPIRO, B., UDACHIN, V., WILLIAMSON, B. J., PURVIS, O. W., TESSALINA, S. G. & WEISS, D. J. 2012. Lacustrine sediments and lichen transplants: two contrasting and complimentary environmental archives of natural and anthropogenic lead in the South Urals, Russia. *Aquatic Sciences*, Published online 30 May: http://dx.doi.org/10.1007/s00027-012-0266-3

STEPANOV, A. M., KABIROV, R. R. ET AL. 1992. Kompleksnaya ecologicheskaya ocenka tekhnogennogo vozdeistviya na ekosistemy yuzhnoi taigi [Integrated Ecological Assessment of Technogenic Impact on Ecosystems of Southern Taiga]. [In Russian]. CEFP, Moscow.

TAYLOR, S. R. & MCLENNAN, S. M. 1985. *The Continental Crust: Its Composition and Evolution. An Examination of the Geochemical Record Preserved in Sedimentary Rocks*. Blackwell, Oxford.

TESSALINA, S. G., ZAYKOV, V. V., ORGEVAL, J. J., AUGE, T. & OMENETO, B. 2001. Mafic–ultramafic hosted massive sulphide deposits in southern Urals (Russia). *In*: PIERSTRZYNSKI, A. (eds) *Mineral Deposits at the Beginning of the 21st Century: Proceedings of the Joint 6th Biennial SGA-SEG Meeting*, Krakow, Poland, 26–29 August 2001. Balkema, Rotterdam, 353–356.

UDACHIN, V., YERSHOV, V. & LANSIART, M. 1998. *Environmental Conditions in the Area of Exploited Massive Sulfide Deposits of the South Ural*. Final Report for TACIS contract FINRUS 9602.

UDACHIN, V., WILLIAMSON, B. J. ET AL. 2003. Assessment of environmental impacts of active smelter operations and abandoned mines in Karabash, Ural mountains of Russia. *Sustainable Development*, **11**, 133–142.

VOSKRESENSKY, G. 2002. The Most Polluted City on the Planet to Breath Easier. [Electronic version.] Eurasian Metals, **3**, (Ecology). http://en.eurasmet.ru/online/2002/3/ecology.php.

WEISS, D. J., KOBER, B. ET AL. 2004. Accurate and precise Pb isotope environmental samples by ratio measurements in MC-ICP-MS. *International Journal of Mass Spectrometry*, **232**, 205–215.

WILLIAMSON, B. J., MIKHAILOVA, I., PURVIS, O. W. & UDACHIN, V. 2004a. SEM-EDX analysis in the source apportionment of particulate matter on *Hypogymnia physodes* lichen transplants around the Cu smelter and former mining town of Karabash, South Urals, Russia. *Science of the Total Environment*, **322**, 139–154.

WILLIAMSON, B. J., UDACHIN, V., PURVIS, O. W., SPIRO, B., CRESSEY, G. & JONES, G. C. 2004b. Characterisation of airborne particulate pollution in the Cu smelter and former mining town of Karabash, south Ural Mountains of Russia. *Environmental Monitoring and Assessment*, **98**, 235–259.

WILLIAMSON, B. J., PURVIS, O. W., MIKHAILOVA, I. N., SPIRO, B. & UDACHIN, V. 2008. The lichen transplant methodology in the source apportionment of metal deposition around a copper smelter in the former mining town of Karabash, Russia. *Environmental Monitoring and Assessment*, **141**, 227–236.

YENISOY-KARAKAS, S. & TUNCEL, S. G. 2004. Geographic patterns of elemental deposition in the Aegean region of Turkey indicated by the lichen, *Xanthoria parietina* (L.) Th. Fr. *Science of the Total Environment*, **329**, 43–60.

YENISOY-KARAKAS, S. & TUNCEL, S. G. 2008. Source apportionment of atmospheric trace element deposition. *Environmental Engineering Science*, **25**, 1263–1271.

ZSCHAU, T., GETTY, S., GRIES, C., AMERON, Y., ZAMBRANO, A. & NASH, T. H. 2003. Historical and current atmospheric deposition to the epilithic lichen *Xanthoparmelia* in Maricopa County, Arizona. *Environmental Pollution*, **125**, 21–30.

Issues and opportunities in urban forensic geology

ALASTAIR RUFFELL[1]*, DUNCAN PIRRIE[2] & MATTHEW R. POWER[3]

[1]*School of Geography, Archaeology & Palaeoecology, Queen's University, Belfast BT7 1NN, UK*

[2]*Helford Geoscience LLP, Menallack Farm, Treverva, Penryn, Cornwall TR10 9BP, UK*

[3]*4669 West Lawn Drive, Burnaby, British Columbia, Canada VSC 3R2*

*Corresponding author (e-mail: a.ruffell@qub.ac.uk)

Abstract: Geological trace evidence including, for example, soil, sand and rock dust has been examined using a wide range of analytical techniques. Whilst such materials are common in rural locations, in urban areas, such geological materials are often perceived to be restricted to parks, recreational areas, gardens and waste ground. However, both geological materials and the wide range of analytical methods used to characterize them are much more applicable to the whole urban environment than is generally realized, with the main differences being the types and amounts of sample analysed and the methods adopted. The range of geological applications can be summarized as those deployed at the broad (decimetres–kilometres) to small (millimetres–decimetres) scale. The broad spatial variation in soil, roadway, water, buildings materials, and wind- or water-borne particles can be contrasted with the variation in urban materials from dwellings to streets or gardens and parks, along with the micro-spatial and stratigraphical variation in each. In addition, geological principles and techniques that have not been used before can be applied to urban materials to provide comparisons of material that were not previously achievable, or to add a further proxy to established methods. The latter point is demonstrated with a case study using X-ray diffraction and QEMSCAN® of a criminal case where building plaster with peculiar qualities could be compared between a suspect's vehicle and plaster present along the escape route from a murder scene.

A common misconception amongst investigating officers and forensic scientists is that the services of a geologist are required only where incidents occur in rural locations or where soil, sediment or rock is recovered (Murray 2011). In this paper we show how the increasing power of geological trace evidence analysis can allow Locard's Exchange Principle to be applied to inorganic particles in the urban and suburban environment. Very-fine-scale resolution trace evidence analysis and the resultant need for multiple samples present the investigator with new possibilities for investigation, yet come with their own pitfalls of interpretation, time- and cost-effectiveness, and contamination (Pye & Croft 2004). None the less, these issues should not daunt us: the preponderance of urban crime, the effectiveness of perpetrator clean-up (Pye & Croft 2004), and the need for more exacting sampling and analysis for the courts mean the forensic geologist should examine what skills they can bring to urban scenes of crime.

This paper is intentionally restricted to a discussion of geological trace evidence (Palenik 2000) in the urban environment and its distribution. Although geological and geographical information has been used in criminalistics for over 100 years (Murray 2011; Bergslien 2012), application to the urban environment has been restricted – partly because of the comparative richness of evidence forthcoming from rural locations, partly because geologists and soil scientists do not commonly analyse urban material, and also because technology has not been sensitive enough to cope with the complex and very small amounts of material common to urban scenes. Forensic geology (Murray 2004, 2011) began with the application of geological microscopy to soil. Our work goes further than a mere transfer of methodology, with a case study that demonstrates how, in the case of industrially manufactured plasterboard, the constituent gypsum could not be discriminated by conventional analysis (visual microscopy) and X-ray diffraction (XRD) analysis. The analysis of such trace evidence requires very sensitive, statistically robust tools (in this case QEMSCAN®) in their analysis.

Urban forensic geology

Over the past 10 years a number of publications, meetings and news items (see, e.g. the summary in Ruffell 2010) have raised the profile of geology (and its cognate subjects) in criminal investigations (Saferstein 2001; Pye & Croft 2004; Ruffell

& McKinley 2005). However, many of the criminal cases reviewed in the works cited above are based on crimes where the scenes of forensic significance are in non-urban environments or they centre on the analysis of soil and sediment, which is commonly less abundant in the concrete- and tarmac-dominated urban environment. This apparent bias towards the examination of rural rather than urban crime scenes is for a number of reasons. First, geologists and soil scientists are more familiar with the 'natural' environment, where their skills are commonly utilized. Second, the urban environment is seen as 'cleaner' and 'harder' for the transfer of geological trace evidence such as soil, sediment or rock. Thirdly, and following on from the above point, *traditional* geoscientific methods of analysis require significant amounts of sample, found most commonly in the non-urban environment (Keaney *et al.* 2009). Lastly, and critically for this work as we wish to dispel this myth, is that non-scientific investigating officers may not realize that geologists can contribute to urban criminal and environmental forensic investigations.

Advances in analysis using non-destructive methods or the analysis of very small samples have allowed the application of geoscience to the urban environment (Pye & Croft 2004). The perception that urban environments lack geological trace evidence is erroneous. Anyone who has seen building excavations within cities knows that the urban environment has soil, sediment, weathered rock and stratigraphy. Similarly, a person who works (and has pale-coloured clothing) in the urban environment knows how much dust occurs in the atmosphere. The visitor to a city cannot help but notice how dirty their clothes become from airborne particulates. In summary, the urban environment is far from 'clean' of transferable materials: the critical aspect is what these materials comprise and how we analyse them.

The preponderance of serious crime in the urban environment (gun crime in many cities, alcohol- or drug-related crime), but also simply reflecting the greater population density, makes this review of how geoscience can assist in criminal investigations timely. The application of the established principles of forensic geoscience to 'mini-rural' scenes such as parks (Fig. 1) is obvious. Such sites can commonly become crime scenes, with, for example, serious sexual assaults within parks or the deposition of murder victims in areas of waste ground awaiting redevelopment. However, serious urban crime also occurs in locations other than parks or waste ground. For example, the petrol station forecourt shown in Figure 2 is nothing like a park or garden and, if a serious crime was committed in this setting, it might appear that the forensic geoscientist would have nothing to offer. However, this is far from correct – the amount of transferable material at the garage, from loose concrete/asphalt, spilt oil, water, car-washing products and foodstuffs, all possibly mixed with mineral and glass fragments, is enough to make the microscopic examination of samples from suspects and scenes very worthwhile. Mapping the distribution of urban materials has never been undertaken with a forensic database in mind: most considerations of the spatial distribution of geological materials are still concerned with soil. However, even from the above consideration of building sites, petrol stations, roadways and parks, it is apparent that spatial variation in surface (dust, soil, debris) and deeper (roadway, pavement, soil) urban materials is likely to be significant.

Types of urban materials that can be analysed

Soil

The analysis of soil in criminal investigations is commonly regarded as being restricted to 'rural' scenes (Murray 2011). Soil is actually defined as any Earth-surface material that can support life. As such, soil is actually far more common in the urban environment than is first apparent. One need only record the locations of vegetation-filled cracks in pavements and roadways (Fig. 3a), in small gardens and parks (Fig. 1), or roadside (Fig. 3b) and window-sill flowerbeds. Likewise, abandoned, vegetation-filled sites are periodically common in the urban situation (Fig. 3c), and, being covert, are often the scenes of criminal activity (Saferstein 2001). The materials of urban construction soon decay through weathering to release minerals; in tandem, the spread of pollen and seeds is so efficient that it is actually rare for ground to remain unweathered, and thus unpopulated by lichen, moss, and, eventually, larger plants (Fig. 3d) and animals. Weathered concrete, brick and stone, with or without the growth of lichen, moss or plants, is a 'sticky' medium for inorganic particles to be captured in, and, thus, transferred to people and vehicles.

Urban soils can be considered in terms of Inman & Rudin's (2002) classification of the materials useful to forensic scientists. Under their first (division of matter) and fourth (the individualization of evidence) classifications, the geographically isolated 'oasis'-like nature of soil and plant communities makes their individual make-up variable from location to location (Fig. 4).

The chances of transfer (second and fifth classifications in Inman & Rudin 2002) appear less in the urban situation. However, areas with soil (flower beds, derelict sites) are often the primary location

Fig. 1. Photograph of the kind of scene where law enforcement officers see 'conventional' forensic geology and geoscience being applied, namely a mini-rural scene such as a park, enclosed by buildings. The advantages of such a location is their relative distinctiveness in terms of comparison to the rest of the town or city, as well as the straightforward application of tried and tested techniques to the location. In many major cities, parks and gardens may have a high preponderance of serious crime (violent robbery, drug-related extortion and killings, and rape) that give forensic geoscientists the opportunity to apply their skills. The disadvantage to such a scene is that not all urban serious crime occurs in such a place and the offender may well be aware of the likelihood that transferable material is present.

Fig. 2. Photograph of a typical urban crime scene – a petrol station. Often open for 24 h a day, sometimes in isolated locations, with few personnel and large quantities of used banknotes, the garage has always been a favourite location for armed robbery. This scene presents many challenges to the urban geoscientist but is not without opportunity: the surrounding area may be geographically distinct and the garage itself contains transferable materials, such as oils, cleaning agents and foodstuffs.

Fig. 3. Photograph of the mini-rural nature of the urban environment. (**a**) Cracks in pavements are abundant, and act as traps for sediment and rubbish that all become transferable and sometimes distinctive material. (**b**) Vegetation is encouraged in some locations – this requires soil, often with an artificial covering (gravel); together they provide a mineralogical profile unlike surrounding soils. (**c**) Nature finds a way. Abandonment allows the establishment of pioneer and successor vegetation (and associated organisms) communities. (**d**) Tangible evidence of the spread of vegetation-related material. Here, tree blossom has collected in the cracks in the paved area outside an urban house. Pollen effectively do the same thing, remaining airborne for longer, increasing the likelihood of transfer but partly homogenizing the environmental signal in the meantime.

Fig. 4. The oasis-like nature of urban vegetation and soils. The advantage of such mini-oases is their isolation, allowing a unique association of plants and soils to develop. The disadvantage of these communities is sampling enough of them to actually allow a comparison between suspect, vehicle, victim and scene, as well as the statistical chances of contact being made.

of criminal behaviour in the first place, as they are less populated and may provide cover from observation. Even away from such 'micro-rural' locations, Locard's Principle still applies, and the urban environment is rich in transferable material – the challenge is to locate and characterize such materials. This leads on to Inman & Rudin's (2002) third classification; the identification of urban soils. This can be problematic as the amounts of material may be small: however, as we shall see with other, non-soil type materials, it is the analytical method that identification relies on, and small samples are no longer an issue for geoforensic analysis.

Roadway and building materials

Fortuitously, main roads, minor roads, tracks, access roads and car parks are variable in make-up by virtue of the construction requirements of each, as well as the different provenance of building materials (Fig. 5). All are made to be hard, durable and, thus, to not provide particles for possible transfer: none the less, the surfaces are made of natural components that erode and decay, as well as having other material added to them, albeit in small amounts. Thus, the problem becomes one of sample size, as opposed to scientific principle. The same can be said of the different buildings that make up the urban environment. In both cases there is variation in space (downtown, industrial, residential) and time (changes in material sources, manufacture methods, construction methods). The movement of vehicles and people causes erosion and alteration (abrasion, chemical changes) of these materials, as well as the addition of material such as tyre rubber, foodstuffs, waste, oil, etc. Erosion creates and recycles loose and, thus, transferable material (Inman & Rudin's 2002 second classification: Locard's Transfer Principle) such as aggregate, tarmac/asphalt and cement (Fig. 6), which may occur loose over surfaces or concentrated in the cracks, joints and gutters that are abundant in the urban environment (Fig. 6a, b). Alteration includes the erosion and abrasion/polishing of surfaces and particles such that adequate identification of controls may allow different uses of roads to be elucidated (Fig. 6c). The end result of both the primary

Fig. 5. (**a**) An unfinished flowerbed against a wall has become a capture zone for leaves, dust, rubbish and waste building materials (builders' sand). At this location, a robbery took place in the car park, from where the photograph was taken, and the gunman ran over this location and into a getaway car. He was apprehended some miles away: his alibi was contradicted by the gunshot residue on his clothing and by the comparison between the soil on his shoes with that from this escape route. (**b**) The spread of building materials by rain and water beyond the limits of a construction site is an irritation to those cleaning streets and walking by but an advantage to the forensic geoscientist, whose scene or related location possesses relatively unaltered, transferable materials such as sand, cement, plaster and insulation materials.

Fig. 6. (**a**) An area designated to be a flowerbed, covered with commercial gravel from a gravel pit no longer extant. Thus, the urban distribution of this type of geological material is limited. (**b**) Even a relatively new paved area has hollows that act as collection points for dust, rubbish and water. (**c**) Loose material eroded from weakened kerbstones, providing a provenance for this transferable material.

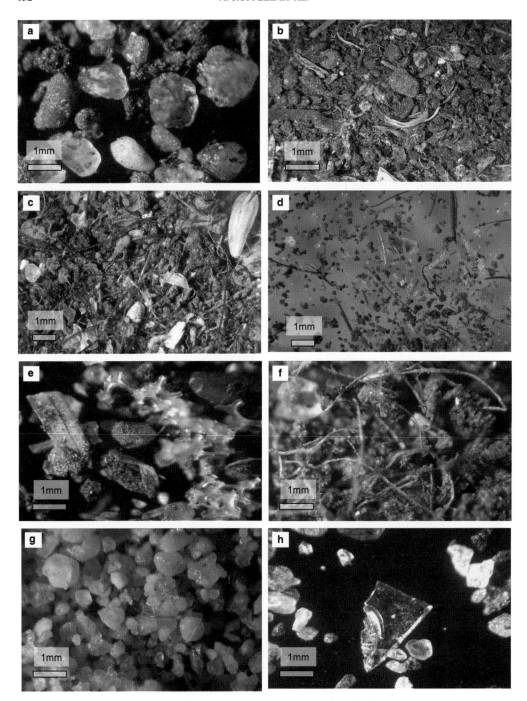

Fig. 7. Photomicrographs of materials found in some of the scenes described, as well as from suspect, witness and victim materials in actual cases. (**a**) Typical urban dust from roadways, a mix of rounded quartz grains, possibly abraded by aeolian activity, and subrounded tarmac grains. (**b**) Typical urban dust from seldom-used roadways, car parks and playgrounds, comprising a mixture of tarmac grains, woody debris, seed husks and paper shreds. (**c**) Typical material from urban gutters and cracks in pavements, comprising tarmac fragments, quartz and carbonate grains, paper, clay, vegetation fibres and seed husks, and fibres and glass. (**d**) Particulates recovered from clothing, worn in a city for 1 day. The mix of clay minerals, asphalt, fibres and glass suggest that the wearer had been in a location with more

variability in construction ingredients and the post-manufacture/construction alteration creates variation in the mineralogical, textural, chemical and biological nature of roadway materials.

Conversely, transport and adherence of construction standards (quality control) creates homogenization of products. Globalization introduces an interesting aspect to the forensic analysis of these kinds of materials. Throughout the late twentieth century, the introduction of international tenders for large-scale building operations (e.g. in the European Union) and the opening up of cheaper material sources (e.g. the far East, India and South America) has resulted, in Europe and North America, in a decrease in the use of primary ingredients close to the construction source (local aggregate, sand, cement) and to some exotic primary ingredients being introduced. This has advantages and disadvantages for the forensic scientist. For instance, changes in construction method and contract tendering and award, as well as ingredients through time, will result in materials and methods being used in a limited number of locations for a specific time period. Therefore, distinctive materials found associated with a suspect or victim may associate that person with a construction location, its transport medium or original source. Equally, prior to these changes in the nature of construction, aggregate, sand, cement and rock dust, combinations may be compared from person to location in much the same way as soil or vegetation is still used.

Wind-blown materials

Urban environments are, by their very nature, akin to the natural arid locations of the Earth. This may seem counter-intuitive for towns and cities located in more humid climatic regions. However, the highly urbanized 'concrete jungle' type of city in a temperate or humid location is constructed with the rapid removal of surface water as a priority. Thus, only short periods of saturation occur. Take, for instance, a period of precipitation on a grass field v. a city street of equal area. The field acts naturally, with percolation and transpiration. The street acts more like a bare rock surface with overland flow to drains, runoff being controlled for health and damage reasons, just as water inputs (taps and hydrants) are controlled. Having established the periodically arid nature of the urban environment, we can suppose that wind-blown materials are common. Types of material include the construction materials discussed above, as well as manufacturing, catering and vehicular (particulates from aircraft and diesel engines). Examination of shoe-treads, clothing, car tyres and undersides that have been limited to the urban environment typically contain sand and asphalt fragments (construction), glass spherules (power station, aircraft, combustion engine emissions), foodstuffs, especially burnt and fried materials, and fibres. The use of wind-blown particles in forensic work has been limited, probably because this type of evidence is 'out of sight, out of mind'; in other words, the particles are too small for easy detection and recovery. However, developments in pollution monitoring, such as advanced filters for the recovery and identification of particulates (e.g. Williamson *et al.* 2013), have shown that urban air is rich in the sorts of trace evidence that will be trapped in clothing, hair and vehicle air filters, and thus provide comparative material between scene, victim and suspect. It seems counter-intuitive that Locard's Principle, developed to describe the contact between people or people and scenes (Locard 1929), can also be applied to contact between people, their clothing and vehicles, and air (or more strictly speaking, the particles suspended in air). A major problem with wind-blown, or aeolian particles, is their homogenization by the very transport medium that makes transfer so widespread. Indeed, it does not take a forensic scientist to observe the 'dirt' one finds adhered to footwear or embedded in one's clothing after a day in the city. What does this comprise? Examination by light microscope or by scanning electron microscopy (SEM) reveals that the types of trapped material will be common to many of the urban environments of the world (Figs 7 & 8). This includes combustion particulates ('soot', 'ash'), loose construction materials, and waste material such as foodstuffs, paper and packaging. The individual make-up of these types of particles will, of course, differ from city to city, town to town, and within different urban environments, and therein lies the discriminatory power of such materials. The major problem is recovery and characterization of such evidence.

Fig. 7. (*Continued*) building material and wind-blown dust than one with a high level of combustion emissions. The sample was prepared by drenching the clothing in filtered water and concentrating the fluid by centrifuge. (**e**) Polymer 'foam', the typical material found in ripped, padded seating or padded clothing, such as car seats, cushioned chairs and insulated coats ('puffer' jackets). (**f**) Fibres in fine-sand aggregates of silt and clay, recovered from the material found in gutters and drains. (**g**) Carbonate sand and silt, a common component of foodstuffs and commonly 'cut' with drugs, such as cocaine. (**h**) Glass shard, recovered from footwear and a common component of urban debris. The thickness, colorants, inclusions and refractive index of different glasses allow comparison.

Fig. 8. Puddles in the urban environment. (**a**) A puddle in the lee of a stairwell, almost impossible to avoid contact with (indeed, footprints can be seen below the puddle). This puddle has abundant paint and rust from the stairwell in it. (**b**) Gutter puddles: these can have a greater longevity than other puddles, by virtue of the greater catchment. Thus, communities of diatoms, testae amoebae and ostracods can develop. (**c**) This doorway to a drinking den in west Belfast has a subtle topographical hollow leading from it, in which abundant debris collects. (**d**) The pedestrian crossing is a topographical hollow where water accumulates and through which people inevitably walk (note the footprints).

Other liquids and waste

The potential for transfer makes liquids of interest. In the urban environment these range from industrial processes, oil on roadways, spilt paint (Fig. 9c) cleaning agents for buildings, and waste from the catering industry and foodstuffs (Fig. 10). Other liquids that are common in the urban environment are associated with some of the other categories of material (see above). These include paint, tarmac (both freshly laid and melted) and fluids with a low water content (e.g. waste oils and sewage sludge).

Summary

Two aspects influence the usefulness of geological and associated materials in forensic work conducted in an urban (or, indeed, semi-urban or industrialized) environment. These are the environment and the materials, which may both be interrelated. The environment (pavements, roadways, parks, gardens, building sites, derelict plots, garage forecourts, etc.) influence some of the particles present, yet the movement of people, vehicles, wind and water also homogenize the loose debris present, and at the same time may introduce exotic material (e.g. volcanic ash, imported construction materials). Thus, challenges and opportunities both exist, which are demonstrated by our case study from a suburban location in which trace amounts of material comprised the only geological evidence available.

Uses of urban geology: a case study of a drug-related 'turf war' murder in Northern Ireland

Background

In order to continue their activities, paramilitary groups in Northern Ireland have often engaged in various legal and illegal money-making ventures.

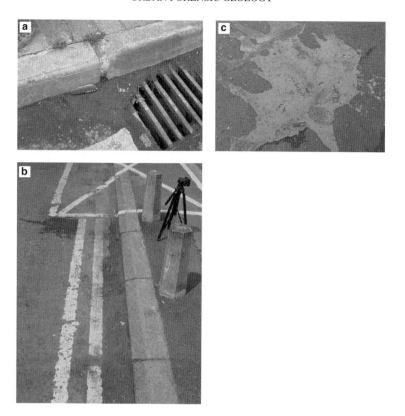

Fig. 9. (a) Drains act as designed collection points not only for water but for fine-grained debris as well. This commonly comprises eroded roadway material (asphalt, quartz and other sand, carbonates) along with waste and vegetation. (b) Subtle topographical hollows that are unconnected with drains, in gutters, also act as sediment collection points through which cars and people pass. (c) An accidental paint spill. In this case, the volume of this gloss paint meant that it remained sticky for many days: a suspect in a rape case nearby to this spill denied having been in the area, yet chemical analysis of the paint spatter on his trousers (his footwear was never recovered) suggested a link to the escape route.

These include property speculation, armed robbery, extortion ('protection'), illegal gaming machines, and drug importation and dealing. One of the problems amongst the loyalist gangs (typically, the Ulster Defence Association (UDA), the Ulster Volunteer Force (UVF), the Loyalist Volunteer Force (LVF) and the Red Hand Commando (RHC)) and groupings was, and still is, the large number of organizations, and their dominance of, and limitation to certain geographical areas, as well as their changing allegiances to one another and to their leaders. With the demise of the nationalist threat (typically, the Irish Republican Army (IRA) and variants, Irish Nationalist Liberation Army (INLA)), inevitably these loyalist groups increasingly found themselves operating against one another in the sort of urban gang warfare familiar to other major cities of the world. Thus, different factions stood to make considerable profit by controlling illegal activities, such as drug dealing, in their rivals' areas. In 1997, two such gangs dominated the drug dealing of loyalist east Belfast. As one gang came to dominate the other, they realized that with the murder of a key rival they could not only diminish that rivalry but also allow take-over of a neighbouring area. In May 1997, a hoax telephone call forced one gang leader to open the gates of his well-fortified suburban house, whereupon two gunmen hired by the rival gang entered his driveway and shot him repeatedly in the head.

It is at this point that the aspects of urban forensic geoscience can be brought in and used in conjunction with the other, geographical, aspects of the case evidence to relate one gunman to the escape route.

The forensic geoscience evidence

Following the fatal shooting, the gunmen's accomplice (in their getaway car) drove away as

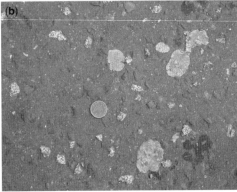

Fig. 10. Waste, trash and rubbish. (**a**) The area beside a waste collection point in an alleyway has abundant spilt material, including liquid and cooked foodstuffs. The location is a classic, dark, urban scene of crime with a complex, chemical profile. (**b**) A problem of only certain cities of the world: discarded chewing gum. In Singapore (for instance), this phenomenon is unheard of, yet in many other places it is common, providing an excellent transfer medium to the soles of footwear.

neighbours were alerted to the noise. The assailants moved from their firing position, into an adjacent property, climbed a locked garden gate by jumping on an adjacent car and landed in a yard that had various waste materials in it. This debris included broken bottles, broken plasterboard, a leaking heating oil tank and abundant leaf litter. Some evidence of the assailants falling or slipping in this location were subsequently noted by scenes of crime officers. From this point onwards, a police tracker dog and handler were able to follow the trail of the assailants as they disposed of some outer clothing and a weapon in the garden of the same house, and then made their way into adjacent farmland

and forest. In this farmland, inspection of a barbed-wire fence along the mapped escape route demonstrated the presence of fresh blood. The tracker dog lost the scent of the escapees at a lane some distance from the scene where it is supposed a getaway car had made a successful pick-up. Police intelligence revealed a number of likely suspects in the shooting, one of whom was arrested. This suspect had abrasions to his body and lacerations to his arm. His blood compared (with a 1 in 100 million chance) to the blood recovered from the barbed-wire fence but this evidence only suggested that the suspect could have been some few hundred metres from the scene. Also recovered at the time of arrest was the suspect's vehicle, which was generally clean inside and out, except for some cream–white powder and fragments on, and in, the tread of a rubber footmat. Preliminary geological examination by X-ray diffraction (Kugler 2003) and visual examination (Fig. 11) revealed this to be gypsum with associated plant fragments and some oil. The total amount of recovered material was less than 0.5 mg. This very small amount of material is entirely typical of urban trace evidence, as well as in cases where the suspect(s) are well versed in clean-up operations. Analysis for comparison with controls (commercial plasterboard samples), the yard plasterboard and associated debris found on the escape route thus required a method that would provide quantitative data from a small sample. Traditional geological analysis, including microscopic examination and analysis by XRD established the presence of gypsum but failed to discriminate any differences in the samples. Consequently, the exhibits were submitted for analysis using automated SEM with linked energy-dispersive spectrometry (EDS) analysis in this case using QEMSCAN® technology.

The use of automated mineral analysis using QEMSCAN® technology has been previously described by Pirrie et al. (2004, 2009, 2013) and further details about the applications of this technology are provided by Pirrie & Rollinson (2011). In this case, all of the samples were measured by QEMSCAN® using the same operating conditions, mode of analysis and pixel spacing. All samples were prepared for QEMSCAN® analysis using the same methodology, and the mineral groupings used in this study are shown in Table 1. Because the phase analysis is based on the acquisition of X-ray spectra, different phases that have the same chemical composition cannot be separately classified. For example, in this study, during automated analysis it is not possible to differentiate between gypsum ($CaSO_4 \cdot 2H_2O$) and anhydrite ($CaSO_4$), although these phases do appear to be both texturally distinct and also anhydrite has a relatively higher backscatter coefficient.

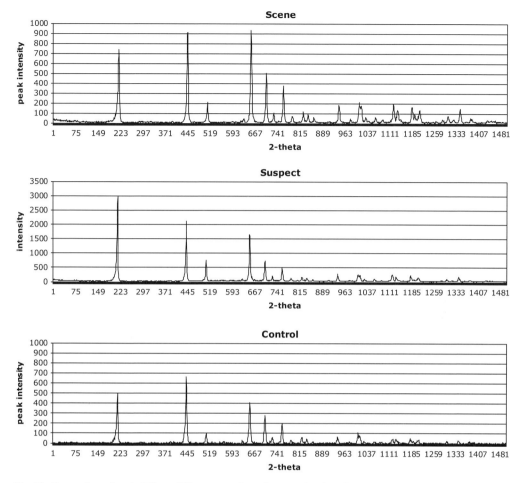

Fig. 11. Comparison of typical X-ray diffractograms from the scene (yard on the escape route), suspect (vehicle floormat) and a control sample (a commercial plaster). Note how the scene and suspect material are comparable but yet so is the control material, providing no further discrimination.

Results

The weight percentage (wt%) mineralogy data for the key samples is summarized in Table 2 and Figures 12 and 13. Despite the dominance of gypsum, these samples can be differentiated on the basis of both the texture of the gypsum particles and also the associated minor/trace phases.

The aim of the investigation was: (a) to test whether the samples recovered from the motor vehicle footmat were comparable with the samples recovered from the scrap plaster in the yard on the escape route; and (b) to consider how distinctive different sources of builders plasterboard are. The particulate material recovered from the footmat of the motor vehicle was found to be comparable mineralogically in terms of both the major (>10%), minor (1–10%) and trace (<1%) phases present with the soil sample recovered from the yard. Although the relative abundance of the phases present in these two samples was variable, there are a number of texturally distinctive phases that co-occur in the two samples. The white particulate material recovered by brushing the footmat from the front offside footwell of the motor vehicle was dominated by gypsum and was interpreted, based on both mineralogy and texture, to be a plaster mineral product. The sample was comparable with two plaster samples recovered from the yard. Two commercial plaster samples (controls) analysed reveal mineralogical and, in some cases, textural differences between these different plasters. The mineralogy and texture of the white particulate material from the footmat of the motor vehicle was comparable with plaster sample recovered from the yard on the escape route and different to

Table 1. *Summary of the mineralogical groupings used to report the automated mineral analysis data*

QEMSCAN® Processed Category	Minerals and phases included within QEMSCAN® category
Gypsum	Gypsum (including plaster) and anhydrite
Quartz	Silica group of minerals (e.g. quartz, opal, cristobalite)
Calcite	Calcite, aragonite. This may also include biogenic carbonate grains
Magnesite	Magnesite
Plagioclase	Albite–anorthite solid solution
Orthoclase	Albite–orthoclase solid solution including sanidine and microcline. This may also include illite
Biotite	Biotite
Muscovite	Muscovite and alteration products, such as sericite
Kaolinite	Kaolinite group minerals, such as halyosite and kaolin
Fe–Al silicates	Chlorite, anthropogenic materials, such as slag and fly ash
Ca–Al silicates	Very fine mixtures of calcium carbonate and calcium silicate minerals interpreted to be anthropogenic cement/concrete
Zircon	Zircon
Fe/Mn oxides/carbonates	Fe-oxide group, such as magnetite, hematite and goethite. This may also include anthropogenic Fe-rich phases, such as welding fumes
Fe sulphides	Pyrite and pyrrhotite
Ti oxide	Rutile and its polymorph anatase and anthropogenic products that use these minerals as filler, such as paints
Apatite	Apatite and Ca phosphates
Ilmenite	Ilmenite
Mg–Fe silicates	Olivine, Fe silicates, may also include fine mixtures of Fe oxide and quartz
Other silicates	Mg–Mn silicates, sphene, epidote
Other	Any other phases not included above

Table 2. *Mineralogical data obtained from QEMSCAN® – abundances of identified major/minor/trace minerals/phases*

Exhibit number*	AR/JWM4-1	AR/JWM-4	AR/G1	AR/G2	AR/G3	AR/Plas1	AR/Plas2
Gypsum	0.93	96.46	95.84	88.96	0.40	91.49	84.64
Quartz	50.06	0.11	0.31	1.39	41.90	1.50	1.95
Calcite	14.79	2.64	3.49	5.78	2.31	2.59	5.30
Magnesite	0.04	0.01	0.00	0.00	0.10	0.01	1.94
Plagioclase	3.74	0.00	0.00	0.20	3.96	0.10	0.18
Orthoclase	4.81	0.06	0.01	0.61	9.18	1.99	0.16
Biotite	0.35	0.00	0.00	0.50	0.79	0.21	0.21
Muscovite	2.03	0.11	0.07	0.55	4.22	0.82	1.38
Kaolinite	2.72	0.05	0.03	0.28	3.14	0.38	0.80
Fe–Al silicates	3.88	0.11	0.02	0.33	3.64	0.16	0.28
Ca–Al silicates	5.71	0.31	0.11	0.97	3.04	0.61	1.36
Zircon	0.02	0.01	0.01	0.05	0.03	0.00	0.03
Fe/Mn oxides/carbonates	8.22	0.04	0.06	0.02	11.18	0.02	1.34
Fe sulphides	0.25	0.00	0.01	0.03	1.36	0.00	0.10
Ti oxides	0.39	0.03	0.01	0.06	13.20	0.02	0.07
Apatite	0.01	0.00	0.00	0.02	0.15	0.00	0.00
Ilmenite	0.65	0.00	0.00	0.00	0.19	0.00	0.00
Mg–Fe silicates	0.62	0.00	0.00	0.12	0.42	0.00	0.06
Other silicates	0.72	0.06	0.02	0.13	0.37	0.10	0.20
Others	0.06	0.00	0.00	0.00	0.41	0.00	0.00

*Numbers refer to samples.
AR/JWM4-1, debris from vehicle footmat; AR/JWM-4, brushing of powder from vehicle footmat; AR/G1 and AR/G2, plasterboard samples from the escape route from the crime scene; AR/G3, soil sample from the escape route from the crime scene; AR/Plas1 and AR/Plas2, plasterboard control samples.

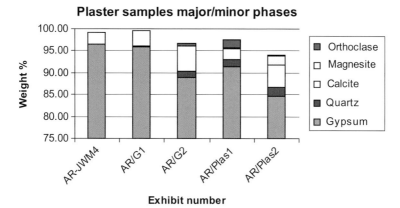

Fig. 12. Major and common minor phases present in the plaster samples from the escape route from the crime (AR/G1 and AR/G2), control plaster samples (AR/Plas1 and AR/Plas2) and the powder sample recovered from the motor vehicle (AR/JWM-4).

the commercial plaster samples. When combined together, the mineralogy of the debris from the footmat and the mineralogy of the material recovered by brushing the footmat are comparable with two different components present in the yard. This suggests that the occupant(s) of the motor vehicle had either been present in the escape route yard or some other place that had exactly the same mineralogical signature in the plaster products present. These materials were transferred to the footwell footmat from the footwear, clothing or body of the person entering the vehicle. Whether more

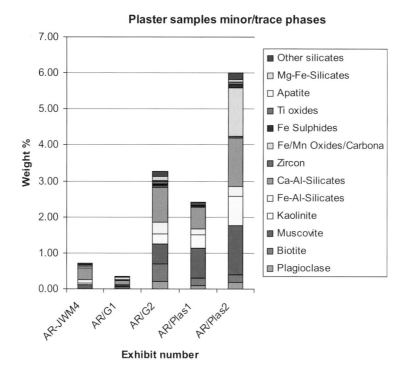

Fig. 13. Minor and trace phases present in the plaster samples from the escape route from the crime (AR/G1 and AR/G2), control plaster samples (AR/Plas1 and AR/Plas2) and the powder sample recovered from the motor vehicle (AR/JWM-4).

material was originally present, we shall never know. However, the micrograms of material that remained make the evidence similar in nature to much of that recovered in urban criminal investigations. The study, therefore, shows how a natural material, processed to become a building product but present in such small quantities, can provide sufficient material for analysis and, thus, comparison.

Discussion

For forensic geoscience to be widely adopted as an investigative tool in the urban environment, several areas of research need to be addressed. The distribution and variation in the types of materials described at the beginning of this work need to be assessed and eventually described in as much detail as possible. This may well start with high-level crime areas, aided by spatial analysis. Geographically defined areas where there are increases in the level of gun crime, drug-related crimes and, in some areas, alcohol-related violence will require mapping and thence control sampling (Hirschfield & Bowers 2001). This appears a proactive strategy but is, in some ways, also reactive and can be adapted as locations and crime types change. Advances in automated analytical methods will speed up this laborious process.

New analytical methods need to be found for the analysis of urban trace evidence. These may be pre-existing techniques used in industry or in environmental monitoring, or they may require development specifically for criminalistics. Most importantly, these methods need to provide statistically robust information from trace amounts of material. Hopefully, any new methods will also be non-destructive, such that other techniques can be used on the same trace evidence, providing the 'multi-proxy' methodology favoured by some for soil evidence (e.g. Bull et al. 2004).

New and established analytical methods need to be tested; this is because of the complexity, and often trace quantities, involved in the analysis of urban geomaterials for criminalistics. The implication of this is that, as far as the legal justice system is concerned, urban geology appears to be in its infancy. The principles of forensic geoscience have been established for over 100 years but the application of these methods has largely been restricted to certain materials or to scenes of crime in rural or similar (parks, gardens) locations. As we have shown, the quantities of trace material and their control by humans, makes the urban environment different to rural locations. The eventual test of the applicability of geology to the urban environment will be in the courts, where the key to success will be the statistical robustness of the data.

This work was inspired by a request from E. Mikhailova (Clemson University), for which we are grateful. We are grateful to the Police Service of Northern Ireland for permission to publish the case study, and for the assistance of staff at the Forensic Science Service of Northern Ireland at the scene and in sampling (J. Logan, W. McCorkhill and J. Holmes).

References

BERGSLIEN, E. 2012. *An Introduction to Forensic Geoscience*. Wiley Blackwell, Chichester.

BULL, P. A., MORGAN, R. M., WILSON, H. E. & DUNKERLEY, H. E. S. 2004. Multi-technique comparison of source and primary transfer soil samples: an experimental investigation'. A Comment. *Science and Justice*, **44**, 173–176.

HIRSCHFIELD, A. & BOWERS, K. 2001. *Mapping and Analysing Crime Data: Lessons from Research and Practice*. Taylor & Francis, London.

INMAN, K. & RUDIN, N. (2002). The origin of evidence. *Forensic Science International*, **126**, 11–16.

KEANEY, A., RUFFELL, A. & MCKINLEY, J. 2009. Geological trace evidence: forensic and legal perspectives. *In*: RITZ, K., DAWSON, L. & MILLER, D. (eds) *Criminal and Environmental Soil Forensics*. Springer, Dordrecht, 221–237.

KUGLER, W. 2003. X-ray diffraction analysis in the forensic science: the last resort in many criminal cases. *Advances in X-ray Analysis*, **46**, 1–16.

LOCARD, E. 1929. The analysis of dust traces. *Revue Internationale de Criminalistique*, **I**, 4–5. (English translation in the *American Journal of Police Science*, **I** (1930), 416–418.)

MURRAY, R. C. 2004. *Evidence from the Earth: Forensic Geology and Criminal Investigation*. Mountain Press Publishing, Missoula, MT.

MURRAY, R. C. 2011. *Evidence from the Earth: Forensic Geology and Criminal Investigation*. Mountain Press Publishing, Missoula, MT.

PALENIK, S. J. 2000. Microscopy. *In*: SIEGEL, J., KNUPFER, G. & SAUKKO, P. (eds) *Encyclopedia of Forensic Science*. Academic Press, New York.

PIRRIE, D. & ROLLINSON, G. K. 2011. Unlocking the application of automated mineralogy. *Geology Today*, **27**, 226–235.

PIRRIE, D., BUTCHER, A. R., POWER, M. R., GOTTLIEB, P. & MILLER, G. L. 2004. Rapid quantitative mineral and phase analysis using automated scanning electron microscopy (QemSCAN); potential applications in forensic geoscience. *In*: PYE, K. & CROFT, D. (eds) *Forensic Geoscience*. Geological Society, London, Special Publications, **232**, 123–136.

PIRRIE, D., POWER, M. R., ROLLINSON, G. K., WILTSHIRE, P. E. J., NEWBERRY, J. & CAMPBELL, H. E. 2009. Automated SEM-EDS (QEMSCAN) mineral analysis in forensic soil investigations; testing instrumental variability. *In*: RITZ, K., DAWSON, L. & MILLER, D. (eds) *Criminal and Environmental Soil Forensics*. Springer, Dordrecht, 411–430.

PIRRIE, D., ROLLINSON, G. K., POWER, M. R. & WEBB, J. 2013. Automated forensic soil mineral analysis; testing the potential of lithotyping. *In*: PIRRIE, D., RUFFELL, A. & DAWSON, L. A. (eds) *Environmental* and *Criminal Geoforensics*. Geological Society, London, Special Publications, **384**. First published online August 7, 2013, http://dx.doi.org/10.1144/SP384.17.

PYE, K. & CROFT, D. J. (eds) 2004. *Forensic Geoscience: Principles, Techniques and Applications*. Geological Society, London, Special Publications, **232**.

RUFFELL, A. 2010. Forensic pedology, forensic geology, forensic geoscience, geoforensics and soil forensics. *Forensic Science International*, **202**, 9–12.

RUFFELL, A. & MCKINLEY, J. 2005. Forensic geoscience: applications of geology, geomorphology and geophysics to criminal investigations. *Earth Science Reviews*, **69**, 235–247.

SAFERSTEIN, R. E. 2001. *Criminalistics: An Introduction to Forensic Science*, 7th edn. Prentice-Hall International, Englewood Cliffs, NJ.

WILLIAMSON, B. J., ROLLINSON, G. & PIRRIE, D. 2013. Automated mineralogical analysis of PM10: new parameters for assessing PM toxicity. *Environmental Science and Technology*, **47**, 5570–5577.

Solid and drift geology in forensic investigations

ALASTAIR RUFFELL

School of Geography, Archaeology and Palaeoecology, Queen's University, Belfast BT7 1NN, UK (e-mail: a.ruffell@qub.ac.uk)

Abstract: Forensic geology, forensic geoscience and geoforensics (see the text for definitions) are largely concerned with searches for buried objects or the use of sediment and soil analysis as evidence. Geophysics, remote sensing, geostatistics, geochemistry and geomorphology are considered part of these disciplines. Traditional outcrop geology is not often considered, with very few publications considering the role of mapped and logged geology. This work comprises a review of the few published works, followed by a range of case studies that demonstrate how traditional, outcrop-scale drift and solid geology can be used in forensic geology. These include: using geology in the search for buried explosives and human remains; the logging of solid and drift geology for analysis of disputed rock quality; and modelling groundwater flow from an illegal landfill site. Geological mapping may assist in limiting a search area or in understanding what failed in a search or forensic investigation. The scale of forensic geology discussed bridges the gap between the two traditions of forensic geology outlined above: that of searching large areas of ground and trace evidence analysis.

Forensic geology (Murray & Tedrow 1975, 1986; Murray 2004, 2011) is concerned primarily with the geological analysis of rocks, sediment and soils in criminal casework. This may include establishing where soil or sediment on a suspect's or victim's shoes or clothing came from, or comparing (and thence excluding) such materials to or from a crime scene. Forensic geology in this sense uses techniques such as petrology, heavy mineral analysis, particle size, soil/sediment colour, various scanning electron microscope technologies and X-ray diffraction, along with the use of microfossils in similar applications. Geological mapping and geophysics as related geological techniques used in searching for buried objects are also included in the definition of forensic geology (as used by Ruffell 2010; Murray 2011). The majority of cases discussed by Murray & Tedrow (1975 and later versions) are where questioned samples (commonly from a suspect's footwear, clothing or vehicles) are compared to alibi locations and the crime scene, or material on the victim, in order to exonerate the innocent or deny the guilty person's alibi. Forensic geoscience first appeared as a distinct application of Earth science techniques to criminal investigations when a conference was organized at the Geological Society of London in 2002 (with an edited book appearing thereafter: Pye & Croft 2004). The organizers had correctly gauged the interest that such a meeting would generate and included far more than the use of mud on someone's shoe being used as exclusionary or associative evidence. Although the various papers within Pye & Croft (2004) do, indeed, have such soil and sediment analysis at their core, geophysics, bedrock geology, microfossils and micro-organisms, geochemistry, statistics and the geological analysis of unusual materials were included, hence the use of 'geoscience' in the title. Simultaneously to the editing of Pye & Croft (2004), Elsevier Europe commissioned Ruffell & McKinley (published a year later in 2005) to write an article on all aspects of the application of geological and geoscience methods to forensic science (Ruffell & McKinley 2005). These authors broadened the scope further, and included remote sensing, GIS (geographic information systems), as well as the varied methods and applications contained in Pye & Croft (2004) in their article. Geoforensics was first used in a publication by Ruffell & McKinley (2008) to encompass all that had gone before (pedology, geology, geoscience) but also included geomorphology, geography, geostatistics, remote sensing and human geography/sociology. Just as Pye & Croft's (2004) *Forensic Geoscience* included more disciplines than in Murray & Tedrow (1975) or Murray (2004), so Ruffell & McKinley (2005) expanded forensic geoscience even further. Thus, through the evolution of these publications, we may observe an increasing number of the subdisciplines of the Earth (and related) sciences being included. This pattern could have continued, as Ruffell & McKinley (2008) alluded to the possible future inclusion of atmospheric science as but one future tool in the forensic arena.

The application of Earth science techniques has thus widened over the past 35 years, from the petrography of soil and sediment, through geophysics and remote sensing to GIS. However, within all these publications, the use of traditional solid and drift geology in three dimensions (geological maps, outcrop photographs, borehole logs and cross-sections) to assist in forensic investigations is somewhat neglected. In this work, the few publications that have considered the use of traditional, outcrop-scale solid and drift geology are reviewed as background to case studies where the understanding of mapped drift and solid geology, and sometimes its subsurface geometry, has assisted in a forensic investigation. Some of the locations described in the case studies share geological attributes; others are different to the rest (see later). In this paper, the known geomorphological, geological, geophysical, hydrological and geochemical features (where relevant) of each location are described in order to demonstrate the types of information that geologists, soil scientists and geomorphologists can provide that may aid a forensic investigation. Not all of the above information is available for each site, yet the same process of desktop study and creating a conceptual geological model (Harrison & Donnelly 2009) can be achieved for each location in order to demonstrate how a better geological understanding assisted in the investigation. To demonstrate how different outcrop geological information may be of forensic assistance, this article intentionally uses a mix of data types in each example, from outcrop photographs, logs, maps and cross-sections. Some case descriptions close with a paragraph that considers any theories as to how geology assisted, concluded or allowed some ideas of what happened to be made.

Previous work

As Murray (2011) showed, while forensic science was in its infancy from the late 1800s to the 1930s, workers such as Hans Gross, Georg Popp, Edmond Locard and Oscar Heinrich were using the analysis of soil, sediment and rock in their cases (see Murray 2011 for details). Gross, Popp and Heinrich all allude to the use of geological maps in their studies, but as an accessory to their analysis of trace evidence. Murray & Tedrow (1975) and Murray (2004, 2011) described how geological maps are useful in considering where a soil or rock sample may have come from, thus excluding areas that are of lesser interest. Donnelly (2003) included a specific section (p. 9) in his overview of search methodology on the usefulness of geological maps in understanding crime scenes, and the provenance of rock debris and soils. Likewise, Ruffell & McKinley (2008) showed how mapped geology can assist the understanding of where hard rocks with thin soils, and soft rocks with thicker soils, may occur, thus limiting the locations where a perpetrator may be able to dig and bury items. Harrison & Donnelly (2009) developed this concept more fully, integrating geology with other aspects of a suspect burial to create a conceptual geological model. This assists our understanding of where a body is buried, and the effect it has on the surrounding subsurface and surface environment. Lee (2004) used mapped geology and cross-sections to show how understanding the geology of a hazardous site (the location of a gas explosion near a landfill but close to a coal mine) aided in locating the source of the methane. The stratigraphy and areal extent of drift geology and soils is also considered by Hanson (2004) in archaeological forensic excavations, and by Lark & Rawlins (2008) in testing the provenance of a soil, respectively. The above publications include elements of what is considered here. Through the following solid and drift geology case studies, I aim to increase awareness of the role of bedrock, and logged and drift (surficial) geology in forensic investigations, hopefully elevating outcrop geology (especially assisting searches) to the status that search methods and trace evidence analysis have in forensic geology.

Solid geology case studies

Case Study 1. The search for buried barrels of explosives, abandoned farm, Cashel, County Tipperary, Ireland

Here the use of a desktop study, and geological mapping of hard and soft rocks is described in order to gain an understanding of the mapped geology, to limit the search area and to collaborate with other investigators in locating buried objects. A farm, outbuildings and adjacent land near Cashel (Co. Tipperary, Republic of Ireland) lay unused for over 15 years following the death of the owner and the disappearance of his family members, most of whom were thought to have emigrated. The land was bought, and planning applications for redevelopment approved and publicized, at which time a member of the public came forth with information that suggested terrorists had hidden a number of plastic barrels filled with explosives in the area or in the farm buildings. The latter were searched with no evidence of the barrels being found. The surrounding lands were also searched, with priority given to the estate and less priority to neighbouring land. A search dog, trained in the detection of explosives, was deployed and did not indicate in the area. At this point a geologist was consulted, primarily to explain the geology of the search area, which had a surprising effect on the operation, as new search locations were identified. Published geological and soil maps were consulted, the borehole database (Geological Survey of Ireland) examined and the geology mapped (Fig. 1). An area to the west of the farm buildings was identified as having thin soil and till on a Carboniferous volcanic succession, with limited areas to dig and bury objects. The central, north–south-orientated ridge of limestone similarly had a thin soil cover, but may have had karstic features, so was considered an area worth searching for voids. The area to the east of the farm was underlain by shale with thick (1–3 m) soft peat and marshy ground above. This area was diggable and considered of high priority. A behavioural analysis from a police psychologist suggested that likely burial locations would be covert, yet close to the house, with one location in particular of interest, it being on soft ground but in a covert location. This 10 × 10 m area was probed for gas venting and the explosives dog was deployed, who reacted strongly in one location. Upon later excavation by army bomb disposal teams, seven barrels filled with nitrate fertilizer and caster sugar mix, some armed with Semtex, were recovered. Other areas were excavated to check on the predictions and no explosives were found. Consideration of the solid geology, its control on drift geology and, thus, diggability (Harrison & Donnelly 2009) assisted in limiting the search when this information was used in conjunction with other law enforcement personnel, saving time and money.

Case Study 2. Hydrogeology of illegally-buried waste, County Tyrone, Northern Ireland

This case relies on understanding the stratigraphy of solid and drift geology, as demonstrated by outcrop photographs and a cross-section incorporating a conceptual geological model of the site. A landowner in Northern Ireland was accused by the local environmental protection agency of burying toxic waste on his land (in an abandoned sandstone quarry) and endangering local water supplies. The latter aspect is critical as the burial and storage of toxic waste is one crime, carrying a specific range of sentences, whilst the contamination of waters of land beyond the owner's land is a second offence, with more severe sentencing guidelines. An excavation by mechanical digger proved the existence of degraded domestic, hospital and other waste in the quarry, with no impermeable lining below. A geological assessment was required in order to understand where the waste was in relation to the solid and drift geology. A desktop study of solid, drift and soil maps, topography and land use (past and present), and boreholes was undertaken prior to a site visit in order to start creating a conceptual geological model. Together, these established that the geology of the landfill site could be split into two units: a lower succession of fractured Carboniferous sandstones (underlying solid geology); and an upper succession of loose sands, peats, and a sand and sandy glacial till (overlying drift geology). The overlying succession dipped to the east and contained an approximately 4 m-thick loose, permeable sand bed that can be seen dipping east at both the southern (Fig. 2a, b) and northern (Fig. 2b) ends of the quarry, and can be seen in the remaining quarry face to the east of the site. Excavations in the permeable sand showed obvious masses (50 cm–1 m thick) of typical domestic waste (black bin bags, degraded foodstuffs). The stratigraphy of the site allowed a cross-section of the likely geometry of the solid and drift geology to be drawn, and a predictive concept of where fluid may flow in the different units (Fig. 3). This allowed targeted water sampling to be undertaken down-dip, excluding other areas and thus reducing costs for the investigating agencies. Polluted water was encountered, and a prosecution for both storing the waste and polluting groundwater followed. A follow-up to the episode is that the offending land owner also lived down-dip of the waste site and had managed to pollute his own (and his neighbours') water supply.

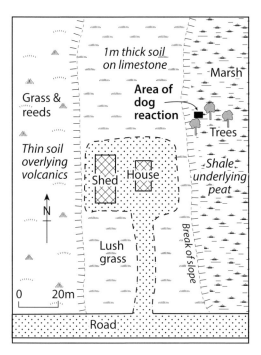

Fig. 1. Map of the location described in Case Study 1. Data from outcrop mapping, field sheets held at the Geological Survey of Ireland and unpublished aerial photographs.

Fig. 2. (a) Photograph of the quarry face described in Case Study 2, with a peat layer above glacial/post-glacial sands. (b) Photograph of the quarry face described in Case Study 2, showing cross-stratified glacial or post-glacial sands. (c) Photograph of an Environment Agency excavation showing waste in direct contact with the permeable sand layer described in Case Study 2. (d) Photograph of the rock units extracted in the quarry discussed in Case Study 3.

Case Study 3. Dispute over taxable quantities of basalt

This case is similar to Case Study 2, where stratigraphy was critical in aiding a forensic investigation. Here, only solid geology was involved, so accurate logging was critical to the work. The case arose because the owner of a basalt quarry in County Antrim (Northern Ireland, UK) was in dispute with Her Majesty's Revenue and Customs (the UK taxation authority) over the quantity of taxable basalt he had sold. Rock that can be used (crushed or uncrushed) for building is of value, and thus subject to tax. Rock, sediment or other soft geological material that is too weak to be crushed or used for building can still be sold or disposed of (say, for the making of rough tracks) but is not subject to tax. A visit by taxation authorities took samples of the material being sold and found it to be of good quality: the thickness and area of the quarry being known, the extracted volume was calculated and the associated tax applied. The quarry owner disputed this, claiming that a substantial proportion (he estimated 50%) of the rock he extracted was of poor quality, and thus he should only be subject to half the tax. An independent assessment was critical in resolving the dispute. A site visit allowed the stratigraphic succession to be defined by inspection and measurement (Fig. 2d); this was compared to geological survey reports of the succession. Four main units (with subunits within) were defined by their dominant rock type (Figs 2d & 4). Visual inspection, hammering and impact measurements with a Schmidt hammer gave a very crude indication of rock quality (Fig. 4). The visit demonstrated minimal recent recovery of quality rock, with good rock having been extracted in a single deep pit (exposing Unit 1) and from the periphery of the working face (exposing Unit 4). The main extraction faces examined comprised 2 m clay, sand and lignite (Unit 2: no quality), 6.6 m rubbly basalt (Unit 3: variable poor quality), 2 m of worked columnar basalt (Unit 4: good quality) and approximately 13 m of columnar basalt worked in the past (NW corner) and in other shallow locations (outside main workings). Of the active faces examined (8–10 m high), 2 m was undoubted waste (20–30%) and 6 m of variable poor quality (over 60%).

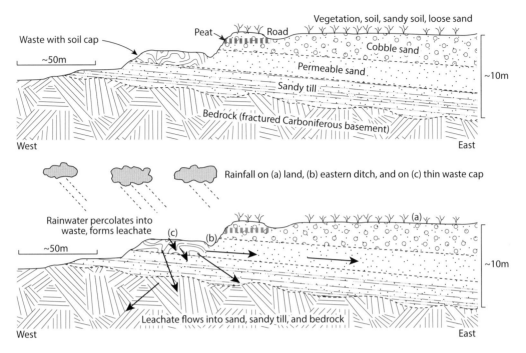

Fig. 3. Cross-section (upper) through the waste and adjacent land, demonstrating the geometry of the solid (lower) and drift (upper) geology. Cross-section (lower) annotated with possible and water/leachate flow, thus creating a conceptual geological model of how the bedrock and drift geology may interact with the waste and rainwater charge.

These observations were outlined to the investigating tribunal and the tax levied was reduced to 50% of the original. The case demonstrates that it was the entire extractable geological succession that needed to be assessed, not just what was being sold.

Drift geology case studies

Case Study 4. Search for the body of a female teenager, Lowland Peat Bog, County Tyrone, Northern Ireland

This case was described (anonymized) in McKinley et al. (2009) and Ruffell & McKinley (2008). The above authors do not mention the facts, nor detail the solid and drift geology, described here. A young woman was last seen in a small town in County Tyrone (UK) in the early hours of August 1994, after having been at a disco in County Donegal in the Irish Republic with two other teenagers and a male, who was acquitted of her murder in 2005 (see McKinley et al. 2009). Following this acquittal, this male's criminal record of rape and murder was released to the public and jury. On the evening that the female went missing, the suspect's car was observed on a nearby road. Later intelligence indicated a location on this road that was known to the suspect. This became the focus of a thorough police investigation, including a search, topographical and geological survey, and cadaver dog deployment: the dog reacted at a number of locations that were excavated, with no body being discovered. The search location is underlain by the Carboniferous (Viséan) Claragh Sandstone Formation (Geological Survey of Northern Ireland 1997), a quartzose, feldspathic and micaceous medium- to coarse-grained, occasionally conglomeratic sandstone. North–south- and NW–SE-orientated faults break up this succession. No other geological unit is known in the area. Three superficial successions were mapped by visual observation and by auger. The search location occurs on a ridge of bedrock (Fig. 5), with a veneer (0–2 m thick) of silty glacial till with a brown earth soil above (c. 30 cm thick). To the immediate east of this ridge, a lowland *Sphagnum* peat bog occurs. This peat is thin (1–2 m) but rapidly thickens to the north to be over 4 m thick (measured by probe, where after it cannot be walked upon). To the west, a meandering lowland river is associated with a 40 m-wide floodplain with fluvial sands, silts and clays. The topography

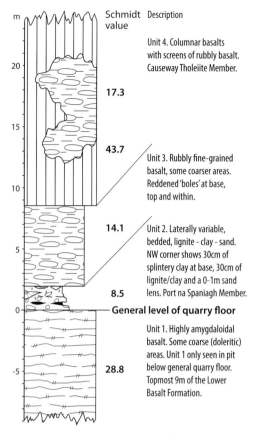

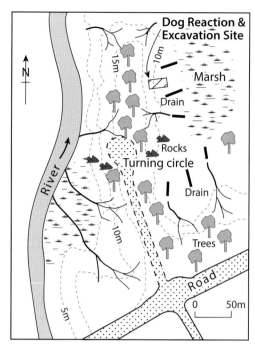

Fig. 4. Log of the quarry discussed in Case Study 3. Units logged relate to those annotated in Figure 2d and described in the text.

Fig. 5. Map of the area described in Case Study 4, showing the track from the road to the turning circle that the suspect's car was observed on. This track runs along a ridge of Carboniferous sandstone with limited diggability, making the areas to the west and east possible areas where ground can be dug. To the west of the ridge, to the river, the area is visible. The area to the east of the ridge to the marsh is covert and diggable, coincident with an area to which the search dog reacted.

of this area was described by Ruffell & McKinley (2008). Topography was used by these authors to demonstrate how to prioritize landscape units for searching and sampling, combining digital elevation models, viewshed analysis and aerial photography. In summary, the area comprises a north–south-orientated ridge, rising to about 15 m above the surrounding floodplain and bog. The ridge is dissected by small (1–2 m deep) valleys and till/soil-filled hollows between upstanding rock. The three drift geology units are each associated with distinct vegetation. The ridge, however, is more complex, with reeds, grass, gorse on the more elevated central area, fringed by silver birch and gorse in the transition areas to the bog to the north, and fields and floodplain to the south. The bog has typical vegetation, dominated by *Sphagnum* and reeds, with both ditches (for drainage) and areas of peat cutting (for fuel). The fields and floodplain comprise grass with some patches of reeds. The dissecting valleys of the ridge drain into the bog and fields: the latter have had drains cut into them in order to improve the ground. Slow flow was observed in the bog drains, while the river to the west is fast-flowing.

Knowledge of the depth to bedrock on all three drift units is incomplete for the whole area, as is the internal structure to these units. On the ridge, drift thickness was limited from a few tens of centimetres to 1 m (measured by auger), making it an unlikely area to bury a body. Likewise, the saturated nature of the peat precludes walking on the surface. The most likely surficial geological unit in which digging could occur comprises the area between the rock-based ridge and impassable peat. Areas where a person could dig in this unit without being observed were identified (Ruffell & McKinley 2008), and a high priority area searched by the police, where a trained cadaver dog indicated. Critically, the thickness of superficial sediment at the search site (where the cadaver dog indicated) and the relationship here between silty till, soil and peat of the bog is unknown: the two must have some form of contact, as the excavated site was in silty till, yet peat could be observed 4–5 m to the north. Similarly, the nature of small-scale (probably shallow) groundwater flow in the search location, as well as from the upland to the bog and fields, is unknown. The main unknown is why the dog reacted at a certain point in the first place: he reacted at a location with disturbed ground (from a previous police dig), close to the peat bog–silty till interface. As far as forming a conceptual geological model (to assist some theorizing on what happened), it is possible that bog gases (possibly methane) from the bog travelled upslope in the silty till to a point where it vented, initially in small amounts. The simpler explanation of the dog reacting to disturbed ground may be applied; however, the coincidence of the reaction at the edge of the bog, in a slightly elevated position, may provide support for the methane hypothesis. The missing person was not found, yet consideration of the geology excluded parts of the area, highlighted others, and, ultimately, saved time and effort for the police.

Case Study 5. Unsuccessful search for the body of a young woman, Lowland Peat Bog, Bangor, County Down

This case was published (anonymized) in Ruffell et al. (2009) but the drift geology, the focus of the description here, was not discussed. A teenager from Bangor (Co. Down, Northern Ireland, UK) went missing after attending a party at a caravan site some 10 miles from her home in February 2005. Ruffell et al. (2009) described how, in late January 2006, a dog-walker observed a shallow depression in rough ground adjacent to a retail park. The location was covert, with easy, yet hidden, access, comprising scrubby woodland adjacent to a small lake. The site thus became the centre of a major police investigation. This began with deployment of a dog trained in the detection of human remains and two GPR (ground-penetrating radar) surveys. The dog indicated at the site, and the GPR profiles were consistent with a cut grave. A full archaeological excavation was completed and no human remains recovered, nor was any evidence of former human remains observed. At this point in the investigation, news of the police dig became public and the origin of the site known. It turned out to be a trial pit to determine the depth of the water table for a nearby building works, dug some months previously. Solid geology has minor impact on the study as the area is of peri-urban nature, on thick (depth unknown) lacustrine peat and glacial deposits. The solid geology comprises deformed greywackes and shales of the Ordovician (Lower Palaeozoic) Gilnahirk Group. These are hard, largely impermeable low-grade metamorphic rocks, presumed (see later on drift geology and trench depths) to be a few metres to tens of metres below the site. Two areas of superficial deposit can be observed on published geological maps of the area, confirmed by field mapping. Beneath raised ground, and including the suspect site, brown sandy soils occur. In lower ground, fringing a nearby lake and in topographically depressed areas, peat occurs (Fig. 6). Both brown sandy soil and peat rest on gravel-rich clay, presumed to be glacial till. Topography in the area is subtle, from the lake, with surrounding waterlogged areas (flat), fringed in a roughly concentric manner by elevated ground (c. 1 m) above the lake, in which the suspect site occurs. Vegetational zones reflect the drift geology. On the flat ground fringing the lake, moss, grasses, sedge and reeds occur amongst pools of water. Pools of water and the lake itself are filled with black rotting vegetation and algae. On the higher ground (around 1 m above lake level, depending on slope), ground flora comprises mosses, grass and clover amongst extensive humus. The canopy comprises alder and birch. Ruffell et al. (2009) reported how, during their excavation

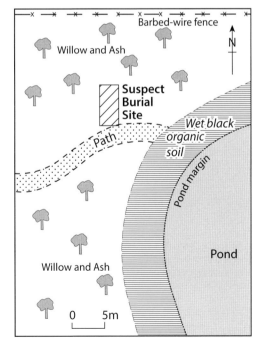

Fig. 6. Map of the area described in Case Study 5. The suspect burial site is approximately 1 m above the level of the pond. The path was possibly used by or created for a mechanical digger.

(intended at recovering any human remains), a water table was encountered at 1 m depth below local ground surface. This level is consistent with the level of the nearby lake below the excavation surface, suggesting a regional water table at this depth. The mapped geometry of the brown sandy soils, till and lacustrine-associated peat is unknown: it is presumed that downslope of the suspect site lacustrine peat overlies the brown sandy soil. The relationship between upland groundwater and surface water in the lake is also not proven but presumed to be in continuity. A simple conceptual geological model shows that the trial pit (the suspect burial site) was dug to the water table, upslope of the peat-fringed lake. One theory is that the scent of decaying vegetation (including methane) travelled along the water table interface, from the lake to the excavation, where it was vented. An alternative explanation is that disturbed ground alone was sufficient for the dog to react. The former explanation is preferred, as the surrounding ground is largely impermeable, with the excavation acting as a discrete vent in this otherwise scent-free area. Thus, the contrast to the dog between this and the undisturbed ground would be greater than simply disturbed ground with no gas emission.

Case Study 6. Search for un-named buried infant(s), Enniskillen, County Fermanagh

This comprises two cases, of almost identical nature, here combined to demonstrate the use of mapped drift geology in the investigations and to protect the identity of two of the perpetrators, who are still alive. In the combined story outlined here, in early 2002, a 55 year-old witness presented herself to police, stating that her mother had recently died and her conscience led her to inform them of a murder in 1960. The event occurred when the witness was a young teenager who developed a relationship with a male temporary worker, stationed nearby. She became pregnant, and upon the birth of her child, her mother (now deceased) suffocated the baby, and buried it in a cloth and leather handbag in fields behind their house. The house had subsequently been rebuilt, but the field remained (apparently) as it had since the 1960s. A geological assessment by desktop study and outcrop mapping was undertaken. This showed the area to be underlain by hard, argillaceous and unkarstified Carboniferous limestones. Augering and surface observations showed the upland area of the field to have a brown sandy loam above bedrock, increasing in thickness from 50 cm (upland) to 1.2 m (mid slope). The lower ground to the north has a 1.5 m drift profile of peat (75 cm) above

brown sandy loam (25–50 cm) above 25–50 cm of calcite-cemented glacial till. The location comprises a south–north, gently sloping field, used for rough pasture (Fig. 7). The higher ground has grass and some isolated reeds on the loam, passing into lower ground with more reeds, rushes and mosses, with reeds on the peat. Rushes occur as small (20–50 cm diameter, 10–20 cm high) tussocks. In between tussocks, the peaty soil was often water-logged. The south–north sloping ground, with permeable brown earths and a shallow water table, that passes to lower ground with water-logged peaty soils suggest a limited shallow groundwater flow above a largely impermeable till and limestone bedrock. The relationship between the upland, brown sandy soil areas and lowland areas with till and brown earth/peat above was not determined by extensive augering as any suspect site could not be disturbed. From the topography and surface exposure of soils, it is presumed that lowland peat lies above the brown soil. This survey showed the limited 'diggability' (Harrison & Donnelly 2008) of over half the suspect area, allowing more time to be spent on geophysical surveying and cadaver dog deployment on the more diggable area. A GPR survey was commissioned that provided a

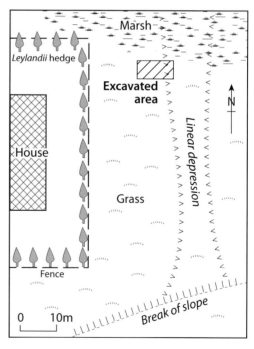

Fig. 7. Map of the area described in Case Study 6. The initial excavation that found buried rubbish was approximately under the label 'Grass'. The burial described was found in the thicker, softer ground labelled 'Excavated area'.

number of anomalies. These were probed, allowing the venting of any subsurface gas. A cadaver dog was then deployed over the area. The dog indicated at three locations, which were also recognized as anomalies on GPR. The area closest to the house was excavated by an archaeological team, who found buried rubbish. At this point intelligence came to the police that explosive devices could also be in the area. A military excavation team took over sometime later, who recovered the handbag in which the baby was buried from a second anomaly in the softer, thicker, more diggable ground further from the house.

Case Study 7. Search for buried neonate infant, West Belfast

In March 2004, a female witness contacted the Police Service of Northern Ireland to state she was raped and abused by a relative when staying in a house on the outskirts of West Belfast, some 20 years earlier. She became pregnant, delivered the child herself and murdered the infant by suffocation. She buried the infant in the dark, in a canvas bag, close to trees, approximately 20 m upslope of the house, using a small hollow patch of ground. On revisiting the scene in 2004 with social care workers, liaison officers from the police and a geologist, the witness failed to recall the location. A cadaver dog was deployed to the area and indicated at some animal burrows made in a hedge (Fig. 8). Excavation of the burrows revealed these to be made by rabbits and forming an extensive network. No remains were found in the burrows or surrounding soil. Redeployment of the dog some months later caused a reaction away from the hedge, 6–7 m downslope. An extensive excavation of this indicated location resulted in recovery of the canvas bag, with neonate bones and black staining inside, buried in the field some 7–8 m from the hedge and closer to the house than suggested by the witness. The dog had, thus, reacted to the empty animal burrows first and also to the burial site, but second. Solid geology comprises over 50 m (thickness) of Palaeogene basaltic lavas with dolerite dyke intrusions (observed in a nearby quarry). These fractured igneous rocks overlie Cretaceous limestones and greensands, in turn above Jurassic shales. Drift geology comprises red sandy glacial till with basalt and limestone cobbles, augered to 50 cm thickness. This till comes to within 5–15 cm of surface red soil, with some black organic remains of vegetation and roots above. There is evidence (terraces, exposed scarps, mounded earth) of extensive landslipping in the area, caused by the solid geology (basalts and limestones) above ductile Jurassic clays. The ground comprises

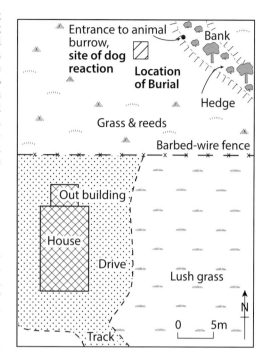

Fig. 8. Map of the area described in Case Study 7. The bank and hedge are approximately 4 m in height above the house, buildings and grassed area. Rabbit burrows extend into the bank, as well as to the SW under the grass and reed area.

moderately steep slopes (10°) with near-horizontal plateaux in between. The search area comprised both, with a steep, grassed field behind the house, and the hedge line forming the boundary to a flatter area, vegetated by scrub blackthorn, horse chestnut and rowan. The hedge is planted blackthorn, with an elevated bank beneath (from studies of the Irish landscape, likely to be of human origin, most likely from stones picked off the fields during seventeenth and eighteenth century attempts at soil improvement). No significant surface water was noted, even after rain. Some evidence of ditch cutting along the hedge lines was noted, although this may be excavation for the hedge banks as opposed to drainage. During excavations, no groundwater was encountered. Deep hydrogeology is not known, but from regional understanding is likely to comprise a fractured aquifer in the basalts and limestone, above a minor aquifer in the greensands, together forming a perched water table above the Jurassic shales, above which a spring line is known from adjacent locations. The obvious unknown is why the dog indicated in two separate locations, following excavation, the initial focus being incorrect. Considering some theories as to what happened, hydrology seems to have little impact here; likewise, small-scale variations in the sandy till, hedge and field vegetation, and animal burrows are unknown. The most likely scenario is that either the bedrock fractures, the bedrock–soil interface, or the soil and rabbit burrows are permeable to gas, allowing scent from the neonate's remains to travel upslope to the animal burrows, where the dog reacted. In summary, a wider consideration of the geology of the area, especially the presence of bedrock fractures but more so the sandy, free-draining soil, may have widened the searchers' perspectives on conditions and caused a more widespread deployment of the cadaver dog.

Conclusions

Previous studies in the use of mapped, logged and photographed solid and drift geology have been used to assist consideration of the likely provenance and distribution of soil, sediment and rock types (see the references cited in the earlier section on 'Previous work'). Geological and soil maps are especially useful in assisting the search for buried materials (Donnelly 2003; Harrison & Donnelly 2009). However, no comprehensive description of the range of uses of solid and drift geology has hitherto been made. The range of cases discussed in this work demonstrate how outcrop descriptions, logging and mapping can variably (and collectively) assist forensic investigations, from the burial of illicit materials and human remains to the correct evaluation of rock quality. Furthermore, geological assessment of the types described here can aid our understanding of why some investigations fail or, better still, simply save time, money and effort in searching large areas when solid and drift geology can focus a search to a smaller area. It is hoped that the mapped and logged solid and drift geology described here will be considered in forensic geology publications along with the uses of geology in trace evidence analysis or in assisting searches.

I am grateful to the many people who have collaborated with me on the criminal cases described. These include, from the various police forces of the United Kingdom and Ireland: J. Shanaghan, J. Gilmore, R. Murray, Sergeant M. Harrison (now Australian Federal Police), M. Grimes, N. Johnson and G. Arbuthnot. From Queen's University Belfast: D. Hughes, J. McKinley, W. Meier-Augenstein (now University of Dundee), A. Keaney, C. Graham, L. Barry, C. Donnelly and E. Murphy. From the Northern Ireland Environment Agency: C. Hunter, P. Parsons, A. Rowney and A. Blacker. Thanks to J. Pringle and P. Doyle for various ideas. I am indebted to L. Donnelly, with whom I presented some of these cases at the British Library in 2010.

References

DONNELLY, L. J. 2003. The applications of forensic geology to help the police solve crimes. *Journal of the European Federation of Geologists*, **16**, 8–12.

GEOLOGICAL SURVEY OF NORTHERN IRELAND 1997. *Northern Ireland Solid Geology*, 2nd edn, 1:250 000. British Geological Survey, Keyworth, Nottingham.

HANSON, I. D. 2004. The importance of stratigraphy in forensic investigation. *In*: PYE, K. & CROFT, D. J. (eds) *Forensic Geoscience: Principles, Techniques and Applications*. Geological Society, London, Special Publications, **232**, 39–47.

HARRISON, M. & DONNELLY, L. J. D. 2009. Locating concealed homicide victims: developing the role of Geoforensics. *In*: RITZ, K., DAWSON, L. & MILLER, D. (eds) *International Soil Forensics*. Elsevier International, Amsterdam, 197–219.

LARK, R. M. & RAWLINS, B. G. 2008. Can we predict the provenance of a soil sample for forensic purposes by reference to a spatial database? *European Journal of Soil Science*, **59**, 100–1006.

LEE, C. 2004. The nature of, and approaches to, teaching forensic geoscience on forensic science and earth science courses. *In*: PYE, K. & CROFT, D. J. (eds) *Forensic Geoscience: Principles, Techniques and Applications*. Geological Society, London, Special Publications, **232**, 301–333.

MCKINLEY, J., RUFFELL, A., HARRISON, M., MEIER-AUGENSTEIN, W., KEMP, H., GRAHAM, C. & BARRY, L. 2009. The importance of spatial thinking in search methodology: case study no body murder enquiry, West of Ireland. *In*: RITZ, K., DAWSON, L. & MILLER, D. (eds) *International Soil Forensics*. Elsevier, Amsterdam, 285–302.

MURRAY, R. C. 2004. *Evidence From the Earth: Forensic Geology and Criminal Investigation*. Mountain Press, Missoula, MT.

MURRAY, R. C. 2011. *Evidence From the Earth: Forensic Geology and Criminal Investigation*. Mountain Press, Missoula, MT.

MURRY, R. C. & TEDROW, J. C. F. 1975. *Forensic Geology: Earth Sciences and Criminal Investigation*. Rutgers University Press, New Brunswick, NJ.

MURRAY, R. C. & TEDROW, J. C. F. 1986. *Forensic Geology: Earth Sciences and Criminal Investigation*. Rutgers University Press, New Brunswick, NJ.

PYE, K. & CROFT, D. J. (eds). 2004. *Forensic Geoscience: Principles, Techniques and Applications*. Geological Society, London, Special Publications, **232**.

RUFFELL, A. 2010. Forensic pedology, forensic geology, forensic geoscience, geoforensics and soil forensics. *Forensic Science International*, **202**, 9–12.

RUFFELL, A. & MCKINLEY, J. 2005. Forensic Geology & Geoscience. *Earth Science Reviews*, **69**, 235–247.

RUFFELL, A. & MCKINLEY, J. 2008. *Geoforensics*. Wiley, Chichester.

RUFFELL, A., DONELLY, C. J., CARVER, N., MURPHY, E., MURRAY, E. & MCCAMBRIDGE, J. 2009. Suspect burial excavation procedure: a cautionary tale. *Forensic Science International*, **175**, 11–16.

Geomorphological and geoforensic interpretation of maps, aerial imagery, conditions of diggability and the colour-coded RAG prioritization system in searches for criminal burials

LAURANCE DONNELLY[1]* & MARK HARRISON[2,3]

[1]IUGS, Initiative on Forensic Geology, Wardell Armstrong International, 2 The Avenue, Leigh, Greater Manchester WN7 1ES, UK

[2]Manager Forensic Operations, Forensic and Data Centres, Australian Federal Police, Canberra, Australia

[3]University of Canberra, Canberra, Australia

*Corresponding author (e-mail: ldonnelly@wardell-armstrong.com)

Abstract: The objective of this paper is to draw attention to the use of air photographs, diggability surveys and the RAG (Red–Amber–Green) prioritization system during police ground searches for burials. The acquisition, analysis and interpretation of aerial imagery by a geologist may provide a useful reconnaissance technique to help delineate and prioritize search areas. A diggability survey may provide information on the ease and efficiency with which the ground may be dug and reinstated by an offender. This is influenced by the depth of the soils, the geology, groundwater, obstructions, the digging implements used, the ability of the offender, the nature of item being buried and the time frames involved. The results of a diggability survey may conveniently be presented as a RAG, map which can help in prioritizing the search. The RAG system appears to have been used independently by geologists, police/law enforcement and the military, and has evolved differently and independently since the early part of the 1900s. These methods have been applied to law enforcement searches for graves and other buried objects as demonstrated by operational case examples.

The geological analysis of high-resolution photographs, the determination of diggability and the use of colour-coded RAG (Red–Amber–Green) maps may often be used by geologists to explore or investigate the ground. These may be required, for instance, as part of mineral exploration, to delineate areas of risk caused by mining and geological hazards or to provide information on the ground conditions ahead of planning, development or construction. The objective of this paper is to draw attention to the possible use of geological maps, air photographs, the determination of diggability and the use of colour-coded RAG maps, which may be used to support some types of law enforcement ground searches.

Search

Geoforensics (known also as Forensic Geology) may be regarded as 'the application of geology to the investigation and detection of crime'. A forensic geologist may assist law enforcement officers by the collection, analysis and interpretation of geological (trace) evidence or by searching the ground for burials.

A 'search' is 'the application and management of systematic procedures and appropriate detection equipment to locate specific targets' (Anon 2006). It may be regarded as 'the skill of looking and the art of finding' (Harrison & Donnelly 2008). In recent years conventional and innovative geological techniques have been applied to law enforcement searches (see for example Donnelly 2009a, b, 2010; Donnelly & Harrison 2010).

The police may search the ground to locate burials, find missing persons, deprive criminals of their resources and ability to commit crime, protect vulnerable targets (person(s) or a venue), or gather intelligence or evidence (Anon 2006). Objects and items recovered from ground searches may include homicide graves, mass burials related to genocide, weapons, firearms, improvised devices, explosives, drugs, stolen items, money and items of value.

Searches for burials may take place in urban, rural and remote locations, in both the terrestrial (land) and marine environments. In order to find a man-made burial or sub-surface criminal hide it is important to first understand what is being searched for, and therefore the development of a conceptual geological model (CGM) of the burial or hide and any associated items is recommended. The search for a burial or sub-surface hide is a pragmatic balance between the minimum acceptable standard of search, equipment and resources, and minimum expenditure of funds required to achieve a high assurance search. For further information on how forensic geology specialists may support the police/law enforcement see Donnelly (2008, 2009a, b, 2010), Donnelly & Harrison (2010), Fenning & Donnelly (2004), Harrison (2008), Harrison & Donnelly (2008, 2009), Pye & Croft (2004), Ruffell & McKinley (2008), Murray (2004, 2005), Murray & Tedrow (1992), Murray & Solebello (2005), Ritz et al. (2008) and Bergslien (2012).

Geological interpretation of imagery to help locate criminal activity

Aerial imagery may be acquired in black and white with variable grey-scale intensities, colour, infrared or the portrayal of different spectra. It may be in hard copy, high-resolution or digital, and may give a vertical or oblique representation of the ground. The type, resolution and quality of the imagery will be dependent upon several factors, such as camera type, satellite imagery sensor, height of aircraft or helicopter, weather conditions, time of day, time of year, topography, geology and the experience of the operator. Stereographic pairs are particularly useful as these may aid identification by artificially exaggerating subtle ground features, which may represent disturbed ground possibly associated with the digging or excavation of a burial or hide.

Aerial imagery requires skill and expertise during each stage of acquisition, analysis and interpretation. The different specialists examining aerial imagery will 'see' different things, which will be a reflection of their training and experience. For example, some military personnel, police officers and geologists are specifically trained in the analysis and interpretation of aerial imagery. It is important, therefore, that the correct specialist is used depending on what is being searched for. The geologist may be able to quickly explain the causes of any ground movements that may be associated with a variety of processes not related to digging, such as natural weathering and erosion of the landscape or past mining activities. These types of disturbances may therefore be disregarded as not being associated with criminal activity (although this is not always the case).

The geological analysis of ground-based and aerial imagery to search the ground is by no means new. A notable example occurred following the arrest of Myra Hindley and Ian Brady (known as the Moors Murderers) in northern England during the 1960s, when several incriminating photographs were found by the police. Distinct landforms and geomorphological features of the Pennine Hills on what initially appeared to be no more than innocent 'holiday snaps' were subsequently shown to contain the locations of their victims' graves (Topping 1989). Another well-known example occurred after the terrorist attack on the World Trade Centre in New York on 11 September 2001. A geologist with field experience working in Afghanistan recognized the rocks that cropped out behind Osama Bin Laden in a video that he broadcast. These were deposits of limestones that cropped out in the southeastern part of Afghanistan (Pinsker 2002). Other examples where aerial imagery or remote sensing have contributed to geoforensic investigations may be found in Abbott (2005), Gomez-Lopez & Patino-Umana (2006), Silvestri et al. (2005), Gripp et al. (2000), Ruffell & McKinley (2008) and Fowler (2004).

Criminal burials and hides in rock

Burials and hides in rock are less common than those in soil. Only soft rocks, such as mudrocks and weathered shales, may offer the opportunity for an excavation to be quickly dug using only hand-held tools. If mechanical excavators are available, for example farm machinery or heavy-duty plant in a quarry, mine or construction site, this may facilitate the digging and subsequent concealment of a victim or criminal/terrorist property in a rock burial.

The method of excavation in rock is determined primarily by geology, mainly the engineering properties of the rock mass, including the mechanical strength of the rock, presence, geometry and type of discontinuities (such as joints, faults, bedding planes and schistocity), groundwater, type and intensity of weathering and erosion.

Typical methods of excavation in rock are digging or ripping (with the use of mechanical excavators) and blasting (with the use of explosives). In forensic searches the use of excavators may need to be considered if, for example, criminal concealment of a homicide victim, drugs or weapons has taken place in a quarry, mine or pipeline trench, or beneath concrete foundations. The authors are aware of only a couple of rare cases and exceptional circumstances where mechanical excavators were used to help search for and recover buried objects. Clearly, the use of such equipment may inadvertently damage forensic evidence and therefore these are not generally used to recover burials where forensic evidence may potentially be obtained.

How easily a rock can be excavated may be determined by a number of engineering geological and geotechnical methods involving assessments of the engineering properties of the rock mass. Further information on these techniques can be found in Pettifer & Fookes (1994), McCann et al. (1990), McCann & Fenning (1995), Scoble & Muftuoglu (1984), Weaver (1975) and Kirsten (1982).

Criminal burials and hides in soil

The sub-surface burial or concealment of a homicide victim, drugs, weapons, explosives or other object/items usually takes place in soil or superficial deposits. There appears to be no universally accepted definition for 'soil', since this term has different meanings depending on the context of its use. For simplicity, the following categories of 'soil' may therefore be considered to exist within the context of near-surface burials and hides as used by murderers, terrorists and organized crime groups:

(1) *Natural soils* – these comprise naturally occurring weathered superficial deposits that have been deposited on the underlying bedrock formed *in situ* by geological and geomorphological processes and have not been significantly changed by the activities of humans. These may include
- *organic soils (peat),*
- *granular soils (silt, sand, gravel, cobbles, boulders),*
- *cohesive soils (silt and clay),*
- *mixtures of the above.*

(2) Artificial (anthropogenic) soils – these have no well-defined landform, boundaries and properties. Their behaviour and physical and chemical characteristics may be highly variable and may be categorized as follows:
- made ground consists of material placed on the ground surface such as domestic, mining, industrial, agricultural and building tipped waste materials or old building foundations;
- worked ground consists of areas excavated by humans, for example engineering and mineral excavations;
- infilled ground comprises excavations that have subsequently been filled in by artificial deposits;
- disturbed ground is complex ground, which may have been subjected to a combination of the above.

Diggability

Diggability is the ease with which soil can be dug and reinstated into an excavation. Diggability is a measure of the ease of excavating a burial because different soil types have different 'strengths'. Diggability assessment tests may be easily performed before the main phase of the search takes place. A pre-search reconnaissance visit to the crime scene or search area is recommended for the diggability survey to be performed (and for other technical, planning, management and logistical constraints to be identified). The 'digger' (offender performing a criminal act of burial) is most likely to choose a site where the soil is thicker (typically over 1 m, or at least the depth of the object being buried), it can be quickly and efficiently dug, then reinstated, and preferably where there is no or little surface expression (such as settlement or colour changes of the displaced soil) to indicate that digging has taken place (provided the offender is forensically aware of this; Table 1). The location where an offender chooses to dig a burial or hide may, consciously or subconsciously, be influenced by several factors, such as the following:

(1) type and thickness of soil/groundwater (i.e. the geology);
(2) concealment from view and low witness potential;

Table 1. *Relative diggability of selected soil/sediment types using a hand-held digging aid (such as a spade, mattock or pick)*

Soil type	Subdivision	Diggability
Very coarse soil	Large boulders	Difficult
	Boulders	Difficult
	Cobbles	Difficult
Coarse soil	Coarse gravel	Difficult
	Medium gravel	Difficult
	Fine gravel	Moderately difficult
	Coarse sand	Easy
	Medium sand	Easy
	Fine sand	Easy
Fine soil	Silt	Easy
	Coarse silt	Easy
	Medium silt	Easy
	Fine silt	Easy
Clay	Dry	Moderately difficult
	Saturated	Easy to moderately difficult
Peat	Fibrous	Very easy
	Amorphous	Very easy
	Humus	Very easy
	Organic clay	Very easy
Artificial (Anthropogenic)	Made ground	Highly variable
	Worked ground	Highly variable
	Infilled ground	Highly variable
	Disturbed	Highly variable

(3) time of day or year;
(4) weather conditions at the time of burial;
(5) places which are known to the offender (this will facilitate navigation, escape routes, aid grave/burial location by recognition of physical features and potentially provide an explanation if compromised during or after the act of burial);
(6) ease of access on foot or by vehicle to the burial/hide location;
(7) diggability and the principle of least effort.

A diggability survey should be performed at the crime scene/search area based on evidence or intelligence to infer that a burial has taken place at a particular location or area. This involves either probing or digging the ground using tools similar to those which the offender is believed to have used. Diggability surveys provide the opportunity to inspect the soil type and structure, and determine the depth of the weathered bedrock interface, groundwater and associated superficial deposits to determine whether it is granular (sand rich), cohesive (clay rich) or organic (peat). These observations are important as they have implications for the efficiency of digging and reinstatement. The diggability of soil may depend on its geological properties, such as intact strength, bulk density, natural water content, depth, weathering, proximity of the underlying bedrock, slope angle, groundwater, surface water, vegetation, stability of the walls upon excavation, bulking and swelling of the soil as well as the method of digging.

There is no generally accepted quantitative measure of diggability in soils for criminal burials. This is a qualitative analysis that can only be determined by *in situ* 'trial and error' testing by the offender (at the time of burial) or forensic geologist (at the time of search). For example, an offender may dig several test pits before choosing a preferred site and thus present the police search team with an initially confusing picture of multiple burials or disturbed ground. When some soils are dug they may increase in bulk, caused by the decrease in density of the displaced soil. Soils with high bulking factors may result in excess material, which the offender may then place on top of the burial. This may form a raised or positive topographic anomaly or become distributed around the burial site. This may cause visible scarring and increase the ease of burial site detection. Field trials in organic peaty soils in the Pennine Hills, UK, showed ideal conditions for the quick and efficient digging and reinstatement of a burial, in just a few minutes. Loose, free-running, granular materials such as sand, when dry, may also offer the opportunity for the quick digging of a burial. However, this will be hindered if the sand becomes wet (increasing its density and causing the collapse of the excavation sidewalls) or if the sand contains increasing numbers of pebbles, cobbles or boulders with increasing depth. Partially consolidated clay, or clayey sidewalls, will require moderate digging but also offers good conditions for the location of a burial and the relative stability of the excavation walls. Wet clays, gravel and large boulders will require considerably more effort during digging. Examples of diggability based on the authors' experiences in the UK Pennines are summarized in Table 2.

As stated above, when soil is excavated it increases in its bulk owing to the decrease in density per unit volume and loss of inter-particulate moisture. This bulking may result in a subtle change in ground elevation or soil distribution around the burial. Subsequent subsidence, or settlement, of the reinstated ground may then cause a depression, which could become filled with standing water. Any mineralogical and geochemical differences in the sub-soil profile should be noted as these differences may be of value during the search for a burial. For example, any excess sub-soil, distributed around a burial, may have characteristics (such as colour, consistency, grain size, structure and mineralogy) that make it distinguishable from the rock, soil or vegetation on the ground surface. The advantages of a diggability survey are summarized in Table 3.

Once the assessment of diggability is complete, a wider geological and geomorphological assessment may be required. This may include, for example, the analysis of surface vegetation to identify anomalous growth, which may be accounted for by a buried human cadaver or the possible generation of leachate plumes from the cadaver. Advanced surveys of this nature benefit from the accompaniment of an environmental geologist, hydrogeologist, palynologist, ecologist or botanist. Before the main phase of the search is implemented, a CGM of the burial is also recommended to be developed. Furthermore, once an item/object/burial has been located, the authors recommend that the forensic geologist should seek the expertise of a forensic archaeologist to provide advice on the state of potential preservation or decomposition and other aspects of search and recovery (Anon 2008).

The British Army produced 'diggability' maps towards the start of the Cold War for potential operations in Germany. These colour-shaded maps showed 'Go/Slow' and 'Go/No' and 'Go' areas and were produced as transparent overlays for use with topographic maps. These were classified as restricted and the system was set up by the Royal Engineers and geologists under their command (Ted Rose pers. comm.).

Table 2. *Summary of the expected geological properties of superficial deposits and their diggability for the excavation of a shallow grave based on the authors' experiences in the UK Pennines*

Type of superficial deposit (soil)	Geological description	Diggability	Suitability for a homicide grave
Peat	Peat accumulates on plateau areas of gently dipping slopes and lowland areas or river valleys. Peat is black to brown, highly variable and may include different types such as fibrous or amorphous, or occur as cohesive clay. High water content, often saturated, high porosity and permeability. Peat may reach 4 m deep, more commonly 0.1–1.5 m deep. Voids may exist in peat caused by sub-surface groundwater flows. Peat is very weak and highly compressible and susceptible to subsidence. Trench supports will be required for deeper excavations as sidewalls may collapse.	Diggable with considerable ease	Highly suitable for the digging of a homicide grave. A shallow burial (<1 m deep) may be excavated and the displaced peat reinstated in less than 5 min, leaving no or little surface evidence of the grave's position. There may be no grave cut detectable in peaty soils. Peat deposits may facilitate the quick disposal of a victim, especially where the deposits are concealed on gully sides.
Alluvium	Alluvium is material deposited by rivers and streams. Usually restricted to the valley floors, although fine sand may be blown beyond incised stream valleys to accumulate in peat, often as discrete lenses. Generally occurs as thin layers (<0.5 m thick, but may be much thicker), loose, non-cohesive and free draining. May contain dense layers with cobbles and pebbles. Laminated alluvium consists of brown to grey, soft, silty clay with peat or sand lenses. Running conditions possible when alluvium is saturated. High water flows may occur into any deep excavations. Supports required in deep excavations for long- and short-term stability.	Diggable with ease when granular (sand rich) or cohesive (clay rich). May not be diggable where alluvium has high proportion of cobbles and boulders.	Generally suitable and diggable with ease where sand deposits have accumulated on valley floors in thicknesses of ≥0.5 m.
Glacial head	Poorly stratified deposits consisting of angular fragments of sandstone, siltstone and shale in a silty–sandy matrix. Found on slopes and valley floors 0.1–0.5 m thick. Formed by solifluction under periglacial conditions.	Usually diggable, but may be difficult if contains boulders	Not generally suitable as these mainly occur as thin deposits, less than approximately 0.3 m thick or exist beneath peat deposits.
Scree	Accumulated material at the toe of slopes. May be diggable if well graded and granular. Properties of the bedrock will vary considerably depending on rock types, degree and intensity of weathering and the groundwater regime.	Not diggable with hand-held tools, may require mechanical excavation	Generally not suitable for the digging and concealment of a homicide grave. Middle and upper slopes, consisting of bedrock, may not be diggable with the use of hand tools and may require mechanical excavation. Some weathered bedrock may possibly be diggable with a spade, but on a slope this would hinder the quick, efficient burial of a victim. What is more, any objects or items buried in scree or weathered rock may become exposed by weathered and erosion, such as the slippage of the slope.

Table 3. *Advantages of a diggability survey*

Characteristic	Details
Soil type and structure	Soils are highly variable and no two crime scenes/search areas will have identical soil types. Diggability surveys enable any 'indicator layers' to be recognized, such as an iron pan or salt layer.
Time	Diggability surveys allow an estimate of the time required for the offender to dig an excavation and reinstate the disturbed soil to conceal the target object. This may be relevant where the investigators have established a 'time window' of opportunity that they believe the offender(s) had to complete the burial.
Resources	Different digging implements will be required to suit different soil conditions. This may also aid investigators in the forensic examination of any seized digging implements to determine if there is a mineralogical association between the implements and the offender, crime scene or victim/item being sought.
Topographic expression	The act of burial may result in the generation of ground surface features to indicate that digging has taken place. These may include scarps, fissures, compression, settlement, excessive soil, changes in soil characteristics and standing water. However, these will require careful consideration to distinguish from biological, geological, geomorphological or anthropogenic processes.

Colour-coded maps and the RAG traffic light system

Geological maps have been produced for a variety of geological and non-geological purposes at a range of scales to give information on the principal soils and rocks and geological structure. However, these are often produced by geologists for geologists, and during a ground search these maps may be difficult to interpret by police and law enforcement officers.

Colour-coded maps have been frequently used to convey geological information to non-geologists and other specialists. These use primary colours to categorize areas of similar geology or to represent the severity, for example, of risks associated with geohazards or certain types of adverse ground conditions.

In a geological, police and military context the Red–Amber–Green system, also known as the 'RAG' or 'traffic light' system, is a convenient method to facilitate the visualization of complex information, or datasets, in a manner that may be quickly and easily interpreted and executed in a search context. Red is often associated with the highest risk, amber moderate risk and green the lowest risk.

Geological RAG maps

It is unclear when colour-coded geological RAG maps were first used in geology. On some geological maps red colours, in planning and development terms, are usually interpreted as signifying caution and a relatively high degree of constraint, whereas green signifies opportunity (for development) and lesser constraint (Smith & Ellison 1999). A red polygon designation on a thematic map signifies 'danger' or 'hazard'. The case of diggability follows the same principle: red polygons are areas with more operational and ground difficulties, such as no or shallow soils. Green signifies thick, more easily diggable, soils. In terms of geological information, areas most likely to yield results should be green (go), whereas red should indicate a lower likelihood (i.e. caution, or a lower priority search).

Colour coding was used on maps (known as thematic maps) developed by the British Geological Survey in the 1980s–1990s to convey information to planners and developers on specific types of ground conditions in the UK, to facilitate decision-making for future urban and rural planning purposes (Richard Ellison pers. comm.; Smith & Ellison 1999).

In Germany, colour-coded maps were also used, including one on 'suitability for construction of various types of ground'. They were classified by colour using 'green and yellow' for ground suitable for construction that would permit bearing pressures of not less than 2.5 bar. 'Orange' indicated moderate ground, whereas 'orange cross-hatching' indicated least favourable ground conditions, where the hydrological conditions necessitated special attention being paid to foundations and allowable bearing pressures. 'Red' corresponded to poor ground requiring costly foundations, comprising generally fill, marshy ground and steep slopes and so on (Dearman 1991; Martin Culshaw pers. comm.).

In the 1980s, during an Andean water capture scheme for diversion to the public water supply of Quito (known as Proyecto Pita–Tambo), the world's tallest volcano, El Cotopaxi (which has been reported to have erupted 34 times since 1534), was located at the centre of the primary watershed-collection area. Geologists involved with this project used a RAG system, associated with a visual and instrumented volcano activity monitoring system (Allen Hatheway pers. comm.). Similarly, colour-coded maps were developed by geologists and volcanologists to delineate high-, moderate- and low-risk zones during the eruption of Nevado del Ruiz in Colombia in 1985 and the Soufriere Hills Volcano on the Caribbean Island of Montserrat (Donnelly 2007; Fig. 1).

In the 1980s, RAG maps were used by geologists in the central USA. Shades of red categorized the expected isoseismal patterns of ground motion intensity. It was proposed that these maps be used by geologists and the military to locate field command posts and traffic management in the event of an earthquake occurring similar to the historical New Madrid Earthquake (Allen Hatheway pers. comm.).

Military RAG maps

US military. In the USA the origin of American RAG usage seems to have been the 'Go'–'No-go' system that developed within the US Geological Survey Military. The geology branch after the Second World War began to develop this system about the time of the beginning of the Vietnam War (i.e. around 1964). This was the simplest advice thought possible to give the harried combat commander, and it meant 'don't commit your forces in this area for you will lose combat advantage'. The same system was employed in the late 1970s for the 1st Infantry Division (known as Big Red 1) in its defence sector of West Germany when the colours 'red' and 'green' were used on the appropriate defensive terrain maps (Allen Hatheway pers. comm.).

British and European military. Simple colour-coding maps were developed by the British Army in the First and Second World Wars for the benefit of (non-geological) military intelligence officers, not for geologists. They were deliberately kept simple in order to enable the key aspects to be considered alongside equally important information concerning other aspects (weather, vegetation, defensive obstacles, enemy deployment, etc.). There were two further considerations for the military products: they had to be printed in a timely fashion, usually within a matter of days, and invariably had to be prepared from diverse and incomplete

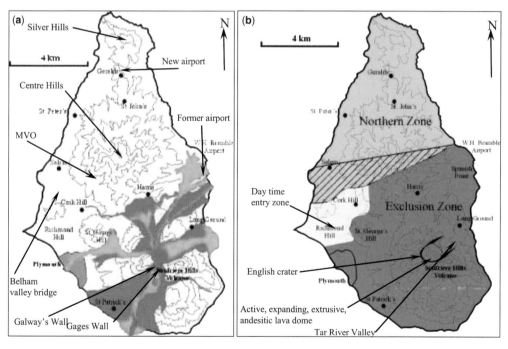

Fig. 1. (**a**) Map of Montserrat showing the approximate location of volcanic and volcaniclastic deposits (green, pyroclastic flow deposits; brown, pyroclastic surge deposits; blue, debris avalanche deposits; purple, lava dome). (**b**) Example of a colour-coded volcanic hazards risk map of Montserrat, in December 1999, showing the location of the active, expanding, extrusive, lava dome in English Crater (*source*: Montserrat Volcano Observatory and British Geological Survey, reproduced in Donnelly *et al.* 2006).

sources of data. The red–amber–green colours (sometimes with variations) were originally based on the colours used for simplified geological maps generated by the Geological Section, Inter-Service Topographical Department unit (British Intelligence) in the Second World War (Rose & Clatworthy 2007). These were originally shown in sequence with amber at the top, followed by green then red. Only three, but sometimes four, colours were used because this enabled simplification of complex geology for non-geologists who required the geological information for military planning purposes. Furthermore, this reduced the costs associated with the production and printing of the maps when they were required in large numbers. Published examples of colour-coded maps date from 1918. These maps portray suitability for dug-outs, essentially based on the 'dampness' of the ground (see for example Institution of Royal Engineers 1922). This work began in 1915 and colour-coded maps may well have first appeared in 1916. The assessment used various shades of two colours, ranging from 'red' for dry (good for dug-outs) to 'green' for wet (bad for dug-outs). Other published examples date from 1941 and the intelligence compilation by the British Army in North Africa regarding cross-country movements (War Office 1949). This was known as 'desert going'. Apparently, a bogus series was produced and leaked to the enemy in advance of the attack on Alamein (August to September 1942). This misled their intelligence appraisal and contributed to the British success, and the turning point of the Second World War. The assessment used four colours: 'blue' for generally impassable, 'green' for reconnaissance essential, 'yellow' for fair going, and 'red' for firm and fast going. Other colour-coded maps were used for terrain analysis for the trafficking of military vehicles and to facilitate the construction of trenches (Rose & Rosenbaum 1993; Rose & Pareyn 1996a, b; Rose & Clatworthy 2007, 2008a, b; Rose 2008, 2009; Willig & Häusler 2012). Other colour-coded maps and the RAG system were also used during the Second World War and Cold War but remain classified (Ted Rose pers. comm.). Further information on the use of British geological maps for excavations and underground facilities on the Western Front during the First World War may be found in Rose & Rosenbaum (2011).

The German military developed a comparable system, and this subsequently provided the basis for the colour-coded maps developed after the Second World War by NATO and later adopted and developed by the Americans. The assessment was quite detailed and complex, and led to the map classification as 'confidential'. The products were therefore rarely used on military exercises because of the difficulties of ensuring sufficient security. They were refined by parallel work undertaken by the Military Engineering Experimental Establishment (MEXE) at Christchurch during the 1950s and 1960s (Griffiths 2009).

British military geologists were deployed on the Western Front (France and Belgium with the British Expeditionary Forces) in 1939, in Egypt, North Africa and the Middle East in 1941 and Italy (including Sicily) in 1943. They provided geological advice, well drilling for potable water supplies, sources of aggregate, terrain intelligence, construction of temporary airfields and general military engineering.

Water supply maps for the Western Front, printed in 1916, are believed to be the first known map produced by a British military geologist in which ground areas were denoted by shades of blue to indicate water resources being 'abundant', 'medium' or 'scarce' and 'good', 'fair' or 'bad' (Rose 2009).

Army maps of Belfast compiled in the early 1980s developed a colour-coded system with 'green', 'orange' and 'grey' zones, corresponding to Nationalist, Loyalist and neutral areas for military patrols (Alastair Ruffell pers. comm.).

In more recent military campaigns in the Falklands War, Middle East and Afghanistan, terrain analysts graded the terrain 'goings' for armoured light vehicles as 'red' (no go), 'amber and green' (go). The terrain analysts also used the RAG system for diggability.

In a military search, 'red' coding is given to the area of most hazard from an explosives or an attack; hence, 'red' is a high-risk search and 'amber' and 'green' comprise lower levels of threat. This system is used on high-risk route searches as well as venue searches to protect a vulnerable person(s) or the venue itself (e.g. where a terrorist might place a bomb in a hotel). This is termed a high-assurance search and the 'red' areas are searched by military trained searchers with specialist search equipment and techniques. Most mapping and imagery to aid this type of search has a key and legend to explain the RAG system (Mick Jenkins, pers. comm.).

Law enforcement and police RAG maps

Colour-coded maps have been used by UK policing since 1985 when they adopted the UK military search prioritization system of red–amber–green. Prior to this there was no recognized system.

The colour coding is a scalable system to easily identify areas vulnerable to attack by criminals and terrorists or identify where it is most likely that evidence is located that has been hidden or discarded. On these maps 'red' denotes the highest vulnerability or highest probability that evidence will be

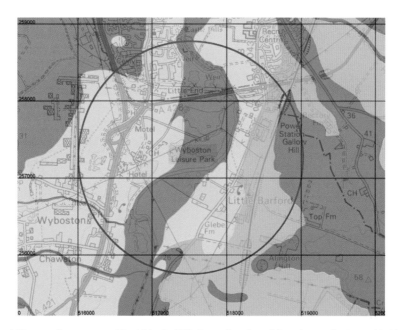

Fig. 2. Diggability map for a suspected burial in the UK. An explanation of the colour codes is provided in Table 4.

found and 'green' the least likely or least vulnerable. This colour coding is used on topographical maps, aerial imagery or structural building plans. The main use for the colour-coding system is to aid the police in determining and allocating appropriate and proportionate search resources to areas that are subject to a high threat of terrorist/criminal attack or in investigations where the terrorists/criminals may have hidden explosives, weapons, drugs or money.

In 2009 the scalable RAG system was further developed to exploit diggability maps based on geological maps published by the British Geological Survey. These diggability maps were primarily developed as a commercial product to provide information for utilities companies that require information at any given location on how diggable the ground is where they wish to place an underground cable or pipe, ensuring they choose the easiest and most cost-effective method and location. The datasets divide the ground into bands of diggability using, for instance, hand tools through to large plant machinery. While it is accepted that this may be a subjective test of diggability, the datasets have been used reliably by utility companies throughout the UK. During a recent police search for a homicide victim, it was requested that the British Geological Survey apply a RAG filter to their dataset, highlighting in 'red' those areas of the ground that could be dug with ease using a pick or shovel. This enabled appropriate levels of resources to be deployed to search the areas denoted 'red' that were most likely to contain a burial and resulted in the successful location of the victim (Fig. 2 and Table 4).

Table 4. *Diggability for selected soils and rocks (see Fig. 2)*

Deposit type	Rating	Diggability description
Alluvium/alluvial floodplain area	Red	Clay and peat soils are diggable. Progress will become more difficult with depth owing to presence of gravel beds. Gravel will be encountered at *c.* 1 m depth
Glacial till	Amber	Loamy to clay soils are diggable at surface, becoming more dense clay with depth. Glacial till deposits can become stiffer at depth where less weathered
Oxford clay (bedrock)	Green	Loamy to clay soils are diggable at surface, but quickly become very heavy clay and mudstone at typically <1 m

Discussion on the application of RAG maps

It is of note that the RAG system as employed by the UK police is at variance with some geologists' interpretation of RAG, in that in policing terms 'red' denotes the area most diggable or where the object is most likely to be located, and 'green' is the least diggable, 'amber' being midway between the two. This became apparent during the search for Anglo Saxon Gold in UK in 2009. The geologist developed a colour-coded RAG map depicting 'green' as the 'go' (i.e. principal search area), whereas the police search manager regarded 'red' as the area of greatest priority.

This variance requires a geologist assisting the UK police to ensure that a clear title and legend accompanies any RAG map produced with the default being to the client's preference, that is the police, law enforcement officer and/or military. In a country such as the UK where the diggability of the ground can be highly variable, the use of diggability RAG maps can be of value, especially in investigations searching for a buried homicide victim or terrorists/organized crime groups who have buried caches of drugs, weaponry and explosives. In these investigations the diggability RAG map can aid the police in prioritizing large search areas or challenging intelligence that indicates a

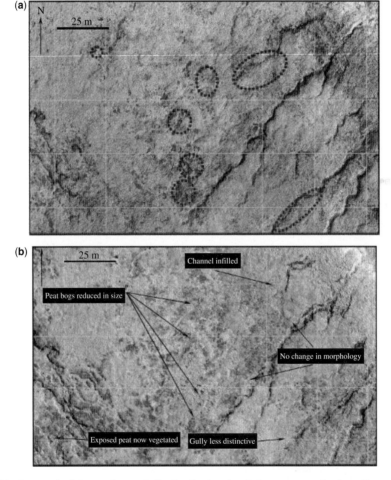

Fig. 3. (a) Air photograph of the search area obtained at the time the victim was reported to have disappeared. (b) Air photograph of the same search area several years later. A geological and geomorphological analysis of the two images shows those parts of the ground where there has been a change in morphology and areas where the ground surface has not changed. Most of the changes were caused by geomorphological processes such as weathering and erosions or by the colonization of peaty soils and rocks by vegetation. These observations may have implications on the search strategy because soils exposed in the past may now have become vegetated.

possible location by providing quantitative and qualitative information. Clearly, if intelligence suggests a specific geographical location and the diggability RAG map suggests that, owing to the ground conditions, digging would be very difficult unless heavy plant machinery was used, then the investigator can reassess the value placed on the intelligence provided by an informant and thus save time and resources on speculative searches.

The 'go/no go' system used by the military also encountered comparable difficulties to those described above with the police. For example, where 'go/no go' information was required for tanks, the green depicted 'go' but tired infantry soldiers interpreted this as woodland and therefore 'good for cover'. Since defensive positions were often occupied during the hours of darkness, this caused problems when dawn broke.

The limited availability of inks during printing in the First World War was also significant since poor colour retention occurred during printing. Some colours also did not reproduce well during photocopying, such as yellow, or became transparent when viewed under artificial light, such as low-power infrared (Mike Rosenbaum pers. comm.).

Colour-coded maps were developed for trench dug-outs in 1916; however, 'red' was used for 'dry' and therefore 'good' conditions and 'green' for wet or 'poor' conditions. This may be the origin of the reversal of the colour scheme, which permeated into the police force when they adopted RAG in the 1980s.

Case examples

In the following examples details of past criminal cases have not been included and/or some operational case details have been purposely changed or made anonymous. In addition, some photographs have been substituted to enable cases to be presented. Furthermore, the RAG maps produced for cases prior to 2009 (i.e. cases 1 and 2) have been changed to 'red' from the original 'green' to denote the areas of high-priority search.

Geological RAG map to search for a spade and unmarked grave

In the 1990s a search was conducted in the UK for a person who had been reported missing several

Fig. 4. Air photograph showing the general layout of a gully that was suspected to contain a homicide grave and a spade. This shows a principal stream, which has incised an otherwise flat to gently tilted peat-covered slope.

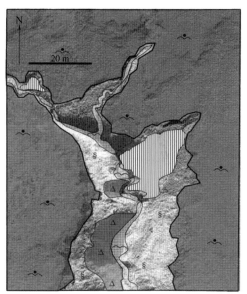

Key to symbols
- Peat
- Alluvium
- Scree
- Glacial head
- Landslide

Fig. 5. Interpretation of air photographs and geological maps published by the British Geological Survey enabled the principal types of superficial deposits to be identified.

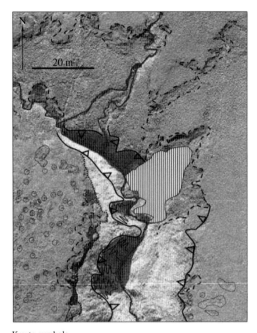

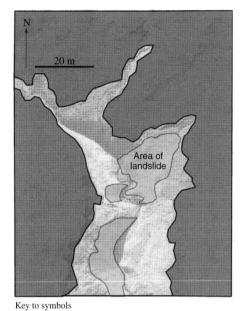

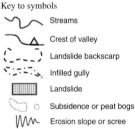

Fig. 6. Geomorphological interpretation of the landscape to show the location of natural (geological) ground disturbances.

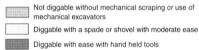

Fig. 7. Diggability RAG map based on the estimated engineering properties of the superficial deposits.

years previously. The objective of the search was to locate a body and a spade. The intelligence suggested that this person may have been murdered and subsequently buried in a shallow grave in a remote upland location. Little additional intelligence was available and the designated search area was considerable, covering at least several square kilometres of moorland and terrain.

Black and white aerial imagery had been taken of the search area at the time the person was originally reported as missing in the 1960s. New photographs were obtained and analysed, paying particular attention to the geomorphology. Comparison of the photographs showed that some significant natural changes had taken place to the ground surface (Fig. 3). This included the reduction in size of peat bogs and the localized natural infilling of incised stream channels. These changes were possibly caused by landslides and increased colonization by vegetation, possibly associated with changes in air quality since the reduction in the burning of fossil fuels in near-by urbanized and industrialized areas. Furthermore, natural geomorphological processes, namely stream erosion, had reduced since the burial had taken place, possibly as a consequence of the reduction in annual rainfall during the intervening years. These observations were significant since areas of peaty soil were exposed at the time the person went missing that would have been diggable with ease at that time, but are now covered with vegetation and therefore more difficult to dig (Fig. 3).

The upper reaches of an incised stream valley was identified as a possible location for the burial. The site measured approximately 100 m long, 50 m wide and 20 m deep (Fig. 4). Aerial imagery and geological maps published by the British Geological Survey were obtained for analysis. The gully had eroded through a plateau, which was covered with a veneer of peat up to 4 m thick. Glacial head deposits (loose angular rocks covering slopes deposited by the thawing of ground ice and the subsequent down-slope movement of groundwater and debris at the end of the last Ice Age) covered the middle and upper slopes. By comparison, exposed sedimentary rocks, mainly shales

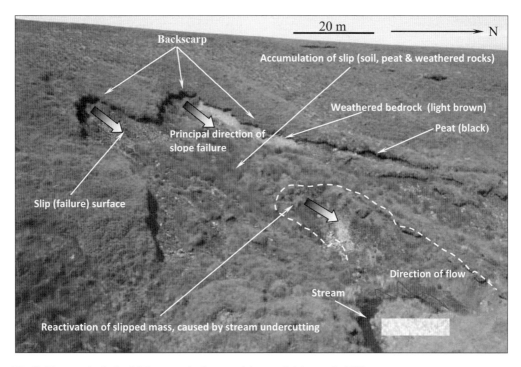

Fig. 8. Photograph of a landslide present in the area of the search (view to the NW).

Fig. 9. Photograph of a landslide that was present in the area of the search (view to the SW).

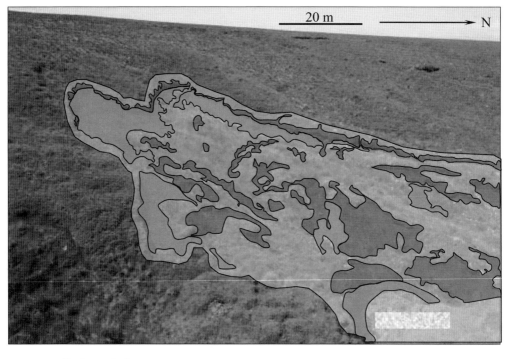

Key to symbols

☐ Not diggable without mechanical scraping or use of mechanical excavators

☐ Diggable with a spade or shovel with moderate ease

☐ Diggable with ease with hand-held tools

Fig. 10. Diggability survey based on the estimated engineering properties of the superficial deposits, verified by *in situ* field observations and the performance of diggability tests (view to the NW).

with some siltstone sequences, were exposed on the side of the valley with scree slopes (accumulations of fallen debris at the toe of the rock slopes) that had been subjected to subsequent removal by stream erosion (Fig. 5).

The crest of the valley (the point where the plateau intersects the upper valley sides) was distinct and numerous ground disturbances were observable on the air photographs. However, the majority of these disturbances were consistent with natural geomorphological processes, which operate on upland slopes, including for example weathering and erosion, mass wasting (soil creep, peat slides and landslides) and peat subsidence (Fig. 6).

Once the types of soils had been determined, their (engineering) properties and behaviour could be estimated (using Table 2). This showed areas of exposed peat, which offered the easiest and most likely soil type for the quick and efficient digging and reinstatement of a burial. Parts of the valley sides and floor were considered to be diggable with moderate ease, but the exposed rock slopes and stream bed were not diggable with the hand-held tool (a spade) believed to have been used by the offender (Fig. 7).

It should be noted that geomorphological analysis and landscape interpretation require different or complementary skills when compared with the engineering geological classification of soils and sediments. In some cases, this may require a competent geomorphologist and/or engineering geologist to be consulted.

The conceptual diggability map was produced during the field reconnaissance visit to the search area (Fig. 7). During the desk study, a landslide on the western flank of the valley had occurred. This was initially envisaged to consist of a mixture or boulders and cobble-sized rocks and it was judged not to be diggable with a spade. However, the field observations enabled the direct inspection of the

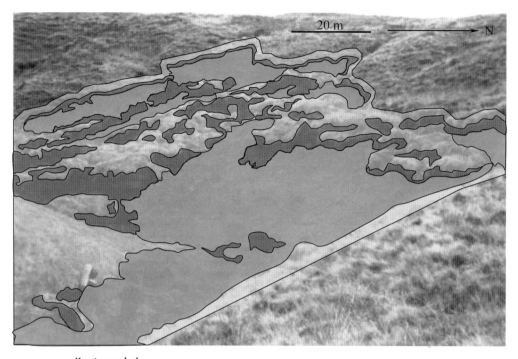

Key to symbols

- Not diggable without mechanical scraping or use of mechanical excavators
- Diggable with a spade or shovel with moderate ease
- Diggable with ease with hand-held tools

Fig. 11. Diggability survey based on the estimated engineering properties of the superficial deposits, verified by *in situ* field observations and the performance of diggability tests (view to the SW).

Fig. 12. Aerial photograph showing the general layout of an area of farmland that was suspected to contain a homicide grave.

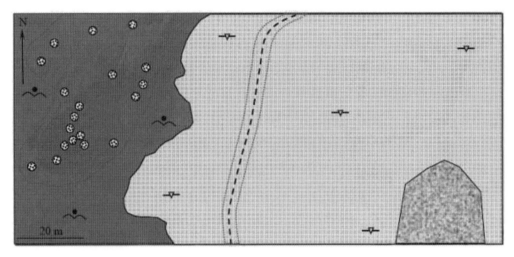

Key to symbols

- Peat
- Glacial till
- Farm and associated buildings
- Approximate outcrop of a coal seam
- Estimated area of possible shallow, abandoned coal and/or ironstone mine workings
- Bell pits associated with coal and/or ironstone mining

Fig. 13. Interpretation of aerial photographs and geological maps (published by the British Geological Survey) enabled the principal types of superficial deposits and areas of past mining to be identified.

landslide, which led to the reclassification of its diggability since pockets of peat were found amongst the displaced debris that were diggable with ease (Figs 8 & 9).

The backscarp and upper failure (slip) surface of the landslide were exposed and consisted of slickenside (striations on the bedrock, caused at the time of failure and indicating the direction of failure), strong, weathered silty sandstone substrate, which was not diggable with a spade, whereas the accumulated slumped mass of peat and glacial head deposits were diggable with ease using a spade, confirmed by *in situ* field tests. This enabled a reinterpretation of the initial conceptual diggability map and the production of preliminary diggability maps for the peat slide. The use of a simple RAG colour-coded system enabled an efficient identification and representation of the primary and secondary search areas (Figs 10 &11).

Following the validation and ground-truthing of the diggability maps, behavioural profiling of the offender, geographical profiling and victimology assessments enabled the identification of the most likely areas where a burial may be located. This included the exposed peat banks along the valley crests and displaced peat within the slipped mass of the peat slide. Resources were therefore focused on these areas and the search found a number of 'items', including a spade.

Interpretation of aerial imagery to help locate a homicide burial

Police had reason to suspect that a person had been murdered and buried on agricultural land in the UK. Aerial photographs were obtained, which showed the general area where the body was believed by the police to be buried (Fig. 12). The

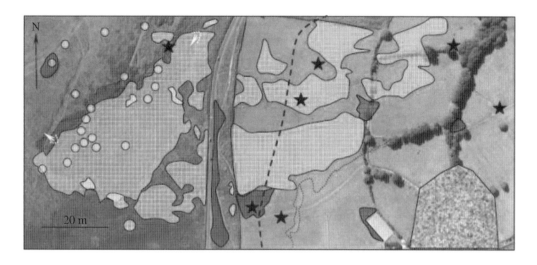

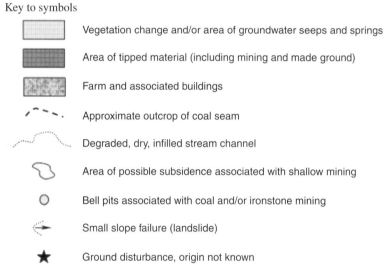

Fig. 14. Geomorphological interpretation of the landscape to show the location of natural (geological) and anthropogenic (man-made) ground disturbances.

police asked whether it was possible to locate the burial site from the aerial photographs. A geological assessment of the photographs was therefore undertaken. Analysis of geological maps for the site, published by the British Geological Survey, showed the ground to consist of Coal Measures strata (alternating sequences of sandstones, shales, siltstones and coal seams) overlain by peat or glacial till (Fig. 13). This enabled the identification of ground disturbances and an assessment of their likely cause. Many of these were explainable as being possibly associated with past mining, landslides, weathering and erosion. For instance, abandoned mine shafts were observable as circular features in the northwestern part of the photograph. These were approximately 1–3 m in diameter, with waste materials tipped concentrically around the shafts. There was no evidence for the recent disturbance of the shafts or the waste. Ground disturbances were also noted along the outcrop of a coal seam, trending roughly north–south in the centre of the photograph, and this was also attributed to past mining activities.

Most of the discoloured vegetation was attributed to groundwater seeps and springs caused by the juxtaposition of permeable and impermeable rocks that crop out in the Coal Measures strata. Other ground disturbances and tipped materials

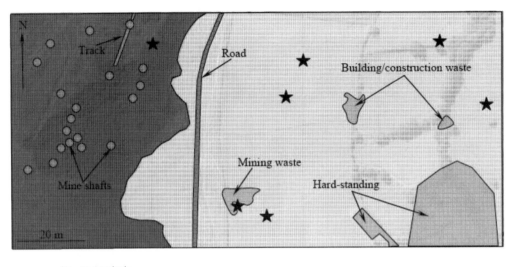

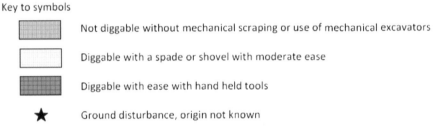

Fig. 15. Diggability survey based on the estimated engineering properties of the superficial deposits and an assessment of past mining and geomorphological process. The star symbols denote ground disturbances which cannot be explained by mining activities, geological or geomorphological processes.

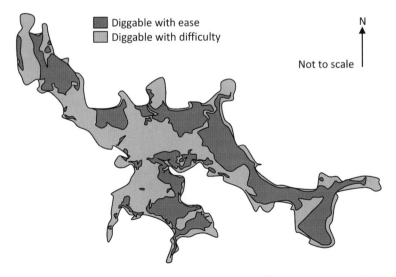

Fig. 16. Diggability map of a remote Scottish island to identify areas most likely to conceal a shallow grave (in red).

Fig. 17. Photos left to right. (**a, b**) Diggability survey being conducted in different sand deposits to determine the ease of digging and the time taken to dig and reinstate a shallow grave. (**c**) Geophysical survey being conducted to search for a homicide grave. (**d**) Positive dog indication at the shallow, unmarked grave.

were identified and interpreted as being possibly associated with the construction of the farm buildings, roads and access tracks. Small landslides and degraded (possibly post-glacial) stream drainage channels were also identified, but there was no evidence to suggest that they been subjected to any recent digging.

There were six ground disturbances that could not be explained by mining, geomorphological process, building and construction works (Fig. 14). An assessment of the diggability of the ground showed that all of these anomalies were diggable with a spade, shovel or similar hand-held implement with moderate ease (Fig. 15). A RAG map allowed the identification of the primary search areas. These locations were identified as a priority and their subsequent exhumation revealed the body of the victim at one of these locations.

Detailed geomorphological analysis coupled with the interpretation of anthropogenic features were both important in this case example, not simply whether the ground could be easily dug. It was the combined interpretation and explanation of all aspects of the landscape that was of value for this police search.

Diggability, geological and RAG maps to aid search for a homicide beach burial

In 2009 an adult male was reported missing on a remote Scottish island. The police treated the disappearance as suspicious and suspected homicide. As the investigation progressed suspects were identified, arrested and interviewed. By analysis of the evidence it was believed that the missing person had been murdered and buried somewhere on the island.

The challenge presented to the police was that the island contained large areas of open ground without trees or other foliage and had many beaches in which traditional methods of over-flights or visual inspection searches of the land would be unlikely to detect a burial. At the early stages of the search a CGM was provided by the geologist to the police search strategist. Furthermore, the collation and analysis of geological maps was recommended and geological advice was also given on likely search strategies and instrumentation.

Diggability surveys were conducted and from these observations, diggability RAG maps were produced (Fig. 16), applying a filter for the use of hand-held tools only. This diggability RAG map then formed the basis for field reconnaissance surveys. This was combined with geological trace evidence collected from a vehicle believed to have been used in the victim's transportation prior to burial, which identified the sand type. From this information a beach was identified as the most probable burial location. This location was further defined by considering the behavioural profile patterns offenders have previously used when faced with similar circumstances, enabling a smaller section of the beach to be chosen for a more detailed search (Fig. 17).

The interpretation of geological maps within a diggability RAG context facilitated the communication of complex data to the police in a readable manner, which greatly assisted the efficiency of prioritizing the search area. The diggability RAG map also enabled the police to discount information provided by the suspects of where the police should search as false and malicious and thereby prevent loss of evidence and prolong the suffering of the victim's relatives. It should be noted that, in addition to geologists, there was additional support provided by a variety of other specialists.

The search was conducted using a team operating near-surface geophysical instruments, thermal imaging and human remains-detecting canine resources. The body of the victim was located buried in sand after a few hours of searching. It was at a depth of less than 1 m and was fully clothed. There were no visual surface topographic features, such as scarring of the sand or any other disturbance to suggest that the burial had taken place.

Conclusions

This paper has demonstrated how the collation and analysis of geological maps, high-resolution aerial imagery, the determination of the diggability of the ground and the production of RAG prioritization maps may facilitate the search for criminal sub-surface concealments.

Aerial photographs, when analysed by a trained geologist/geomorphologist, may enable the identification of ground disturbances. These may subsequently be explained as being caused by natural geomorphological processes (e.g. landslides or stream erosion), anthropogenic activities (e.g. mining, farming, building or construction) or associated with the possible digging of a burial.

Where a sub-surface concealment may have been dug in the past (years to decades), the geological analysis of aerial imagery can verify whether the landscape and ground conditions may have changed since the concealment took place, although this may be more difficult in urbanized and densely populated areas.

In a law enforcement context diggability can be considered as a measure of the ease by which a burial may be dug and the displaced soil reinstated. This will depend on several factors, such as geological characteristics of the burial site and in particular the associated engineering properties of the soils, sediments or superficial deposits. Most burials are in soil owing to its relative ease of digging. However, the diggability of the soil is dependent on whether the material is primarily granular, cohesive, organic or anthropogenic, and the presence of groundwater. Diggability will also be influenced by the type of digging implement used by the offender(s) (such as a spade, pick, mattock or mechanical excavator).

Colour-coded maps using the RAG system may be convenient in some cases for determining and readily communicating higher and lower priority search areas, ensuring that search resources are deployed in the most efficient and effective manner. To reduce the possibility for confusion, all colour-coded maps must have a title, key and legend to explain the RAG system employed, clarifying which colours denote the most and the least diggable areas. It is not advocated that the use of a simple RAG map is appropriate in all cases, but in some situations, it may be of value as a basic means of communication between the geologist and police officer. However, the RAG map must be used in conjunction with all other search techniques (e.g. behavioural profiling) and not in isolation.

It is considered that the CGM is a more useful tool to assist with the design of some ground searches. Law enforcement officers would benefit from being made aware of how RAG, diggability maps and CGM are developed and the decision process involved, and understanding their advantages and limitations and that a CGM is a model, to be tested and updated as more geological information becomes available during the search. A geologist therefore may, in some cases, positively influence search strategies and their implementation by allowing search assets and resources to become more focused at locations where burials, hides and concealed objects may be located.

The authors would like to acknowledge the continued assistance provided by the following: M. Culshaw (formerly, British Geological Survey), M. Jenkins (formerly UK military), A. Ruffell (Queens University Belfast), R. Ellison, R. Lawley, K. Northmore and J. Trick

(British Geological Survey). The authors are particularly grateful to T. Rose, A. Hatheway and M. Rosenbaum for helpful communication during the production of this paper; and to IUGS Initiative on Forensic Geology and the Geological Society, London, Forensic Geoscience Group.

References

ABBOTT, D. M. 2005. Investigating mining fraud. *Geotimes*, January, 30–32.

ANON 2006. *In*: HARRISON, M., HEDGES, C., SIMS, C. (eds) *Practice Advice on Search Management and Procedures*. National Police Improvement Agency, Bedfordshire, UK.

ANON 2008. *In*: JONES, D. M. (ed.) *Geophysical Survey in Archaeological Field Evaluation*. English Heritage Publishing, Swindon.

BERGSLIEN, E. 2012. *An Introduction to Forensic Geoscience*. Wiley, New York.

DEARMAN, W. R. 1991. *Engineering Geological Mapping*. Butterworth-Heinmann, Oxford.

DONNELLY, L. J. 2007. Engineering geology of landslides on the volcanic island of Montserrat, West Indies. *Quarterly Journal of Engineering Geology & Hydrogeology*, **40**, 267–292.

DONNELLY, L. J. 2008. Communication in geology: a personal perspective and lessons from volcanic, mining, exploration, geotechnical, Police and geoforensic investigations. *In*: LIVERMAN, D. G. E., PEREIRA, C. P. & MARKER, B. (eds) *Communicating Environmental Geoscience*. Geological Society, London, Special Publications, **305**, 107–121.

DONNELLY, L. J. 2009a. Geoforensics in Colombia. *Police Professional. Investigative Practice Journal*, May, 1–2.

DONNELLY, L. J. 2009b. The geological search for a homicide grave. *The Investigator*, July/August, 42–49.

DONNELLY, L. J. 2010. The role of geoforensics in policing and law enforcement. Emergency Global Barclay Media Limited, Custom Print, Manchester, January, 19–22.

DONNELLY, L. J. & HARRISON, M. 2010. Development of geoforensic strategy & methodology to search the ground for an unmarked burial or concealed object. Emergency Global Barclay Media Limited, Custom Print, Manchester, July, 30–35.

DONNELLY, L. J., JONES, L., PALMER, M. & DILKES, C. 2006. Engineering geological and geotechnical aspects of the Soufrière Hills volcanic eruption, Montserrat. *In*: CULSHAW, M. G., REEVES, H., SPINK, T. & JEFFERSON, I. (eds) *Engineering Geology for Tomorrow's Cities*. 10th IAEG International Congress, Nottingham, 6–10 September, Theme 4, paper 114. Geological Society, London.

FENNING, P. J. & DONNELLY, L. J. 2004. Geophysical techniques for forensic investigations. *In*: PYE, K. & CROFT, D. (eds) *Forensic Geoscience: Principles, Techniques and Applications*. Geological Society, London, Special Publications, **232**, 11–20.

FOWLER, M. J. F. 2004. Archaeology through the keyhole: the serendipity effect of aerial reconnaissance revisited. *Interdisciplinary Science Reviews*, **29**, 118–134.

GOMEZ-LOPEZ, A. M. & PATINO-UMANA, A. 2006. Whose is missing? Problems in the application of forensic archaeology and anthropology in Colombia's conflict. *In*: FERLLINI, R. (ed.) *Forensic Archaeology & Human Rights*. Charles C. Thomas, Springfield, IL.

GRIFFITHS, J. S. 2009. Reply to discussion by R. J. G. Edwards on the paper 'Some limitations in the interpretation of vertical stereo photographic images for landslide investigations' by A. B. Hart, J. S. Griffiths & A. E, Mather. *Quarterly Journal of Engineering Geology & Hydrogeology*, **42**, 261–264.

GRIPP, W. M., GRIPP, R. W. & MORRISON, R. 2000. Application of aerial photography in environmental forensic investigations. *Environmental Forensics*, **1**, 121–129.

HARRISON, M. 2008. How can geologists aid the understanding of a crime scene? *Police Professional Investigative Practice Journal*, 25 September, 22.

HARRISON, M. & DONNELLY, L. J. 2008. Buried homicide victims: applied geoforensics in search to locate strategies. *The Journal of Homicide and Major Incident Investigations* (Produced on behalf of the Association of Chief Police Officers Homicide Working Group, by the National Policing Improvement Agency).

HARRISON, M. & DONNELLY, L. J. 2009. Locating concealed homicide victims; developing the role of Geoforensics. *In*: RITZ, K., DAWSON, L. & MILLER, D. (eds) *Criminal and Environmental Soil Forensics*. Springer, Berlin, 197–219.

INSTITUTION OF ROYAL ENGINEERS 1922. *Geological Work on the Western Front*. Institution of Royal Engineers, Plate IV.

KIRSTEN, H. A. D. 1982. A classification system for excavation in natural materials. *The Civil Engineers in South Africa*, **24**, 293–308.

MCCANN, D. M. & FENNING, P. J. 1995. Estimation of rippability and excavation conditions from seismic velocity measurements. *In*: EDDLESTON, M., WALTHALL, S., CRIPPS, J. C. & CULSHAW, M. G. (eds) *Engineering Geology of Construction*. Geological Society, London, Special Publications, **110**, 335–343.

MCCANN, D. M., CULSHAW, M. G. & NORTHMORE, K. J. 1990. Rock mass assessment from seismic measurements. *In*: BELL, F. G., CULSHAW, M. G., CRIPPS, J. C. & COFFEY, J. R. (eds) *Field Testing in Engineering Geology*. Geological Society, London, Special Publications, **6**, 257–266.

MURRAY, R. C. 2004. *Evidence from the Earth*. Mountain Press, Missoula, MT.

MURRAY, R. C. 2005. Collecting crime evidence from the earth. *Geotimes*, January.

MURRAY, R. C. & SOLEBELLO, L. P. 2005. Forensic examination of soil. *In*: SAFERSTEIN, R. (ed.) *Forensic Science Handbook*, 2nd edn., **1**. Prentice Hall, Englewood Cliffs, NJ.

MURRAY, R. C. & TEDROW, J. C. F. 1992. *Forensic Geology*. Prentice Hall, Englewood Cliffs, NJ.

PETTIFER, G. S. & FOOKES, P. G. 1994. A revision of the graphical method for assessing the excavatability of rock. *Quarterly Journal of Engineering Geology*, **27**, 145–164.

PINSKER, L. M. 2002. Geology adventures in Afghanistan. *Geotimes*, We Feature, February, http://www.geotimes.org/feb02/Feature_Shroderside.html (accessed 13 March 2003).

PYE, K. & CROFT, D. (eds) 2004. *Forensic Geoscience: Principles, Techniques and Applications.* Geological Society, London, Special Publications, **232**.

RITZ, K., DAWSON, L. & MILLER, D. (eds) 2008. *Criminal and Environmental Soil Forensics.* Springer, Berlin, 197–219.

ROSE, E. P. F. 2004. The contribution of geologists to the development of emergency groundwater supplies by the British Army. In: MATHER, J. D. (ed.) *200 Years of British Hydrogeology.* Geological Society, London, Special Publications, **225**, 159–182.

ROSE, E. P. F. 2008. British military geological terrain evaluation for Operation Overlord: the allied invasion of Normandy, in June 1944. In: NATHANIAL, C. P., ABRAHART, R. J. & BRADSHAW, R. P. (eds) *Military Geography & Geology: History & Technology.* Land Quality Management Press, Nottingham.

ROSE, E. P. F. 2009. Credit Due to the Few: British Field Force Geologists of World War II. *Abstracts of the papers presented at the Open Meeting of the History of Geology Group*, Geological Society, London, 23.

ROSE, E. P. F. & CLATWORTHY, J. C. 2007. Specialist maps of the Geological Section, Inter-Service Topographical Department: aids to British Military planning during World War II. *Geographical Journal*, **44**, 13–43.

ROSE, E. P. F. & CLATWORTHY, J. C. 2008a. Fred Shotton: a hero of military applications of geology during World War II. *Quarterly Journal of Engineering Geology & Hydrogeology*, **41**, 171–188.

ROSE, E. P. F. & CLATWORTHY, J. C. 2008b. Terrain evaluation for Allied Military operations in Europe and the Far East during World War II: 'secret' British reports and specialist maps generated by the Geological Section, Inter-service Topographical Department. *Quarterly Journal of Engineering Geology & Hydrogeology*, **41**, 237–256.

ROSE, E. P. F. & PAREYN, C. 1996a. Roles of Sapper geologists in the liberation of Normandy, 1944, Part 1. Operational planning, beaches and airfields. *Royal Engineers Journal*, **110**, 36–42.

ROSE, E. P. F. & PAREYN, C. 1996b. Roles of Sapper geologists in the liberation of Normandy, 1944, Part 2. Quarries, water supply, bombings and cross country movements. *Royal Engineers Journal*, **110**, 138–144.

ROSE, E. P. F. & ROSENBAUM, M. S. 1993. British military geologists: through the Second World War to the end of the Cold War. *Proceedings of the Geologists' Association*, **104**, 95–108.

ROSE, E. P. F. & ROSENBAUM, M. S. 2011. British geological maps that guided excavation of military dug-outs in Belgium during World War I. *Quarterly Journal of Engineering Geology and Hydrogeology*, **44**, 293–306.

RUFFELL, A. R. & MCKINLEY, J. 2008. *Geoforensics.* Wiley-Blackwell, Chichester.

SCOBLE, M. J. & MUFTUOGLU, Y. V. 1984. Derivation of diggability index for coal measure excavations. In: *Proceedings of the 27th US Symposium on Rock Mechanics, Tuscoloosa*, AIME, New York, 443–448.

SILVESTRI, S., OMRI, S. & ROSSELLI, R. 2005. The use of remote sensing to map illegal dumps in the Veneto Plain. In: *Proceedings Sardinia 2005, Tenth International waste Management and Landfill Symposium, Santa Margherita di Pula, Cagliari, Italy, 3–7 October 2005*. CISA Environmental Sanitary Engineering Centre, Italy.

SMITH, A. & ELLISON, R. A. 1999. Applied geological maps for planning and development. A review of examples from England and Wales. *Quarterly Journal of Engineering Geology & Hydrogeology*, **32**, S1–S44.

TOPPING, P. 1989. *Topping, The Autobiography of the Police Chief in the Moors Murder Case.* Angus & Robertson, London.

WAR OFFICE 1949. *Application of Geology.* War Office, London, chapter 6 section 53, Restricted.

WEAVER, J. M. 1975. Geological factors significant in the assessment of rippability. *The Civil Engineer in South Africa*, **17**, 313–316.

WILLIG, D. & HÄUSLER, H. 2012. *Aspects of German Military Geology and Groundwater Development in World War II.* Geological Society, London, Special Publications, **362**, 187–202.

The design and implementation of a high-assurance forensic geology and police search following the discovery of the Staffordshire (Anglo Saxon) Gold Hoard

LAURANCE DONNELLY

IUGS, Initiative on Forensic Geology, Wardell Armstrong International, 2 The Avenue, Leigh, Greater Manchester WN7 1ES, UK (e-mail: ldonnelly@wardell-armstrong.com)

Abstract: In 2009 a hoard of gold and silver objects was found in a ploughed field in Staffordshire, by a member of the public using a metal detector. The site was subjected to a detailed archaeological excavation and approximately 3940 items in total were found. Archaeologists interpreted the find as belonging to the Anglo Saxon age (seventh century AD) and probably comprising the military hilts and fixings from swords, helmets, shield, clothing and possibly books, chests and what is now thought to be a cross from the cover of a bible. Archaeologists considered that all hoard-related material that was recoverable at that time had been retrieved from the excavation. To confirm this, a forensic geology and police search was commissioned. This search provided a high level of assurance and was able to confirm that the original archaeological dig was likely to have found all/most of the buried gold that was reasonably and practicably recoverable at that time and buried in the top soil to a depth of 280 mm. In 2012, further items of interest were found in this field. These may have been buried at deeper levels or beyond the original excavation and were possibly brought to the surface by ploughing.

On 5 July 2009, an amateur metal detectorist found a collection of gold artefacts buried at a shallow depth in a recently ploughed field, at Ogley Hey in the Lichfield district of Staffordshire. Numerous additional similar items were found (by the same metal detectorist) in the days that followed. In accordance with the Portable Antiquities Scheme and the Treasure Act (1996), these were reported to the Finds Liaison Officer for Staffordshire, who in turn informed the Historic Environment Team at Staffordshire Council.

The hoard was identified by archaeologists as being the military hilts and fittings from swords, helmets, shield, clothing and/or books and chests, the possible cover decoration cross from an early medieval bible and a bishop's pectoral cross. An archaeological survey of the field was subsequently conducted by Birmingham Archaeology. This included topographical and geophysical surveys and the digging of the single area in an open excavation. Approximately 1600 pieces were initially recovered in total. The hoard contained about 5 kg of gold, along with 1.5 kg of silver and some copper items. Many of these were aesthetically beautiful and contained fine, intricate carvings, decorations and Latin inscriptions giving verses from the Bible. This was interpreted by the archaeologists as a unique find, belonging to the Anglo Saxon period. There are no other finds of this type in terms of the quality of preservation, age and quantity. It was also potentially the most important find ever relating to the Mercian Kingdom, which prevailed in central England during the seventh century AD. It was estimated to be of vast national and international, historical and cultural significance.

Given the significance of this find, Staffordshire Council contacted Staffordshire Police for advice on how the site may be secured and searched to help confirm that all the items reasonably recoverable had been found or to recover any additional gold items that may still remain buried beneath the ground surface at a shallow depth. The police in turn called upon the services of the Home Office Scientific Development Branch (HOSDB; now the Centre for Applied Science and Technology, CAST). On 10 September 2009 HOSDB contacted the author for advice on what geological techniques and methods may be deployed to search the field to confirm that all of the gold items had been recovered by the archaeologists. The anticipated issuing of the coroner's report was envisaged to make the finds public knowledge, potentially attracting other metal detectorists and 'treasure hunters' to the field. Since the time frames were very short, it was therefore desirable for the search to be conducted immediately.

The hoard was later declared as treasure at a coroner's inquest held at Cannock on 24 September 2009. The objectives of the geological investigation were to rapidly design and deploy a ground search and to recover any additional gold artefacts or to prove, so far as possible, the absence of any further gold items buried at a shallow depth beneath the ground in the immediate vicinity of

From: PIRRIE, D., RUFFELL, A. & DAWSON, L. A. (eds) 2013. *Environmental and Criminal Geoforensics*.
Geological Society, London, Special Publications, **384**, 195–208.
First published online July 22, 2013, http://dx.doi.org/10.1144/SP384.8
© The Geological Society of London 2013. Publishing disclaimer: www.geolsoc.org.uk/pub_ethics

the location where the Anglo Saxon Gold Hoard items were found.

Geophysical surveys

Geophysics is routinely used by geologists, archaeologists, the military and the police during ground searches for burials to help locate, for example, graves, weapons and fire-arms, explosive devices and items of value. A variety of instruments are available, including conventional metal detectors, ground penetrating radar, and magnetic, electromagnetic, resistivity and conductivity techniques. Less common methods available for use in forensic investigations include microgravity and seismic methods. Geophysical investigations may be airborne or ground-based, the latter normally being most suitable for forensic ground searches. Although geophysical surveys tend to be non-invasive, their deployment may potentially cause ground disturbance by the operators and therefore this has to be carefully assessed if other forensic evidence may potentially exist. The choice of geophysical instruments to be used should only be determined following a detailed analysis of the geology of the burial and the physical properties of the target. It is recommended that this is undertaken by the development of a conceptual geological model (CGM). Once the choice of instrumentation has been determined, a suitable control site should be established to test that the equipment is fully operational, determine the instruments detection limits and resolution and provide the opportunity for the operator to become familiarized with the instrument(s). The outer limits of the search area may then be defined before search lanes (or search sectors) are then cordoned off using GPS and high-visibility tape or rope. Geophysics will not normally identify the type of buried item, but instead provides a location and depth of the physical anomaly. Geophysical surveys have a resolution and detection limit and this must be determined and understood. Following the positive identification of a geophysical anomaly (sometime known as a 'hit'), invasive works and recovery are normally conducted by a forensic archaeologist, working closely with the forensic geologist. Further technical and more scientific information on the use of geophysics in searches is beyond the scope of this paper but may be found in Fenning & Donnelly (2004), Ruffell & McKinley (2008), Hunter & Cox (2005), Pringle et al. (2012), Linford (2004) and Walkington (2010).

Location

The field is located in the parish of Ogley Hey, south of Hammerwich, approximately 6 km to the SW of Lichfield and 3.5 km to the west of Wall, in the Lichfield district of Staffordshire. The field is bounded to the north by the A5 (Watling Street), on its eastern side by the M6 Toll and to the west by the B4155 (Haney Road). The field where the hoard was found is approximately 5.5 ha in size. It was under a crop of hay stubble at the time of the find and the search, but in the past has been used for a variety of crops, including potatoes, wheat, barley and carrots. The field boundary consists of hedges and wire fences and posts, the main access gate is to the south and no transmission lines or other utilities cross the site. According to the farmer, between c. 1939 and 1950 the field had had butchers' waste spread on it. Also, potato tops had been burnt across the field at intervals. This variable past land use may influence the usage and suitability of some geophysical techniques and preclude some methods altogether (Fig. 1).

Archaeological investigations

The find

The find was reported to be a scatter of Anglo Saxon objects dating to the seventh century AD.

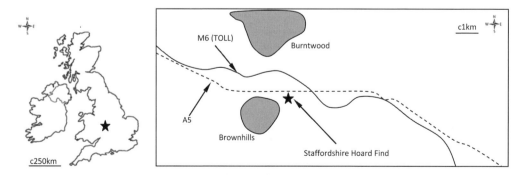

Fig. 1. Location of the Staffordshire (Anglo-Saxon) Gold Hoard.

Following micro-examination of the 26 soil blocks lifted by the detectorist at Birmingham Museum, the final total for the hoard reached 3940 items of treasure. This included over 300 complete pieces recovered and some items 'cemented' together. Most of the items were made from gold, many inlaid with garnets and enamel, although some were made from silver or copper. The complete objects were 20–100 mm in length; the sizes of the fragments varied considerably but were generally <20 mm long. The items were exceptionally well-preserved, although many displayed historic damage probably inflicted when they were stripped from their mounts, which may have included swords and daggers, scabbards, books, armour and possibly furniture. The field had been recently ploughed and all of the artefacts were found in the ploughed soil, which was less than approximately 200–300 mm thick, but thicker in places. There was little evidence of plough damage on the objects, despite the disturbance they were exposed to during ploughing. No pit or feature was found directly associated with the hoard; instead all elements were recovered from within the topsoil. It is hypothesized that, after its burial in a single location on the ridge, a process of physical weathering (including mass wasting and soil creep) had removed soil from the ridge, eroding the feature until the burial site was further degraded during ploughing, bringing the objects into the topsoil and closer to the ground surface. The function of the hoard still remains the focus of conjecture; it may have been a deliberate burial to hide valuable items with the intention to return and reclaim or possibly a ritual burial or safe keeping (Fig. 2).

The Treasure Act (1996)

All searches for treasure must obtain the correct permissions to detect on land and the agreement of the landowner as to what will happen if anything is found. Any finds in Britain that are more than 300 years old and with precious metal content >10% are regarded as 'treasure', along with any other items found associated with them. Treasure hoards and finds belong to the Crown. However, any museums interested in the items may purchase them and pay the finder and the landowner their commercial value (usually split 50:50 between the finder and the landowner). The Treasure Valuation Committee and an independent body of experts assess the value of the finds.

The Staffordshire Hoard is, to date, a unique find for this period. Most hoards come from one or two context: burials or votive offerings. This hoard may be different. The information contained within the hoard shines a light onto a poorly understood but extremely important period in British history and can tell us much about art, material science, trade, warfare, the use of symbols and the role of hierarchy and gift-giving during the early medieval period. The Staffordshire Hoard was declared as treasure at a coroner's inquest held at Cannock on 24 September 2009. The hoard was subsequently sold in November 2009 for £3.3 million.

Overview of initial archaeological investigations

Following the discovery of the gold items a meeting was held on 21 July 2009 at Birmingham Museum and Art Gallery attended by representatives from the Portable Antiquities Scheme and the Historic Environment Team of Staffordshire County Council. It was agreed, with the landowner's and finder's permission, to excavate a 1 m² area in the vicinity of the finds to recover any additional objects and to determine the archaeological context of the finds.

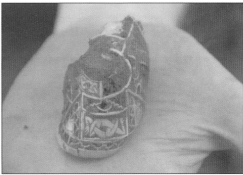

Fig. 2. Oval gold 'brooch' encrusted and inlaid with garnets, found at a shallow depth, just below the soil cover, which was one of the first items to be discovered (© Staffordshire County Council 2009).

The initial excavation took place on 23 July 2009 and large quantities of gold objects were found and recovered. With the agreement of English Nature, a more extensive archaeological excavation was commissioned. The excavation was therefore extended with the support of Birmingham Archaeology and many more gold items were found within the top soil. Further excavations continued for about 2.5 weeks, starting on 28 July 2009. In total an area approximately 9 × 13 m was excavated by a team led by Birmingham Archaeology and Staffordshire County Council. The investigation consisted of the stripping and sieving of the top soil (Fig. 3).

Topographic and magnetic geophysical surveys were conducted by Birmingham Archaeology. The results revealed the presence of a curvilinear 'ditch-like' feature located to the south and west of the find. This was investigated by the digging of an exploratory trench, but no items were found there and there was no evidence to help explain the nature of this geophysical anomaly. Numerous other geophysical anomalies were observed and subsequently investigated in the field, but these were found to be 'background noise', mainly relating to discarded iron objects of no value (perhaps relating to past and recent farming and including old screws and ferrous objects fallen from farming equipment). The archaeological investigation was terminated on 24 August 2009 following a reduced quantity of finds being recovered. The final area of excavation was approximately 12 × 13 m (Figs 4 & 5).

Geoforensic search strategy

In order to deliver a high assurance ground search, the three following strategic phases were adopted:

- Phase 1 – pre-search (briefing, objectives, reconnaissance and a rapid desk study);
- Phase 2 – the actual search; and
- Phase 3 – post-search (debriefing, exit strategy and reporting).

Phase I: pre-search (briefing, objectives, reconnaissance and desk study)

Briefing. On Friday 11 September 2009 an initial briefing meeting was held with the police, Home Office and all interested stakeholders. The background and context of the hoard were discussed and a search of the field was required immediately to confirm that no other gold and silver items were present in the area searched by the archaeologists. The geological search of this field at the place where the gold items were found took place on Monday 14 and Tuesday 15 September 2009.

Search objectives. While the archaeologists were confident that they had recovered everything that it was possible to recover at that point in time, confirmation was required. The objective of the geological search was to design, deploy and manage a systematic operational procedure, combined with appropriate detection equipment, to locate and recover any further gold and silver items that may have not been found in the original archaeological excavation. The search was conducted so far as possible with the understanding of the depth and resolution limitations of the available detection equipment, geophysical instruments and associated technology.

Geology. According to the British Geological Survey's (1919; 1971) published map (*old series,*

Fig. 3. The original archaeological trench, in July 2009 (© Staffordshire County Council 2009).

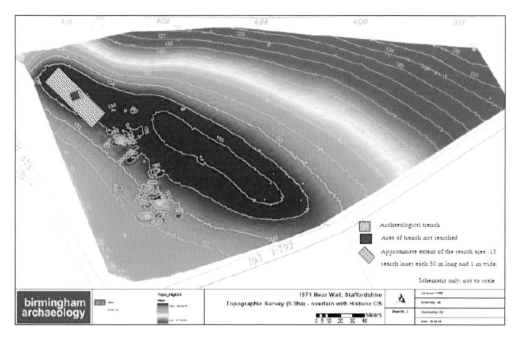

Fig. 4. Topographic survey showing the location of the hoard on the top of a NW–SE trending ridge (modified after original geophysical and surveying work undertaken by Birmingham Archaeology, © Birmingham University 2009).

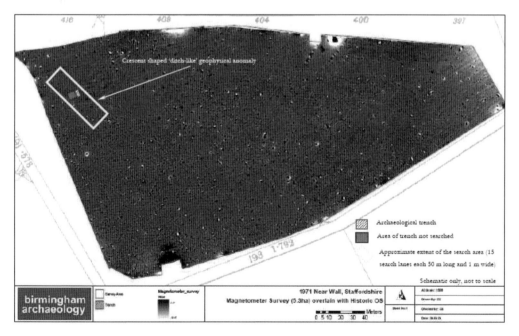

Fig. 5. Geophysical (magnetic) survey showing the location of a crescent-shaped magnetic anomaly in close proximity to the hoard (modified after original geophysical and surveying work undertaken by Birmingham Archaeology, © Birmingham University 2009).

one inch to one mile scale, sheet 154, Lichfield, solid 1926, reprinted 1964 and drift 1926, reprinted 1971), the field where the Anglo Saxon hoard was found is located approximately 2 km to the east of the South Staffordshire Coalfield, which terminates in the east against the north–south-trending Eastern Boundary Fault.

The bedrock comprises Triassic rocks that probably dip gently to the east and rest unconformably on Carboniferous (Westphalian) Coal Measures. The rocks belong to the middle Triassic Sherwood Sandstone Group (known previously as the Upper Mottled (Bunter) Sandstone), which has an extensive outcrop in the Midlands. This is a succession of sandstones, mudstones, marls and siltstones with occasional beds of gypsum and other evaporites. The sandstones are generally red to orange/buff coloured, often cross-bedded (although parallel laminations are also common) and may contain a variety of coarse, well-rounded and high-sphericity pebbles and cobbles. The sandstones are usually poorly cemented, friable and possibly unconsolidated near the ground surface, owing to weathering and erosion, where they grade into superficial deposits and soils.

There is no natural exposure of rock in the field, as the bedrock is overlain by Quaternary deposits and soils. Glacial, periglacial and/or other superficial deposits are distributed unevenly in the region, and they may vary considerably in thickness, type, composition, physical and chemical properties and geotechnical characteristics. These 'drift' deposits may also be variable across the field, but in general probably represent undifferentiated glacial till, deposited during the retreat of the glaciers in the closing stages of the ice age, during the Pleistocene.

The hoard was found on the top of a NW–SE-trending ridge, the site being approximately 280 m long, 80 m wide and 15–20 m high. The ridge is conspicuous, distinct and rises above the general elevation of an otherwise flat to gently undulating plain that dominates this part of central England. The ridge is asymmetrical, dipping more steeply to the west and more gently to the east. Part-way along the crest of the ridge there is a subtle reduction in its height. This may be associated with natural geomorphological processes or it potentially may be related to an anthropogenic excavation or 'ditch-like' feature that has been recorded on a magnetic survey by Birmingham Archaeology. The ridge is probably glacial or periglacial in origin and is composed of red to orange-brown sand and gravel, with cobbles and boulders.

The Sherwood Sandstone Group is a major aquifer, with both intergranular and fracture flow characteristics. Similarly, the granular superficial deposits are likely to have high permeability with the possible presence of perched water tables and fluctuating water levels depending on seasonal rainfall intensity.

Since the burial of the gold items, in the seventh century AD, the site has been subjected to natural geomorphological processes of weathering and erosion. This may have included processes such as mass wasting, including soil creep and sheet wash erosion during times of prolonged heavy rainfall, or frost or snow melt. Rates of soil creep are not currently known for this particular location, although accumulation of displaced soil on the up-slope sides of field boundary fences, to at least 300 mm high, suggests that such processes may be significant in transporting and redistributing soil down-slope. These processes, in association with farming activities (such as ploughing), may have contributed to the displacement and/or exposure of some of the gold objects. Past edition Ordnance Survey maps were also collated and studied, and discussions with the farmer were held to better understand the recent land use changes in the region and to the field.

Reconnaissance site visit. A 'walk-over' survey was undertaken on Friday 11 September 2009 to inspect the geology, observe the ground conditions, better understand active geomorphological processes, assess the topography, obtain an appreciation of the search area and identify any technical, logistical or access constraints (like wire fences, which may influence the choice of geophysical techniques; Fig. 6).

Diggability. During the reconnaissance 'walk-over' survey, *in situ* diggability tests were performed before the main phase of the search, to provide the opportunity to inspect the soil and weathered bedrock and associated superficial deposits on which the gold hoard had been buried (Fig. 7). The Quaternary deposits were unconsolidated, dry, moderately dense and proved to be resistant and difficult to excavate with hand-held implements such as a pick, mattock, shovel or spade. The diggability of the ground was highly variable. On the crest of the ridge, where compacted, granular soils with siliceous and highly spherical pebbles, cobbles and occasional boulders occurred, it was possible to dig an excavation with a spade to a maximum depth of 280 mm, when either boulders (or bedrock) prevented further digging. On the middle and lower parts of the slope it was possible to dig to deeper parts of the soil profile, as the soils were increasingly finer, less dense and contained fewer pebbles, cobbles and boulders.

The shallow depth where the gold objects were found may be related to the presence of dry, compacted, denser soils, creating relatively difficult

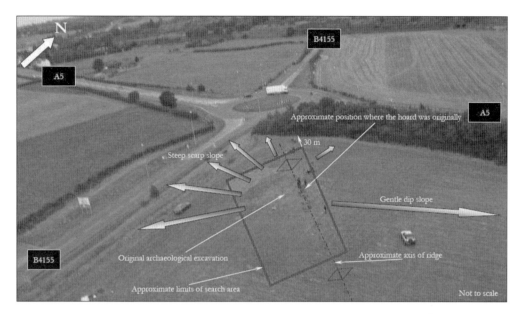

Fig. 6. Oblique air photograph to show the location of the search (air photograph supplied courtesy of Staffordshire Police 2009).

digging conditions. It is possible (although clearly somewhat speculative) that those who 'hid' and buried the gold hoard in the seventh century were faced with similar difficulties if they used hand-held implements to dig the excavations. This may also explain why the gold was buried at such a shallow depth.

Conceptual geological model. The development of a CGM for police searches was first applied to searches for burials in the Pennines in the middle part of the 1990s (Donnelly 2003; Harrison & Donnelly 2008, 2009). The basic concept of the CGM is that it provides an understanding of the target objects. It also enables the optimum choice of search assets to be decided to help deliver a high-assurance search. The CGM may not be applicable in all search cases. However, this has been found to be a particularly useful method when conducting ground searches for burials.

A rapid and basic desk study was started but not completed before the design and implementation of the search owing to the critical short time constraints that were imposed. This therefore introduced an element of vulnerability into the search. However, a reconnaissance site visit and walk-over survey was conducted with the Police PolSA (the force Police Search Adviser), HOSDB (now CAST) and the Staffordshire County Council archaeologist. This, combined with an analysis of British Geological Survey and Ordnance Survey published maps, enabled a general appreciation of the geology and ground conditions and a CGM to be developed. A search was designed to detect gold and associated metal objects, less than approximately 100 mm in size, buried at a shallow depth of no more than about 280 mm, in a dry, red-brown, coarse grained, granular soil, with cobbles and boulders (Fig. 8).

RAG map. The RAG system was deployed. Red indicated the high priority to be searched first, amber moderate priority, to be searched second, and green lowest priority, furthest away from the archaeological excavation, to be searched last. It was notable that the initial RAG map produced by the author denoted green as the high priority search area and red as the area of least interest. This caused initial confusion and so the colour-coded system was reversed to suit the police requirements.

Search team and resources. The search teams consisted of a forensic geologist and police search adviser (PolSA), HOSDB (now CAST) geophysical instrumentation specialists, Staffordshire Police Tactical Planning Unit and police search adviser, a security co-ordinator and Staffordshire Council archaeologist. Field support was also provided by Staffordshire Council, English Nature and Birmingham Archaeology. The finder and landowner were also present as observers during parts of the search.

Fig. 7. The determination of diggability and examples of the soil types. Following separation by sieving, the soils comprised sand and gravel, with cobbles, pebbles and occasional boulders. Cohesive clay soil and finer soils were found on the middle to lower slopes of the ridge. In general, the soils were progressively finer, less granular and more cohesive (increased clay content) on the lower parts of the ridge, possibly associated with soil creep and mass wasting).

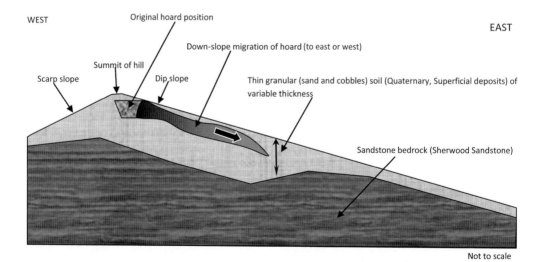

Fig. 8. Conceptual geological model showing the location of the Staffordshire (Anglo Saxon) Gold Hoard. The detectables are non-ferrous metal, about 300 mm deep, in a moderately dense, sandy soil with cobbles, pebbles and boulders.

Pre-search considerations and logistics. The field was directly visible from the A5 and other surrounding roads, farms and areas of public access. Therefore, one of the principal constraints was the risk of the search being compromised by either the press/media or passing members of the public. An appropriate and agreed 'cover story' was therefore devised to give any inquiring persons a consistent message so that the search could continue unhindered.

Defining the search (target) area. Defining the search area was one of the most difficult aspects of the search, and attracted much debate and discussion. The position of the hoard and the initial archaeological excavation was located towards the northwestern corner of the field. The time and resources were not available for the entire 5.5 ha field to be searched. This would also be impracticable, expensive and taken longer than the time available.

Consideration was originally given to the possibility for the search to be extended from the original area of the find, in a southeasterly direction along the axial crest of the ridge. This was based on one object of historical value (but of possible Roman age) being found on Friday 11 September 2009 several tens of metres from the hoard along the crest of the hill.

What is more, the burial of 'loot' or objects of value rarely takes place randomly. Instead they may be consciously or sub-consciously located near a fixed physical feature (such as a tree, rock or the top of a hill) to facilitate their possible future recovery. However, since the majority of gold objects were scattered with an area of approximately 9×13 m it was decided to concentrate the search on the western part of the crest, as it was envisaged that a combination of ploughing and mass wasting (soil creep) since the time the hoard was buried (i.e. in the past 1300 years) may have caused the gradual down-slope displacement of the objects. The outer limits of the search zone were therefore defined on this basis during the initial site reconnaissance. The search area was defined on what was achievable during the two days available to implement the search. However, it was recognized by the search team that the possibility for finds existed further downslope and beyond the imposed search limits.

Choice of detection instrumentation. The aim of the search was to prove the presence or absence of gold, silver and copper objects, buried no deeper than about 280 mm in granular soil. Therefore, ferrous and non-ferrous metal detecting (magnetic and electromagnetic) geophysical instruments were considered to be the most suitable to give a high-level of assurance. These types of instruments and search methodology are typically applied during counter-terrorism and homicide grave detection where minimal metal components are targeted, for example related to either explosive devices or clothing on human remains, respectively. The two principal instruments used were as follows.

Fluxgate gradient magnetometer. This detects subtle changes in the Earth's magnetic field caused by metal objects buried in the ground. This instrument has the capability to detect metal at a shallow depth or near-surface. It will only detect ferrous metals and will not detect gold, silver or alloys such as brass. It has been used successfully on numerous police and military searches in the UK and internationally. This method is particularly useful for detecting 'background geophysical noise', that is, small pieces of ferrous iron of no historical (Anglo Saxon) interest. Equally, if there were any iron-containing helmets, swords or shields buried at a shallow depth, they may potentially be detectable by this instrument.

Electromagnetic minimal metal detector. This is based on the 'continuous wave' method but works with an 'exciter coil' and two symmetrical 'receiver coils' in gradient arrangement. It is suitable for detecting all metals, whether ferrous or non-ferrous. Originally, it was designed for the detection of antipersonnel mines containing minimal metal components. This instrument is able to detect small pieces of gold, silver, copper and other metals, including ferrous and alloys, at a shallow depth, to no more than approximately 280 mm.

The advantages of the above two instruments include their ease of operation (trained police search officers became familiar in their use following a brief introduction and instruction) and the accurate localization of the objects to be detected; even where there is minimal metal content, the detection of separate objects in close proximity to one another is possible. This allowed rapid and 'real-time' interpretation and investigation. A disadvantage is the absence of any recorded data for subsequent analysis, processing and recording.

Additional instrumentation used to investigate positive geophysical anomalies and to assist during the recovery of metal objects buried in the ground was as follows.

Heliflux magnetic locator GA-52B. This is a basic magnetometer capable of detecting ferrous objects to a detection depth limit of 300 mm.

MD4 metal detector. This instrument is capable of detecting all metal objects to a detection depth limit of approximately 150 mm, but, as with the electromagnetic instrument, it does not discriminate between different metals.

Soil probes. In addition, 1.2 m-long soil probes (window samplers) were used to investigate the base and walls of excavations. However, they were only capable of penetrating approximately 0.5 m into the compacted, desiccated and dense soils.

Phase 2: search

Search dates. The search team began at 08:00 h Monday 14 September 2009 and finished at 17:00 h on Tuesday 15 September 2009 when the search was completed. Staffordshire County Council commissioned Birmingham Archaeology to spend a further two days investigating any remaining positive geophysical indications. No further archaeological finds were recovered and this additional invasive work was terminated on Thursday 17 September 2009.

Geophysical control measures. Control measures were undertaken to ensure the suitability of the geophysical detection equipment and to identify the maximum detectable depth of the gold. The control site also enabled the testing of the instruments throughout the survey, in case of instrument failure, or the existence of too many 'false-positives'. The geophysical instruments were tested on the control site at the start and end of each day, to check that they had remained operational.

Before the start of the main geophysical survey a location was chosen 10 m to the NW of the search area, along the crest of the ridge, in an area of similar geology. In the presence of a representative from the British Museum, some of the original gold objects were buried in the soil 5 m apart, to depths of 50, 100, 150, 200, 250 and 280 mm. Gold buried at deeper levels was not detectable.

The tests performed confirmed that the maximum detectable depth of the gold was at least 280 mm below ground level with the electromagnetic instruments, but, as expected, this was not detected using the magnetometer. A small metal knife made with a steel blade buried at 280 mm below ground level was detected with ease using both the electromagnetic instrument and magnetometer.

Search strategy and methodology. There were four main stages of the search, as follows (Figs 9 & 10).

Stage A: marking the search lanes. Fifteen search lanes were established each measuring 50 m long and 1 m wide. The position where the hoard was found was located in the centre of search lane 1. The survey lanes were marked using high-visibility white string and plastic pegs. The lanes were progressively marked in a westerly, down-slope direction.

Stage B: geophysical surveys. Each lane was systematically swept, at a rate of about 0.3 m s^{-1} with the magnetometer followed by the electromagnetic instrument. All positive indications were marked with a red (ferrous) or yellow (non-ferrous) plastic peg or flags or white spray and the instrument operator continued with the search.

Fig. 9. Search methodology and search lanes (not to scale).

Fig. 10. (**a**) Establishment of search lanes, 50 m long and 1 m wide. (**b**) Positive geophysical indications (known also as anomalies or 'hits') marked with paint and flags to denote which instrument generated each anomaly. (**c**) Verification of geophysical anomalies. (**d**) Digging, sieving and dividing the soil to investigate the cause of a positive geophysical anomaly. (**e**) Post-recovery investigation of a positive geophysical anomaly. (**f**) Archaeologists identify metallic objects recovered by the geophysical survey.

Stage C: intrusive investigation. Each positive indication was investigated by a designated police 'recovery team' consisting of highly trained and experienced police search team officers, from Staffordshire Police Tactical Planning Unit, under supervision of the force PolSA and forensic geologist. This involved the sweeping of the location with the instruments to verify the geophysical anomaly followed by the digging, dividing, sieving and scanning of the displaced soil to recover the target object that caused the geophysical anomaly.

Stage D: recovery and identification of objects. Any metal objects recovered were identified by the forensic geologist (author) if geological and/or archaeologists if anthropogenic.

Stage E: probing. When considered necessary, the base and walls of the excavations were augered using soil probes to a variable depth depending on the density of the soil, to the point of refusal, to confirm the absence of any artefacts or associated items of archaeological interest.

Stage F: geophysical clearance. The excavation and displaced soil were scanned with the geophysical instruments to confirm that there were no additional metal objects.

Stage G: reinstatement. The excavations were in-filled and reinstated with the displaced soil and restored by compaction to as close as possible its original state.

Phase 3: post-search

Objects recovered. No further gold or other objects of potential monetary value associated with the hoard were found in the areas investigated by the search team. The items found included numerous metal objects of no archaeological significance or value. These were mainly ferrous items and discarded alloys that consisted of, for example, bolts, unidentifiable metal, parts from old tools and a knife blade. Some metal objects 3 mm in size were found 150 mm below ground level and occasionally some objects were recovered from a depth of 300 mm along with one copper (or brass) tube, approximately 100 mm long and less than 5 mm wide and one copper (or brass) flake about 5 mm^2.

Recording of the search. Field records were kept for each search sector, showing the locations and boundaries of each search lane, ground conditions, weather, personnel and assets deployed, start and end times, details of any instrumentation errors, non-search personnel who visited the site and GPS coordinates of potentially important objects found.

GPS traverses were conducted around and across the site to provide topographic sections. Numerous digital photographs and video footage were taken to record each stage of the search. Additional imagery was provided from the police drone, which was operated remotely from the field command unit.

Search limitations and levels of assurance. The search was originally designed to investigate 15 search lanes, each 50 m long and 1 m wide. Traditionally, searches begin slowly and increase in rate once the search team has become familiar with the search methodology and instrumentation. However, progress was slower than expected because of a greater than expected number of positive anomalies detected by the instruments, which required subsequent invasive investigations. These were caused by background geophysical noise, for example, discarded ferrous metal of no significance and on a few occasions some of the sandstone cobbles which contained sufficient iron minerals to trigger a positive response, requiring calibration of the instruments. Furthermore, poor diggability of the ground was encountered because it was dry (in places desiccated), dense and extremely difficult to dig with shovels and picks, particularly along the crest of the ridge, where pebbles, cobbles and boulders were more abundant.

The geoforensic search gave a high level of assurance and confidence of the absence of any additional gold objects within the area searched, to a depth of approximately 280 mm. Therefore, the objectives of the search had been met and it could be concluded that the archaeologists had conducted a thorough initial recovery of the gold items, within the upper parts of the soil profile.

The possible existence of additional Anglo Saxon gold items in this field was recognized and discussed by the search teams. This was speculated to exist within the area searched but at deeper stratigraphic levels (greater than 280 mm) within the soil profiles or superficial deposits, or beyond the confines of the defined search area and in particular along the ridge crest or on the adjacent slopes.

The 2012 finds

In November 2012, a further 91 gold and silver items of interest were discovered by a team from Archaeology Warwickshire, Staffordshire Council and English Heritage. These were located in the same field where the 2009 hoard was found. On 4 January 2013, the South Staffordshire Coroner ruled at an inquest that 81 of the items were treasure and part of or associated with the original find. Ten of the items were rejected but may still be of some interest. Interestingly, the items found in 2012 were discovered immediately following the ploughing of the field. This is consistent with the original find, three years earlier. The resolution of the geophysical surveys test work conducted in 2009 was limited to 280 mm. A working hypothesis by the archaeologists suggests that the gold items may have been buried at deeper ground levels or beyond the original search area. These items were possibly brought to the surface by ploughing. This suggests future archaeological and geological searches could be implemented immediately following the future ploughing of this field.

Conclusions

In September 2009, gold artefacts of Anglo Saxon age were found by an amateur metal detectorist in a recently ploughed field near Lichfield, in Staffordshire. A comprehensive and thorough archaeological excavation was carried out and numerous other items were found and recovered, resulting in 3940 items of treasure, including 5 kg of gold and 1.5 kg of silver. Confirmation was required by the archaeologists to verify that no additional gold items had been left in the ground, immediately below the ground surface (in the upper soil layers) at the site of the archaeological excavation. A forensic geology and police search was designed and deployed in association with HOSDB (now CAST) and Staffordshire Police.

The forensic geology and police search strategy included magnetic and electromagnetic geophysical instruments and highly trained and experienced police search teams. This proved, as far as possible, given the technological limitations of the instrumentation, the absence of any additional gold or associated metals of value to a depth of approximately 280 mm, within a search area 50 m long and 15 m wide, at and surrounding the excavation where the gold finds were made. Many metal objects were found during this search but they were of no archaeological value, apart from two small copper (or brass) items. All of the gold and associated objects therefore seemed to have been removed during the initial stripping of the uppermost top soil by the archaeologists.

The forensic geology and police search did not recover any further gold or associated items of value. However, the aim of this search was not to locate further gold items since the archaeologists were confident that they had recovered everything that was reasonably and practically recoverable from just beneath the ground surface at that point in time. The archaeologists therefore required an authoritative opinion on whether this was the case. The forensic geology and police search provided high assurance that no further gold had been left in the ground immediately below the ground surface, in the area searched, to a depth of approximately 280 mm. This was important since the archaeologists expected the pending coroner's report to potentially attract amateur treasure hunters to the field in search of any remaining gold items.

Forensic geology and police ground searches are not necessarily always aimed at recovering items. Sometimes, a defensive or protective search may be commissioned to prove the absence of a suspected burial and this was the case with the Staffordshire Hoard. By comparison, an offensive or detective search will be to locate a known grave or burial.

In 2012, a further 91 items were discovered in this field, of which 81 were subsequently regarded by the coroner as treasure. It may be, depending on the distribution plot from the current work, that these items lay off the ridge or further downslope from the original excavation and search area, but could still be associated with the hoard. The 2012 finds may also have been located at a depth beyond the 280 mm detection limit of the geophysical instruments and been brought up closer to the ground surface by more recent ploughing.

Both the 2009 and the 2012 finds occurred after ploughing and previous searches of the field recovered nothing. As with all forensic geology and police ground searches, if there is a change in the geological and/or environmental conditions or any new information and/or intelligence is provided, then repeated ground searches are recommended to be conducted.

Archaeological, geological and forensic searches of this type, particularly those involving geophysics and trained search teams can give a high level of assurance, but not a 100% guarantee that complete coverage has been attained. There remains a possibility that additional gold and silver items still remain in the field, including at the site of the original archaeological excavations, but at a greater depth. There also remains a possibility for other gold and silver items to exist at deeper levels in the soil profile in other parts of this field. This is somewhat speculative but nevertheless further searches are recommended to take place immediately following ploughing to recover any additional items of historical value.

This investigation was undertaken in association with several organizations and individuals and the author gratefully acknowledges their help and support. These included T. Herbert, M. Ferguson and G. Hewings (Home Office Scientific Development Branch, HOSDB, now the Centre for Applied Science and Technology, CAST); J. Overend (PolSA), S. Thompson (Inspector) and Staffordshire Police Tactical Planning Unit; S. Dean and I. Wykes (Staffordshire County Council); E. Macey (Birmingham Archaeology), Professor S. Esmonde Cleary and Birmingham University; British Museum; English Nature; H. Siddle (Halcrow) and J. Pringle (University of Keele). Some of the figures in this paper have been published courtesy of Staffordshire Council and Birmingham University. The author is particularly grateful to S. Dean for his advice and support on the production of this paper, IUGS, Initiative on Forensic Geology and Geological Society, London, Forensic Geoscience Group.

References

BRITISH GEOLOGICAL SURVEY 1919. *Memoir for the country around Lichfield*. British Geological Survey, NERC.

BRITISH GEOLOGICAL SURVEY 1971. *Geological map, old series, one inch to one mile scale, sheet 154, Lichfield*. British Geological Survey, NERC.

DONNELLY, L. J. 2003. The applications of forensic geology to help the police solve crimes. European Geologist. *Journal of the European Federation of Geologists*, **16**, 8–12.

FENNING, P. J. & DONNELLY, L. J. 2004. Geophysical techniques for forensic investigations. *In*: PYE, K. & CROFT, D. (eds) *Forensic Geoscience: Principles, Techniques and Applications*. Geological Society, London, Special Publications, **232**, 11–20.

HARRISON, M. & DONNELLY, L. J. 2008. Buried Homicide Victims: Applied geoforensics in search to locate strategies. *The Journal of Homicide and Major Incident Investigations*, **4**, 71–86 (produced on behalf of the Association of Chief Police Officers Homicide Working Group, by the National Policing Improvement Agency).

HARRISON, M. & DONNELLY, L. J. 2009. Locating concealed homicide victims; developing the role of geoforensics. *In*: RITZ, K., DAWSON, L. & MILLER, D. (eds) *Criminal and Environmental Soil Forensics*. Springer, Berlin, 197–219.

HUNTER, J. & COX, M. 2005. *Forensic Archaeology Advances in Theory and Practice*. Routledge, Abingdon.

LINFORD, N. 2004. Magnetic ghosts: mineral magnetic measurements on Roman and Anglo-Saxon graves. *Archaeological Prospection*, **11**, 167–180.

PRINGLE, J. K., RUFFELL, A. ET AL. 2012. The use of geoscience methods for terrestrial forensic searches. *Earth Science Reviews*, **114**, 108–123.

RUFFELL, A. & MCKINLEY, J. 2008. *Geoforensics*. Wiley-Blackwell, Chichester.

WALKINGTON, H. 2010. Soil science applications in archaeological contexts: a review of key challenges. *Earth-Science Reviews*, **103**, 122–134.

Detection of fatty acids in the lateral extent of the cadaver decomposition island

MELINA LARIZZA[1] & SHARI L. FORBES[1,2]*

[1]University of Ontario Institute of Technology, 2000 Simcoe Street, North, Oshawa, Ontario, Canada L1H 7K4

[2]University of Technology Sydney, PO Box 123, Broadway, Sydney, NSW 2007, Australia

*Corresponding author (e-mail: shari.forbes@uts.edu.au)

Abstract: Identifying biomarkers of decomposition may prove to be an important area of environmental and criminal forensics research. Biomarkers released during the decomposition process can be detected in soil as a means of confirming the presence of a decomposition site in the case of relocated or scavenged remains. This study was conducted to characterize the fatty acid profile in soil containing decomposition fluid and to determine the lateral extent of fatty acid release in the cadaver decomposition island (CDI). Owing to practical and ethical restrictions, the study utilized pig carcasses as human analogues to investigate postmortem decomposition on a soil surface. Soil samples were collected from directly beneath the carcasses and at increasing distances from the carcasses within the CDI. Fatty acids were extracted with chloroform, derivatized with a silylating agent and analysed using gas chromatography–mass spectrometry (GC-MS). Saturated and unsaturated fatty acids were detected including myristic (C14:0), palmitic (C16:0), palmitoleic (C16:1), stearic (C18:0) and oleic (C18:1) acids. Fatty acids were detected up to 50 cm in the lateral extent of the CDI at significantly higher levels in decomposition soil than in the control soil. The results indicate that fatty acid analysis of decomposition soil could be used to confirm the location of a decomposition site.

The decomposition process of human or animal remains is inherently affected by the environment in which it occurs. Cadaveric material can be rapidly introduced to the surrounding environment through putrefaction and/or insect activity (Putman 1978; Dent et al. 2004), thus forming a cadaver decomposition island (CDI; Carter et al. 2007). The CDI forms as decomposition fluid leaches from decomposing remains into the soil environment (Carter et al. 2007) and is apparent as a dark stain around the remains, often associated with the death of nearby vegetation. Decomposed remains provide a localized pulse of concentrated nutrients into the soil (Benninger et al. 2008), having the potential to positively or adversely affect biological communities (e.g. microorganisms), depending on the specific energy requirements of the community (Carter et al. 2007). Decomposition fluid is composed of nutrients and decomposition by-products (Swann et al. 2010a, b) and supplies the soil with an abundance of organic matter and inorganic compounds (Carter et al. 2007). CDIs are generally associated with a proliferation in microbial biomass and activity as they are hubs for carbon and nutrient flow (Carter et al. 2007; Benninger et al. 2008).

Soil is a complex matrix comprised of mineral particles of varying sizes (e.g. gravel, sand, silt and clay) and organic material (Fitzpatrick 2008). Traditionally in forensic science, soil is analysed as a passive medium for the purpose of comparison of soil material to a crime scene (Benninger et al. 2008; Fitzpatrick 2008; Tibbett & Carter 2008). However, Tibbett & Carter (2008) assert that soil is a dynamic medium rapidly influenced by environmental change. The analysis of the nutrients and biomarkers in soil has the potential to provide information about the decomposing remains and/or the decomposition site.

Increasingly, studies are being conducted on the interaction of soil and the decomposition of human and animal remains. Vass et al. (2002) analysed volatile fatty acids, anions and cations in soil solutions produced from soft tissue and bone decomposition of human remains. Soil samples collected from grave sites were analysed by Forbes et al. (2002) to characterize the composition of adipocere formed by human remains. Benninger et al. (2008) analysed nitrogen- and phosphorus-based by-products in soil collected within the CDI of decomposing pig carcasses placed on a soil surface. Significant increases in soil pH, total nitrogen, soil-extractable phosphorus and lipid phosphorus were detected throughout the decomposition process. Similarly, Van Belle et al. (2009) identified significant increases in ninhydrin-reactive nitrogen in grave soils of buried pig carcasses compared with control soils. All of these biomarkers demonstrated some potential for estimating the postmortem interval (PMI) of decomposed remains or confirming a decomposition site.

Fatty acids derived from adipose tissue also have the potential to be used as biomarkers of decomposition in soil. Human adipose tissue is typically composed of 5–30% water, 2–3% proteins and 60–85% lipids or fats, by weight (Dent et al. 2004). Triglycerides (neutral fat) comprise 90–99% of the lipids in adipose tissue (Dent et al. 2004) and are composed of a glycerol molecule bound to three free fatty acids (Fiedler & Graw 2003; Dent et al. 2004). Monounsaturated oleic acid (C18:1) is cited as the most abundant fatty acid in adipose tissue, followed by polyunsaturated linoleic acid (C18:2), monounsaturated palmitoleic acid (C16:1) and the saturated palmitic acid (C16:0) (Pfeiffer et al. 1998; Dent et al. 2004; Janaway et al. 2009). A similar fatty acid composition has also been described for pig adipose tissues (Davenel et al. 1999).

At death, neutral fats are hydrolysed by intrinsic lipases (Fiedler & Graw 2003; Dent et al. 2004; Notter et al. 2009), yielding a mixture of unsaturated and saturated fatty acids (Forbes et al. 2004). Subsequent to the hydrolysis of lipids, fatty acids can undergo oxidation or hydrogenation depending on oxygen availability. Both processes are carried out by microorganisms associated with the decomposition process. In aerobic conditions, microorganisms (bacteria and fungi) can carry out oxidation reactions to produce aldehydes and ketones (Fiedler & Graw 2003; Dent et al. 2004; Notter et al. 2009; Mohan Kumar et al. 2009). Conversely, in an anaerobic environment, unsaturated fatty acids can undergo hydrogenation by bacterial enzymes to yield saturated fatty acids (Fiedler & Graw 2003; Dent et al. 2004; Mohan Kumar et al. 2009; Notter et al. 2009). A single-step β-oxidation, in which a two-carbon (C_2) unit is lost from the fatty acid chain, can also be carried out by bacterial enzymes (Bereuter et al. 1996; Fiedler & Graw 2003). Chemical analyses of decomposition fluid collected from human and pig remains has confirmed the presence of short- and long-chain fatty acids as well as other compounds implicated in lipid and protein degradation (Cabirol et al. 1998; Swann et al. 2010a, b). Fatty acids typically identified in decomposition soils include the saturated fatty acids myristic (C14:0), palmitic (C16:0) and stearic (C18:0) acid, as well as the unsaturated fatty acids palmitoleic (C16:1), oleic (C18:1) and occasionally linoleic (C18:2) acid (Forbes et al. 2002, 2003, 2004; Vane & Trick 2005).

The analysis of fatty acids in decomposition soil is often conducted with respect to adipocere formation. Adipocere is a soap-like substance that can develop following the postmortem conversion of adipose tissue from decomposing remains into a lipid mixture under anaerobic conditions (Forbes et al. 2003). Adipocere may take weeks or years to develop (Forbes et al. 2003) as anaerobic processes occur at a slower rate than aerobic processes (Janaway et al. 2009). The analysis of fatty acids in soil can be conducted over extended time periods, as fatty acids are relatively stable in soil. Long-chain fatty acids have been detected in 12-year-old grave soil samples collected from exhumations of Australian cemeteries (Forbes et al. 2003). Vane & Trick (2005) detected long-chain fatty acids in soil samples recovered from burial sites of cattle and pig carcasses from the Foot and Mouth epidemic of 1967.

The identification of fatty acid biomarkers of decomposition has the potential to provide valuable information about the decomposing remains and/or the decomposition site. The aim of this study was to characterize fatty acid profiles in sandy soil containing decomposition fluid and to determine the lateral extent of fatty acid release in the CDI.

Materials and methods

Research site

The research was conducted in collaboration with the Royal Canadian Mounted Police (RCMP). Research trials were carried out at the Technical and Protective Operations Facility (TPOF), an RCMP facility in Ottawa, Ontario, Canada. Ottawa is part of the Great Lakes–St Lawrence Forest, which encompasses a mix of coniferous and deciduous trees and is located at latitude 45°22′N and longitude 75°43′W, 114 m above sea-level (http://ottawa.ca/city_hall/glance/index_en.html). TPOF is enclosed by barbed wire fencing and measures approximately 3.5 km². The taphonomic experimental area was approximately 0.004 km² (1 acre) of open land lacking shrubs and trees. Two research trials were conducted during the summer months of 2008 (June to August) and 2009 (June to July).

An Onset HOBO® Pro V2 External Temperature and Relative Humidity Data Logger was placed at the field site along with a precipitation gauge. The data logger was programmed to record the ambient temperature and relative humidity every hour for the duration of the study and the amount of rainfall was recorded using a standard rain gauge each sampling day. The data logger sensor was placed in a radiation shield prior to placement at the field site to eliminate the effect of solar radiation on the ambient temperature readings. Data were retrieved from the data logger using HOBOware® Pro Data Logging Software Version 2.X. Temperature was converted to Accumulated Degree Days (ADD). ADD was used as a temporal measurement in this study. It is the sum of the average daily temperature for a given time period and was calculated using temperature data collected over the duration of the trials.

This study involved the analysis of the decomposition of domestic pig carcasses (*Sus scrofa*) on a soil surface containing vegetation. The use of human bodies is ideal for conducting decomposition studies. However, owing to practical and ethical restrictions, pig carcasses have been cited as appropriate analogues for human decomposition attributable to their similarities to an adult human torso (Catts & Goff 1992; Anderson & VanLaerhoven 1996; Schoenly *et al.* 2006).

Trial 1 – summer 2008

Trial 1 was conducted to characterize the fatty acid profile in soil containing decomposition fluid. A reference soil sample was collected in a 20 ml glass vial from the area on which each pig was to be deposited (experimental site) prior to placement. A control site, an area on the soil surface lacking the presence of a pig carcass, was established 2 m away from each carcass. The site possessed similar topographic features to those of the surface on which each corresponding pig carcass was deposited. A control soil sample was also collected for each experimental site.

Three 30 kg, antibiotic-free pigs were obtained from the Experimental Farm in Ottawa, Ontario. All carcasses were killed by a barbiturate overdose. To prevent scavenging, the carcasses were wrapped in chicken wire with a mesh size of 2 cm. The carcasses were deposited on the soil surface approximately 20 m apart within hours of their death.

Soil samples were collected daily from the day of deposition of the carcasses until the removal of soft tissue was almost complete (14 days). Subsequently, samples were collected once a week until the carcass was skeletonized and only desiccated tissue and bones remained. Soil samples were collected beneath each carcass by lifting the chicken wire that encased the carcass and collecting approximately 10 g of soil from the top 2 cm. Soils samples were collected beneath the torso of each carcass from an area previously undisturbed by sampling. Control soil samples were also collected each sampling day. All soil samples were placed in glass vials and stored in a −20 °C freezer, until further analysis. In total, approximately 150 soil samples were collected from experimental and control sites in Trial 1.

Trial 2 – summer 2009

Trial 2 was conducted to confirm the fatty acid profile detected in Trial 1 and to determine the lateral extent of fatty acid release in the CDI. A reference soil sample was collected using a stainless steel soil core sampler with an internal diameter of 2 cm that extended 5 cm into the soil from the area on which each pig was to be deposited (experimental site) prior to placement. A control site was established 2 m away from each pig carcass and possessed similar topographic features to the surface on which the corresponding pig was placed. A control soil sample was also collected for each experimental site.

Three 50 kg, antibiotic-free pigs acquired from the Experimental Farm were killed by means of a barbiturate overdose. Each carcass was wrapped in chicken wire with a mesh size of 2 cm and placed on the soil surface containing vegetation approximately 20 m apart within hours of their death. Soil samples were collected each sampling day. Samples were collected daily from the commencement of the study (day of deposition of carcasses) until the removal of soft tissue was almost complete (14 days). Subsequently, samples were collected three times a week for a further two weeks.

Core soil samples were collected for each carcass on each sampling day. A soil core sampler that extended 5 cm into the soil was used at 10 cm intervals commencing at the soil–carcass boundary (0 cm) up to a distance of 50 cm from the carcass (visible perimeter of the CDI). This sampling method aimed at characterizing the CDI at regular intervals throughout the decomposition process. Samples were always collected along a gradient but were randomly sampled around each carcass and collected from undisturbed soil. The top 2 cm of the core soil sample was separated from the bottom 3–5 cm and placed in separate bags. The separation of the soil layers allowed for the independent characterization of each layer. A core soil sample was collected at the same time from the control site of each carcass. All soil samples were placed in Ziploc® Brand Plastic Bags and stored in a −20 °C freezer until further analysis. In total, approximately 420 soil samples were collected from experimental and control sites in Trial 2.

Decomposition stages

Decomposition stages were monitored visually based on stages described by Payne (1965) for the decomposition of pig carcasses. Stages were described in terms of the physical appearance of the pig carcasses and included fresh, bloat, active decay, advanced decay and dry stages (Payne 1965).

Physicochemical properties of soil

Three representative soil samples (10 g of reference soil) were dried in an aluminium container in an oven at 80 °C overnight and stored in a glass vial. The samples were sent to the Agriculture and Food Laboratory at the University of Guelph in Guelph, Ontario, where soil characterization tests were performed to identify the soil texture and determine the electrical conductivity of the soil.

The moisture content and pH of soil samples were determined at the University of Ontario Institute of Technology. Soil samples were thawed, weighed and dried at 80 °C overnight. Soil samples were subsequently reweighed to determine moisture content. Distilled water was added to dried soil samples in a 1:5 soil to water ratio and pH was measured using a Mettler Toledo Seven-Easy pH Instrument, calibrated with solutions of pH 4, 7 and 10.

Fatty acid analysis

Two-hundred milligrams of dried soil were placed in a disposable screw-top culture tube and 1.75 ml of chloroform (Fisher Scientific, Ottawa, Canada) was added to the tube. Samples were sonicated at 55 °C for 1 h then mixed by pulse vortexing every 30 min. Samples were centrifuged at 8000 rpm for 5 min. The chloroform layer was drawn off and transferred to a disposable culture tube. Prior to transfer, appropriate dilutions were performed, as necessary, based on the colour of the chloroform layer. Preliminary studies indicated high levels of fatty acids in samples in which the chloroform layer possessed a yellow to dark brown colour. Samples that required dilution were typically diluted 10, 100 or 1000 times. Subsequently, 0.050 ml (50 µl) of 100 µg ml^{-1} internal standard, nonadecanoic acid (C19:0; Fluka Analytical, Seezle, Germany), and 0.200 ml (200 µl) of *N, O-bis*(trimethylsilyl)-trifluoroacetamide (BSTFA; Fluka Analytical, Seezle, Germany) were added to the samples and incubated at 70 °C for 15 min. Upon cooling, samples were transferred to 2 ml vials and either immediately analysed by gas chromatography–mass spectrometry (GC-MS) or stored in a −20 °C freezer pending analysis by GC-MS.

All analyses were performed on a Varian 450-GC Gas Chromatograph coupled with a Varian 240-MS Ion Trap Mass Spectrometer. Samples were injected into the gas chromatograph with a Varian PAL Autosampler. A 1 µl aliquot of the sample was analysed on a VF-23ms (Varian Inc., Mississauga, Canada) high-polarity fused silica capillary column (30 m × 0.25 mm × 0.25 µm). Helium was used as the carrier gas at a constant column flow of 1.0 ml min^{-1}. The initial column temperature was 50 °C, where it was held for 2 min. The temperature was then increased at a rate of 15 °C min^{-1} to 150 °C and held for 1 min. The temperature was again increased at a rate of 10 °C min^{-1} to 220 °C, where it was held for 3.33 min. The total run time was 20 min. All samples were injected in the splitless mode. The analysis was carried out in total ion count mode. The ion trap mass spectrometer employed electron ionization. It was held at 250 °C and activated at 6 min from the commencement of the run to circumvent the detection of the solvent and excess derivatizing agent. The target total ion count was 20 000 counts and the scan time was 0.50 s/scan. The scan range analysed was 50–450 amu.

Peaks corresponding to trimethylsilyl fatty acid esters were identified based on comparison of their retention time to the fatty acid standards and mass spectra against the NIST MS Search 2.0 Library. The GC-MS output provides the peak area for the compound of interest, a measure of the abundance of the compound. The peak area ratio was calculated by dividing the peak area of the fatty acid of interest by the peak area of the internal standard (area/IS area). The peak area ratio was used as a measure of the relative abundance of the fatty acids extracted from the soil to allow for statistical analysis. Statistical analyses were performed using VSN International Genstat 12th Edition. Two-way analysis of variance tests were performed on the fatty acid data.

CDI perimeter measurements

The 'Observed CDI Perimeter' was measured using a metre stick. The CDI was made visible by the decomposition fluid that leached into the soil environment, forming a dark stain around the remains and killing nearby vegetation; the perimeter of the dark area was defined as the 'Observed CDI Perimeter' (Fig. 1). The 'Maximum CDI Perimeter Detected' was defined as the furthest distance from the carcasses at which fatty acids were chemically detected above control levels in the decomposition soil by GC-MS analysis.

Results

Climatic conditions

The average daily temperature for Trial 1 was 20.0 °C, with an average daily maximum and minimum of 26.8 °C and 14.7 °C, respectively. The total precipitation was 201.5 mm of rain. The average daily temperature for Trial 2 was 20.5 °C with an average daily maximum and minimum of

Fig. 1. Formation of the cadaver decomposition island (CDI). The CDI has formed around the skeletonized remains and is apparent as a dark stain and death of nearby vegetation. The perimeter of the CDI was measured as the 'Observed CDI Perimeter'.

28.0 °C and 15.1 °C, respectively. The total precipitation was 79.5 mm of rain.

Decomposition stages

The fresh stage was characterized by minimal visible changes and was observed from 0.0 ADD until the onset of the bloat stage in each trial. The bloat stage was characterized by the complete distension of the carcasses and putrefactive colour changes; this stage was observed from 58.9 ADD until 113.8 ADD in Trial 1 and from 56.3 ADD until 95.5 ADD in Trial 2. The deflation of the carcasses and the corresponding leaching of decomposition fluid from the remains initiated the active decay stage. Extensive invertebrate activity was also associated with the remains at this stage. Active decay was observed from 107.4 ADD until 176.0 ADD in Trial 1 and from 113.8 ADD until 173.6 ADD in Trial 2. The advanced decay stage was characterized by a lack of invertebrate activity and soft tissue associated with the remains; this stage was observed from 192.4 ADD until 396.6 ADD in Trial 1 and from 192.4 ADD until 398.6 ADD in Trial 2. Only bones, cartilage and dry skin remained in the dry stage. This stage was observed at 416.6 ADD in Trial 1 and at 419.5 ADD in Trial 2 and can persist indefinitely. The range of ADD values observed for each stage was comparable for both trials.

Physicochemical properties of soil

Soil characterization tests conducted at the University of Guelph indicated that the soil collected at the research site was composed of 0.0% gravel, 0.0% very coarse sand, 0.2% coarse sand, 1.9% medium sand, 61.1% fine sand, 27.8% very fine sand, 5.9% silt and 3.1% clay. The soil texture was determined to be fine sand and the electrical conductivity was measured at 0.194 mS cm^{-1}.

The moisture content of the soil was estimated to be approximately 20–50% over the course of both trials. An increase in moisture content did not necessarily coincide with the leaching of decomposition fluid into the soil or formation of the CDI. The soil pH was measured for control and experimental samples during both trials. In general, the pH of experimental soil samples was significantly less than the pH of control soil samples in both trials ($p < 0.05$; Fig. 2a, b). The greatest difference in pH values between control and experimental soil samples was observed from 123.0 ADD to 254.5 ADD in Trial 1 and from 156.5 ADD to 223.8 ADD in Trial 2. This period coincided with the active decay stage and thus the leaching of decomposition fluid from the carcasses into the soil.

Fatty acid analysis

A fatty acid profile was characterized from the soil samples collected during Trial 1 to identify the expected fatty acid composition of soils containing decomposition fluid. Fatty acids were identified by GC-MS at the following retention times: 10.53 min for myristic acid (C14:0), 11.95 min for palmitic acid (C16:0), 12.24 min for palmitoleic acid (C16:1), 13.29 min for stearic acid (C18:0) and 13.52 min for oleic acid (C18:1). The internal standard used in this study was nonadecanoic acid (C19:0) with a retention time of 13.90 min.

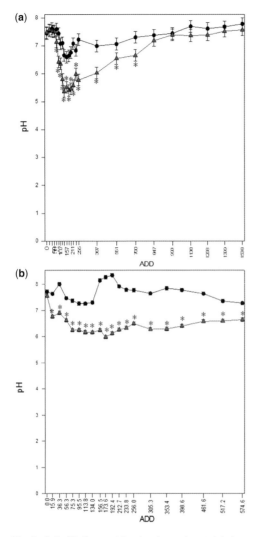

Fig. 2. Soil pH of control (-●-) and experimental (-▲-) soil collected within the cadaver decomposition island at the Technical and Protective Operations Facility in Ottawa, Ontario, Canada during (**a**) Trial 1 and (**b**) Trial 2. * Statistical difference, $p < 0.05$.

Table 1. *Mean relative percent composition of fatty acids detected in Trials 1 and 2*

Fatty acid	Trial 1 (relative percentage)	Trial 2 (relative percentage)
Myristic acid (C14:0)	2.6	2.2
Palmitic acid (C16:0)	60.5	48.5
Palmitoleic acid (C16:1)	0.6	0.9
Stearic acid (C18:0)	29.2	29.7
Oleic acid (C18:1)	7.1	18.8

Additional peaks detected were identified as siloxane-containing compounds, suggesting that the peaks were probably generated by excess derivatizing agent (BSTFA), artefacts produced by the derivatizing agent and/or derivatization of non-specific compounds extracted from the soil matrix (Little 1999).

Table 1 demonstrates the mean relative percentage composition of fatty acids detected over the course of both trials. Saturated fatty acids, palmitic and stearic acids, were the most abundant fatty acids followed by the unsaturated fatty acid, oleic acid. Myristic and palmitoleic acids were the least abundant fatty acids detected throughout the trials.

In addition to characterizing the fatty acid profile, the sampling regime for Trial 2 was aimed at characterizing the lateral extent of the CDI. The top 2 cm of the core soil sample was separated from the bottom 3–5 cm. Preliminary results indicated that the level of fatty acids in the bottom 3–5 cm was negligible in comparison to that in the top soil layer (data not shown). Hence, fatty acid analysis was conducted on the top 2 cm layer only for each sample.

Fatty acids were detected throughout the lateral extent of the visible CDI (up to 50 cm). Figure 3 depicts the relative fatty acid content of experimental samples with respect to the distances at which the samples were collected (irrespective of ADD values). The level of palmitic acid was significantly higher ($p < 0.001$) than all other fatty acids at 0 cm, 10 cm and 20 cm. Stearic acid was the second most abundant fatty acid detected. With the exception of palmitic acid, the level of stearic acid was significantly ($p < 0.001$) higher than all other fatty acids at 0 and 10 cm. The third most abundant fatty acid, oleic acid, was significantly ($p < 0.001$) higher than myristic and palmitoleic acids from 0 to 20 cm. The levels of myristic and palmitoleic acids were negligible when compared with the levels of palmitic, stearic and oleic acids.

In general, there was a successive decrease in the level of each fatty acid with an increase in distance from the pig carcasses. The levels of palmitic, stearic and oleic acids were significantly ($p < 0.001$) higher at 0 and 10 cm than at all distances located further away. Furthermore, the palmitic acid content was significantly ($p < 0.001$) greater at 20 and 30 cm than at 40 and 50 cm. The level of stearic acid was significantly ($p < 0.001$) higher at 20 cm than at 40 and 50 cm, whereas the level of oleic acid at 20 cm was only significantly higher than that at 50 cm. There was no significant

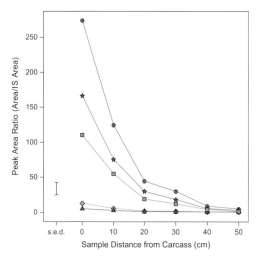

Fig. 3. Relative fatty acid content with respect to the distances at which experimental soil samples were collected within the CDI at the Technical and Protective Operations Facility in Ottawa, Ontario, Canada during Trial 2. The fatty acids identified were myristic (◆), palmitic (●), palmitoleic (▲), stearic (★) and oleic acids (■). Analysis of variance, $F_{\text{Distance}(5, 1770)} = 58.95$, $p < 0.001$, $F_{\text{Fatty acid}(4, 1770)} = 41.85$, $p < 0.001$, $F_{\text{Int}(20, 1770)} = 11.51$, $p < 0.001$, bar = standard error of differences of means.

Table 2. *Perimeter measurements of the cadaver decomposition island (CDI)*

ADD	Observed CDI perimeter (cm)	Maximum CDI perimeter detected (cm)
0	–	–
15.9	–	–
36.3	–	–
56.3	0	–
75.3	0	–
95.5	10	0
113.8	20	10
134.0	20	20
156.5	50	50
173.6	50	30
192.4	50	50
212.7	50	50
233.8	50	20
256.0	50	10
305.3	50	50
353.4	50	10
398.6	50	50
461.6	50	50
517.2	30	10
574.6	30	20

difference between the levels of myristic and palmitoleic acids at any distance.

CDI perimeter measurements

Table 2 demonstrates the 'Observed CDI Perimeter' and the 'Maximum CDI Perimeter Detected' at each ADD for Trial 2. The perimeter measurements were dissimilar across most ADD. The 'Observed CDI Perimeter' increased from 0 cm until 156.5 ADD where the maximum CDI perimeter was observed at 50 cm from the carcasses. The 'Observed CDI Perimeter' remained at 50 cm until 517.2 ADD when it appeared to reduce to 30 cm. Conversely, no distinct pattern was apparent for the 'Maximum CDI Perimeter Detected' measurements. The maximum CDI perimeter was detected at 0, 10, 20, 30 and 50 cm across varying ADD; however, it was not detected at 40 cm at any ADD value.

Discussion

The effect of soil pH on the decomposition process is not well understood (Tibbett & Carter 2008) and may depend on soil type as well as intrinsic and extrinsic factors. The pH of experimental soil samples collected beneath the carcasses in both trials was significantly lower overall than control soil samples. The pH of experimental soil samples collected at fixed intervals from the carcass in Trial 2 was not examined in this study. Towne (2000) determined that the pH of soil beneath ungulate carcasses on the soil surface was also significantly lower than that of the surrounding soil. Soil pH may decrease in the early stages of decomposition as a result of the fermentative processes carried out by anaerobic bacteria in the gastrointestinal tract and soil (Gill-King 1997; Forbes & Dadour 2009) and subsequently increase owing to proteolysis (Rodriguez & Bass 1985). The decreased pH values observed in this study could have been related to the leaching of decomposition fluid, which contains acidic by-products, such as volatile and long-chain fatty acids (Vass et al. 1992; Swann et al. 2010a, b).

In gas chromatography, peak area is proportional to the amount of compound that is present. Peak area ratio was the unit of concentration of fatty acids in this study and was calculated by dividing the peak area of the fatty acid of interest by the peak area of the internal standard. Peak area ratio is a relative measurement of abundance that standardizes the GC-MS output of the fatty acid of interest to that of the internal standard. The use of this type of measurement afforded the ability to provide a quantitative measure for statistical analysis, while accounting for slight variations in instrument responses from sample to sample. It should be noted that the quantitative analysis conducted was based on the assumption that the solvent extraction efficiency, the stability of the derivatized products and the GC-MS detector responses were equivalent for all fatty acids.

Soil analysis from Trial 1 and Trial 2 characterized a fatty acid profile containing the following fatty acids in order of abundance: palmitic acid (C16:0), stearic acid (C18:0), oleic acid (C18:1), myristic acid (C14:0) and palmitoleic acid (C16:1). These fatty acids have also been detected by other researchers who conducted fatty acid analysis of soil containing decomposition fluid from buried remains (Forbes et al. 2002, 2003, 2004; Vane & Trick 2005) as well as decomposition fluid in the absence of a soil matrix (Swann et al. 2010a, b).

Fatty acids became detectable in the soil upon the leaching of decomposition fluid from the carcasses into the surrounding environment. Although oleic acid is the most abundant fatty acid in adipose tissue (Hirsch et al. 1960; Pfeiffer et al. 1998; Davenel et al. 1999; Dent et al. 2004; Monziols et al. 2007; Janaway et al. 2009), it was not the most abundant fatty acid detected over the course of both trials. Furthermore, linoleic acid, another widespread fatty acid in adipose tissue, was not detected in any of the decomposition soil samples. In this study, palmitic acid and stearic acid were the most abundant fatty acids detected in decomposition soil overall. The increased levels of these fatty acids can be attributed to the anaerobic putrefactive processes that occurred within the carcasses. Oleic and linoleic acids were most likely converted to stearic acid while palmitoleic acid may have been converted to palmitic acid through the process of hydrogenation. A single-step β-oxidation of stearic acid to palmitic acid may have also contributed to the increased levels of palmitic acid observed.

The sampling method in Trial 2 was designed to characterize the CDI at regular intervals throughout the decomposition process. In general, the most abundant levels of each fatty acid were detected at 0 cm and successively decreased with increasing distance from the carcasses. The increased levels of fatty acids at 0 cm were expected as the carcasses were in direct contact with the soil at this distance. A decrease in fatty acid levels was expected to coincide with an increase in sample distance from the carcasses; however it was not known to what extent fatty acids could be detected from the carcass. It can be speculated that the soil matrix played a diluting role with respect to the fatty acid levels; the highest levels of fatty acids entering the soil would be found closest to their source (i.e. the carcass) and as decomposition fluid dispersed outwards, fewer fatty acids would be available to enter the soil matrix with increasing distance from the carcass.

Soil matrix effects may have also impacted the fatty acid extraction efficiency, reducing the levels of fatty acids detected. Free fatty acids are polar compounds and therefore can be difficult to extract by organic solvents, such as chloroform. The ionization of fatty acids in alkaline conditions can decrease the extractability of fatty acids from alkaline soils. The possibility exists that the extractability of fatty acids was decreased in the more alkaline soil conditions observed throughout this study, in particular the control soil samples depicted in Figure 2. Acidic soil conditions observed over the course of decomposition may have contributed to an increase in fatty acid extractability as non-ionized forms of fatty acid are more readily extractable from the soil matrix. This may help to explain the reduced concentration of fatty acids further from the carcass (as shown in Fig. 3), although pH was not measured at these distances to confirm their increased alkalinity. The extraction efficiency may have also been reduced by the affinity of fatty acids to other soil particles and organic material.

The methodology developed for fatty acid analysis in this study was an adaptation of the methodology developed by Forbes et al. (2003) and Notter et al. (2008). The choice and quantity of solvent was optimized for the soil type investigated in this study but may require optimization for use in other soil types. The use of mixtures of polar solvents (i.e. chloroform–methanol or dichloromethane–acetone) or the addition of an acid to the solvent may aid in increasing extraction efficiency. An increase in the volume of solvent per unit weight of soil may also assist with improving extraction efficiency. Improved extraction efficiency may result in increased levels of fatty acids being detected further from the carcass, particularly in different soil types.

The 'Observed CDI Perimeter' was physically measured at the research site whereas the 'Maximum CDI Perimeter Detected' was chemically determined through GC-MS analysis. The measurements of the 'Observed CDI Perimeter' did not correspond to the 'Maximum CDI Perimeter Detected'. The CDI is noted visually as a dark stain surrounding the decomposing remains (Carter et al. 2007). However, the exact chemical nature of the CDI has yet to be elucidated. The highest fatty acid levels were detected at the soil–carcass boundary (0 cm) and therefore soil sample collection should be carried out closest to the decomposing remains to detect the highest levels of fatty acids. In cases where remains have been relocated, for example, soil samples should be collected at the site where the remains are suspected to have been placed to increase the likelihood of fatty acid detection. The maximum CDI perimeter was not detected at 40 cm at any ADD value, whereas it was detected at 50 cm. Furthermore, the moisture content of the decomposition soil within the CDI did not differ significantly from control soil samples. These results, along with the lack of distinct patterns observed for the 'Maximum CDI Perimeter Detected', emphasize the dynamic nature of the chemical composition of the CDI and the need for further investigation into its composition.

Although the extent of fatty acid release in the CDI has yet to be determined conclusively, several studies have been conducted on other decomposition products (e.g. soil nutrients) detected along a gradient extending from the centre of animal carcasses. Melis et al. (2007) detected higher concentrations of calcium in soil at the centre of bison carcasses, decreasing towards the periphery (up to 4 m), 3 and 6 years after death. The delayed appearance of calcium in soil can be attributed to the slow release of calcium through bone decomposition (Melis et al. 2007). A higher concentration of nitrate was detected at carcass centres 1 year after death; however, a distinct pattern of variation was not detected along the gradient (Melis et al. 2007). No difference in nitrate levels was detected in carcasses older than 1 year. These results indicate that nitrogen cycling is rapid and that nitrogen by-products may be quickly taken up by plants and microorganisms (Melis et al. 2007) or affected by volatilization processes (Towne 2000; Carter et al. 2007).

Nitrogen in the form of protein, protein catabolic products, amines and ammonium can react with ninhydrin to form a coloured product (Van Belle et al. 2009). Previous research has indicated that ninhydrin reactive nitrogen (NRN) can be used to identify decomposition soil (Carter et al. 2009). In a study conducted to investigate the lateral extent of NRN in decomposition soil, Van Belle et al. (2009) determined that there was a higher concentration of NRN at the centre of the grave of buried pig remains than in the periphery. The lateral extent of NRN was most extensive during the most active stages of decomposition, probably when decomposition fluid was purged from the remains into the surrounding soil environment (Van Belle et al. 2009).

This study and the comparative studies presented reiterate the notion that decomposition produces a localized pulse of nutrients and biomarkers into the surrounding soil environment. Several nutrients have been measured and detected along a gradient from the carcass through to the periphery of the CDI. Nutrient responses and biomarkers of decomposition have proven to be an important tool in forensic science as a means of confirming the original deposition site in the case of the relocation of remains and scavenged remains. For example, Carter et al. (2009) were able to reconstruct a death scene where human remains had been scattered in four distinct areas using NRN. Continued research is required to further elucidate the chemical composition of the CDI in varying environmental conditions. A combined approach, using several nutrients and/or biomarkers known to be released during decomposition (e.g. fatty acids, NRN, phosphorus), should be utilized as a confirmatory test for the presence of a decomposition site in the case of the relocation of remains and scavenged remains. Such confirmatory tests can provide law enforcement agencies with information about the decomposition site that can be used to reconstruct the series of events.

Conclusion

The purpose of this study was to characterize the fatty acid profiles produced by decomposing pig carcasses in a sandy soil environment and to determine the lateral extent of fatty acid distribution in the cadaver decomposition island. Saturated and unsaturated fatty acids typical of carcass decomposition were detected in significantly higher concentrations in the decomposition soil samples when compared with the control soil samples. These fatty acids were successfully detected throughout the lateral extent of the CDI up to 50 cm, having important implications for the consideration of soil sampling for research purposes as well as for environmental and criminal investigations. The increased fatty acid content of decomposition soil associated with pig decomposition has the potential to be used in forensic investigations as a means of confirming a decomposition site in the case of the relocation of remains and scavenged remains.

We would like to thank the Royal Canadian Mounted Police for volunteering their facility for the research project and Sergeant Diane Cockle for aiding in the organization of the project and her insightful advice throughout. We would also like to extend our gratitude to Dr R. Christopher O'Brien for his assistance with statistical analysis as well as the members of the Decomposition Chemistry Laboratory at the University of Ontario Institute of Technology for their insightful discussions and assistance in completing this study.

References

ANDERSON, G. S. & VANLAERHOVEN, S. L. 1996. Initial studies on insect succession on carrion in southwestern

British Columbia. *Journal of Forensic Sciences*, **41**, 617–625.

BENNINGER, L. A., CARTER, D. O. & FORBES, S. L. 2008. The biochemical alteration of soil beneath a decomposing carcass. *Forensic Science International*, **180**, 70–75.

BEREUTER, T. L., LORBEER, E., REITER, C., SEIDLER, H. & UNTERDORFER, H. 1996. Post-mortem alterations of human lipids – part I: evaluation of adipocere formation and mummification by desiccation. *In*: SPINDLER, K., WILFING, H., RASTBICHLER-ZISSERNIG, E., ZUR NEDDEN, D. & NOTHDURFTER, H. (eds) *Human Mummies: A Global Survey of their Status and the Techniques of Conservation*. Springer, Innsbruck, 265–274.

CABIROL, N., POMMIER, M. T., GUEUX, M. & PAYEN, G. 1998. Comparison of lipid composition in two types of human putrefactive liquid. *Forensic Science International*, **94**, 47–55.

CARTER, D. O., YELLOWLEES, D. & TIBBETT, M. 2007. Cadaver decomposition in terrestrial ecosystems. *Naturwissenschaften*, **94**, 12–24.

CARTER, D. O., FILIPPI, J., HIGLEY, L. G., HUNTINGTON, T. E., OKOYE, M. I., SCRIVEN, M. & BLIEMEISTER, J. 2009. Using ninhydrin to reconstruct a disturbed outdoor death scene. *Journal of Forensic Identification*, **59**, 190–195.

CATTS, E. P. & GOFF, M. L. 1992. Forensic entomology in criminal investigations. *Annual Review of Entomology*, **37**, 253–272.

DAVENEL, A., RIAUBLANC, A., MARCHAL, P. & GANDEMER, G. 1999. Quality of pig adipose tissue: relationship between solid fat content and lipid composition. *Meat Science*, **51**, 73–79.

DENT, B. B., FORBES, S. L. & STUART, B. H. 2004. Review of human decomposition processes in soil. *Environmental Geology*, **45**, 576–585.

FIEDLER, S. & GRAW, M. 2003. Decomposition of buried corpses, with special reference to the formation of adipocere. *Naturwissenschaften*, **90**, 291–300.

FITZPATRICK, R. W. 2008. Nature, distribution, and origin of soil materials in the forensic comparison of soils. *In*: TIBBETT, M. & CARTER, D. O. (eds) *Soil Analysis in Forensic Taphonomy*. CRC Press, Boca Raton, FL, 1–28.

FORBES, S. L. & DADOUR, I. 2009. The soil environment and forensic entomology. *In*: BYRD, J. H. & CASTNER, J. L. (eds) *Forensic Entomology: The Utility of Arthropods in Legal Investigations*. CRC Press, Boca Raton, FL, 407–426.

FORBES, S. L., STUART, B. H. & DENT, B. B. 2002. The identification of adipocere in grave soils. *Forensic Science International*, **127**, 225–230.

FORBES, S. L., KEEGAN, J., STUART, B. H. & DENT, B. B. 2003. A gas chromatography-mass spectrometry method for the detection of adipocere in grave soils. *European Journal of Lipid Science and Technology*, **105**, 761–768.

FORBES, S. L., STUART, B. H., DADOUR, I. R. & DENT, B. B. 2004. A preliminary investigation of the stages of adipocere formation. *Journal of Forensic Sciences*, **49**, 566–574.

GILL-KING, H. 1997. Chemical and ultrastructural aspects of decomposition. *In*: HAGLUND, W. D. & SORG, M. H. (eds) *Forensic Taphonomy: The Postmortem Fate of Human Remains*. CRC Press, New York, 93–108.

HIRSCH, J., FARQUHAR, J. W., AHRENS, E. H., PETERSON, M. L. & STOFFEL, W. 1960. Studies of adipose tissue in man – a microtechnic for sampling and analysis. *American Journal of Clinical Nutrition*, **8**, 499–511.

JANAWAY, R. C., PERCIVAL, S. L. & WILSON, A. S. 2009. Decomposition of human remains. *In*: PERCIVAL, S. L. (ed.) *Microbiology and Aging: Clinical Manifestations*. Springer Science and Business Media, New York, 313–334.

LITTLE, J. L. 1999. Artifacts in trimethylsilyl derivatization reactions and ways to avoid them. *Journal of Chromatography A*, **844**, 1–22.

MELIS, C., SELVA, N., TEURLINGS, I., SKARPE, C., LINNELL, J. D. C. & ANDERSEN, R. 2007. Soil and vegetation nutrient response to bison carcasses in Bialowieza Primeval Forest, Poland. *Ecological Research*, **22**, 807–813.

MOHAN KUMAR, T. S., MONTEIRO, F. N., BHAGAVATH, P. & BAKKANNAVAR, S. M. 2009. Early adipocere formation: a case report and review of literature. *Journal of Forensic and Legal Medicine*, **16**, 475–7.

MONZIOLS, M., BONNEAU, M., DAVENEL, A. & KOUBA, M. 2007. Comparison of the lipid content and fatty acid composition of intermuscular and subcutaneous adipose tissues in pig carcasses. *Meat Science*, **76**, 54–60.

NOTTER, D. J., STUART, B. H., DENT, B. B. & KEEGAN, J. 2008. Solid-phase extraction in combination with GC/MS for the quantification of free fatty acids in adipocere. *European Journal of Lipid Science and Technology*, **110**, 73–80.

NOTTER, S. J., STUART, B. H., ROWE, R. & LANGLOIS, N. 2009. The initial changes of fat deposits during the decomposition of human and pig remains. *Journal of Forensic Sciences*, **54**, 195–201.

PAYNE, J. A. 1965. A summer carrion study of the baby *Sus scrofa* Linnaeus. *Ecology*, **46**, 592–602.

PFEIFFER, S., MILNE, S. & STEVENSON, R. M. 1998. The natural decomposition of adipocere. *Journal of Forensic Sciences*, **43**, 368–370.

PUTMAN, R. J. 1978. Flow of energy and organic matter from a carcase during decomposition: decomposition of small mammal carrion in temperate systems 2. *Oikos*, **31**, 58–68.

RODRIGUEZ, W. C. & BASS, W. M. 1985. Decomposition of buried bodies and methods that may aid in their location. *Journal of Forensic Sciences*, **30**, 836–852.

SCHOENLY, K. G., HASKELL, N. H., MILLS, D. K., BIEME-NDI, C., LARSEN, K. & LEE, Y. 2006. Recreating death's acre in the school yard: using pig carcasses as model corpses. *The American Biology Teacher*, **68**, 402–410.

SWANN, L., CHIDLOW, G. E., FORBES, S. L. & LEWIS, S. W. 2010a. Preliminary studies into the characterization of chemical markers of decomposition for geoforensics. *Journal of Forensic Sciences*, **55**, 308–314.

SWANN, L., FORBES, S. L. & LEWIS, S. W. 2010b. Observations of the temporal variation in chemical content of decomposition fluid: a preliminary study using pigs as a model system. *Australian Journal of Forensic Sciences*, **42**, 199–210.

TIBBETT, M. & CARTER, D. O. 2008. *Soil Analysis in Forensic Taphonomy: Chemical and Biological Effects of Buried Human Remains*. CRC Press, Boca Raton, FL.

TOWNE, E. G. 2000. Prairie vegetation and soil nutrient responses to ungulate carcasses. *Oecologia*, **122**, 232–239.

VAN BELLE, L. E., CARTER, D. O. & FORBES, S. L. 2009. Measurement of ninhydrin reactive nitrogen influx into gravesoil during aboveground and belowground carcass (*Sus domesticus*) decomposition. *Forensic Science International*, **193**, 37–41.

VANE, C. H. & TRICK, J. K. 2005. Evidence of adipocere in a burial pit from the foot and mouth epidemic of 1967 using gas chromatography–mass spectrometry. *Forensic Science International*, **154**, 19–23.

VASS, A. A., BASS, W. M., WOLT, J. D., FOSS, J. E. & AMMONS, J. T. 1992. Time since death determinations of human cadavers using soil solution. *Journal of Forensic Sciences* **37**, 1236–53.

VASS, A. A., BARSHICK, S. A., SEGA, G., CATON, J., SKEEN, J. T., LOVE, J. C. & SYNSTELIEN, J. A. 2002. Decomposition chemistry of human remains: a new methodology for determining the postmortem interval. *Journal of Forensic Sciences* **47**, 542–53.

Ground penetrating radar use in three contrasting soil textures in southern Ontario

A. C. LOWE[1], D. V. BERESFORD[2], D. O. CARTER[3], F. GASPARI[1], R. C. O'BRIEN[4] & S. L. FORBES[1,5]*

[1]*Faculty of Science, University of Ontario Institute of Technology, 2000 Simcoe Street North, Oshawa, Ontario, Canada L1H 7K4*

[2]*Department of Biology, Trent University, 1600 West Bank Drive, Peterborough, Ontario, Canada K9J 7B8*

[3]*Department of Entomology, University of Nebraska–Lincoln, 302 Biochemistry Hall, Lincoln, NB 68583-0816, USA*

[4]*Forensic Science Institute, University of Central Oklahoma, 100 North University Drive, Box 203, Edmond, OK 73034, USA*

[5]*Centre for Forensic Science, University of Technology Sydney, PO Box 123, Broadway, Sydney, NSW 2007, Australia*

*Corresponding author (e-mail: Shari.forbes@uts.edu.au)

Abstract: Ground penetrating radar (GPR) is a non-invasive, geophysical tool that can be used for the identification of clandestine graves. GPR operates by detecting density differences in soil by the transmission of high frequency electromagnetic waves from an antenna. Domestic pig (*Sus scrofa domesticus*) carcasses were clothed in 100% cotton t-shirts and 50% cotton/50% polyester briefs, and buried at a consistent depth at three field sites of contrasting soil texture (silty clay loam, fine sand and fine sandy loam) in southern Ontario. GPR was used to detect and monitor the graves for a period of 14 months post-burial. Analysis of collected data revealed that GPR had applicability in the identification of clandestine graves in silty clay loam and fine sandy loam soils, but was not suitable for detection in the fine sandy soil studied. The results of this research have applicability within forensic investigations involving decomposing remains by aiding in the location of clandestine graves in loam soils in southern Ontario through the use of GPR.

The ability to identify the location of a clandestine grave is of importance to forensic investigators in order to identify the victim and further advance the criminal investigation. Traditional methods of locating a clandestine grave site include observations of foliage, soil changes, and determination of soil conductivity, temperature and pH (Rodriguez & Bass 1985; Ruffell *et al.* 2009). Immediate changes in the foliage over a freshly dug grave may be evident, as the disturbance of the soil reduces plant growth (Rodriguez & Bass 1985). However, an older grave of approximately one year or more may have an increased amount of foliage on the grave and in the surrounding area owing to the organic nutrients that are released from a decomposing body into the soil (Rodriguez & Bass 1985). Soil sinking or compaction as decomposition proceeds, such as when the chest cavity collapses, can also be apparent. Traditional methods used to locate a grave are presumptive and cannot determine with certainty if a body is located at that site.

One method of locating buried anomalies is through the use of ground penetrating radar (GPR). GPR is a non-invasive, geophysical tool that can be used for the location of unexploded ordnance, buried utility lines, landfill debris, mineral resources and artefacts at prehistoric sites (Rodriguez & Bass 1985; Miller 1996; Neubauer 2001). Law enforcement search teams and military organizations have used GPR to search for buried organic remains (Miller 1996; Ruffell 2005; Ruffell & McKinley 2008; Ruffell *et al.* 2009). GPR is increasingly used in the search for forensic evidence because of its non-destructive nature and because it can be used in combination with other non-invasive methods to locate areas for further testing (Schultz *et al.* 2006). Other methods used for locating clandestine graves and human remains include magnetometry, electrical resistivity, probing, cadaver dogs and geochemical sampling (Owsley 1995; Nobes 2000; Buck 2003; Schultz *et al.* 2006; Ruffell *et al.* 2009).

From: Pirrie, D., Ruffell, A. & Dawson, L. A. (eds) 2013. *Environmental and Criminal Geoforensics.*
Geological Society, London, Special Publications, **384**, 221–228.
First published online July 11, 2013, http://dx.doi.org/10.1144/SP384.12
© The Geological Society of London 2013. Publishing disclaimer: www.geolsoc.org.uk/pub_ethics

The underlying physics of GPR involves the transmission and reflection of high-frequency electromagnetic (EM) waves into the ground from an antenna, and reflection back to the surface with detection by the receiving antenna (Ruffell 2005; Ruffell & McKinley 2008; Ruffell et al. 2009). The antenna transmits the EM waves, which are reflected when changes in the electrical properties of the ground are detected, such as the difference between buried human remains and the surrounding soil texture (Davis et al. 2000). The electrical properties of soils will vary depending on the amount of moisture held by soil particles. For example, sands typically have a low electrical conductivity, while silts and clays have medium and high electrical conductivities, respectively. The electrical conductivity of a soil correlates strongly to its particle size and texture (Grisso et al. 2009).

Common GPR models use antennae of 300, 500 or 900 MHz centre frequency (Miller 1996; Davis et al. 2000; Schultz et al. 2006). A short pulse antenna (900 MHz) is effective with near-surface targets (≤ 0.5 m), such as buried ordnance (Miller 1996). A 500 MHz antenna is useful for depth investigations of 0.5–3.5 m, which includes most of the items of interest in a forensic investigation (Miller 1996). A long pulse antenna (300 MHz) is effective for subsurface imaging of depths greater than 3.5 m and up to 9.0 m, such as water-tables (Miller 1996; Davis et al. 2000). Overall, a decrease in antenna frequency will increase the depth of investigation, while decreasing the vertical resolution of the subsurface (Schultz et al. 2006). Alternatively, an increase in antenna frequency will decrease the depth of investigation, while increasing the resolution of subsurface objects (Schultz et al. 2006). An antenna in the frequency range of 500 MHz is ideally suited to forensic investigations, as it provides a suitable compromise between depth of penetration and resolution of subsurface features.

Controlled forensic studies using GPR provide training for operators and determination of soil properties and environmental conditions that are applicable to the use of the radar and detection of burial location. Operator experience can be a limiting factor of GPR use in a forensic setting, and therefore, research conducted in a known setting is necessary to interpret the data collected during a criminal investigation (Schultz et al. 2006). Experienced GPR operators may overlook a body when conducting a survey if transects are not collected over a grid or line pattern that utilizes appropriate spacing (Schultz et al. 2006). Davis et al. (2000) and Neubauer (2001) suggest applying archaeological GPR parameters to forensic cases using transects separated by 0.5 m or less. The use of control graves, which consist of only disturbed backfill, is also important to demonstrate that hyperbolic anomalies are primarily the result of a decomposing cadaver.

Ground penetrating radar has proven useful in the search for historical burial grounds. Ruffell et al. (2009) used GPR for the location and assessment of an unmarked, historical burial ground in NW Ireland believed to contain decedents of the Great Famine of 1845–1851. Soils in the area comprised post-glacial sands, glacial till and Carboniferous sandstones (Ruffell et al. 2009). Prior to GPR use, 84 possible burials were located based upon historical records, aerial photographs and landscape interpretation (Ruffell et al. 2009). The target area (area of suspected burials) was analysed using GPR with three different antenna frequencies: 100, 200 and 400 MHz (Ruffell et al. 2009). After data interpretation, it was determined that the 400 MHz antenna centre frequency was the most appropriate antenna to use, as the locations of over 300 possible burials were obtained using this antenna. In contrast, the 100 MHz antenna gave only an indication of some possible burials, whereas the 200 MHz antenna detected 210 possible burials (Ruffell et al. 2009).

Soil properties and environmental conditions can enhance, limit or impair GPR performance. Research has shown that GPR yields reliable results in sandy soils (typically low moisture and conductivity; Schultz et al. 2006), permafrost (Davis et al. 2000), glaciers and concrete/pavement (Ruffell et al. 2009). The use of GPR is often difficult in clay soils (high moisture and conductivity; Schultz et al. 2006; Schultz 2008) and after periods of heavy rain (Ruffell et al. 2009). Clays demonstrate a high adsorptive capacity for water and exchangeable cations causing high attenuation losses. As a result, the penetration depth of GPR in clay soils is restricted, often penetrating less than 1 m in wet clays (Doolittle et al. 2007).

Schultz et al. (2006) found that pig carcasses buried in sandy soils could be detected using GPR for 21.5 months, while exhibiting variable decomposition stages, including complete skeletonization. However, there was a weak contrast between the skeleton and the surrounding soil (Schultz et al. 2006). Difficulties imaging the carcasses during the later stages of decomposition were experienced in clay soils. During the first 6 months of burial, the graves and carcasses were generally detectable (Schultz et al. 2006). However, as the disturbed ground became more compact over the duration of the study, the response became increasingly difficult to interpret, even though the carcasses had undergone little decomposition (Schultz et al. 2006). Despite the fact that carcasses buried in clay were difficult to detect, Schultz et al. (2006) found that it was possible to image disruptions or breaks in the clay horizon that were the result of soil disturbance from the presence of the grave and carcass. However, detecting clandestine graves based solely on soil features may not be possible, as the response from the disturbed soil of the grave will be reduced over time (Schultz *et al.* 2006).

A more recent study in sandy loam soil (Pringle *et al.* 2012) demonstrated that a wrapped or clothed victim in a shallow burial can be located using medium dominant frequency (110–450 MHz) GPR antennae because the wrapping produces a good reflective contrast. An unclothed 'naked' victim could also be located initially, but after 18 months' burial, the remains attenuated a large proportion of the signal, making it difficult to locate the clandestine graves using GPR. Resistivity surveys were recommended for clay-rich soils owing to the possibility of a highly conductive leachate being retained in the soil from the decomposing body and the poor penetration depths typically experienced by GPR in these soil types. However, GPR was recommended over resistivity surveys in the sandy loam soil owing to its ease of data processing.

The applicability of GPR to forensic investigations involving homicide victims buried in clandestine graves has been demonstrated by controlled research in the USA and UK (Schultz *et al.* 2006; Schultz 2008; Pringle *et al.* 2012). The research consisted of burying pig carcasses as human body analogues, and subsequently detecting and monitoring the carcasses for a period of time post-burial. The current study involved the burial of clothed, domestic pig carcasses (*Sus scrofa domesticus*) in a range of contrasting soil textures (silty clay loam, fine sand and fine sandy loam) at three field sites in southern Ontario, Canada. GPR was used to detect the graves over a range of post-burial intervals representing the first large-scale study to investigate the applicability of GPR to forensic investigations in Canada.

Materials and methods

Site locations

Field experiments, which consisted of burying and subsequently exhuming domestic pig (*Sus scrofa domesticus*) carcasses in contrasting soil textures, were conducted over a 14-month period. The domestic pig is commonly used as a model for human decomposition in forensic research (Schoenly *et al.* 2006; Notter *et al.* 2009). This is due to the ethical restrictions on using human bodies for research (Notter *et al.* 2009), their similar internal anatomy, fat distribution, size of chest cavity, lack of heavy fur and omnivorous diet, suggesting a similar gut fauna (Schoenly *et al.* 2006).

Three field site locations within southern Ontario, Canada were selected for GPR data collection based upon soil texture: 'Nashville', a grazing field located in Nobleton, Ontario; 'Springwater', a commercial gravel pit located in Springwater, Ontario; and 'Dummer', a grazing field located in Douro-Dummer Township, Ontario. Analysis of control soil samples collected from each site to determine soil texture and electrical conductivity was performed by the University of Guelph Laboratory Services – Agriculture and Food Laboratory.

The Nashville field site (43°54′08″N, 79°41′10″W) soil texture was silty clay loam, with the following components: gravel 0.0%, sand 19.1%, silt 53.4% and clay 27.5%. The electrical conductivity was 7.5 mS m^{-1} and the soil moisture content varied between 20 and 30% throughout the study. Annual temperatures in the region range from −32.8 to 40.6 °C, with a daily mean temperature of 9.2 °C. Average annual rainfall in the region is 709.8 mm with 834 mm of precipitation and 133.1 cm of snowfall (www.climate.weatheroffice.ec.gc.ca).

The Springwater field site (44°22′48″N, 79°45′80″W) soil texture was fine sand, with the following components: gravel 0.0%, sand 97.6%, silt 1.2% and clay 1.2%. The electrical conductivity was 5.9 mS m^{-1} and the soil moisture content varied between 2 and 6% throughout the study. Annual temperatures in the region range from −35 to 36 °C, with a daily mean temperature of 6.7 °C. Average annual rainfall in the region is 700.2 mm, with 938.5 mm of precipitation and 238.4 cm of snowfall (www.climate.weatheroffice.ec.gc.ca).

The Dummer field site (44°18′00″N, 78°19′00″W) soil texture was fine sandy loam, with the following components: gravel 0.0%, sand 59.9%, silt 35.2% and clay 4.9%. The electrical conductivity was 39.5 mS m^{-1} and the soil moisture content varied between 15 and 18% throughout the study. Annual temperatures in the region range from −35.5 to 36.5 °C, with a daily mean temperature of 6.6 °C. Average annual rainfall in the region is 715.3 mm, with 869.6 mm of precipitation and 165 cm of snowfall (www.climate.weatheroffice.ec.gc.ca).

Burial parameters

A total of 45 pig carcasses were buried across the three field sites. Burial formations at the Nashville and Dummer sites were in the shape of a cross (Fig. 1). This grave arrangement was used for ease of data collection for GPR and other geophysical surveys (data not included in this study). At the Springwater site, burials were arranged in two parallel lines owing to a space constraint and potential safety hazards to researchers. Burial occurred

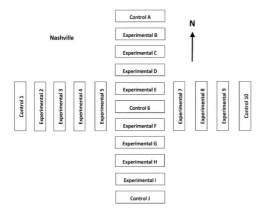

Fig. 1. Nashville field site burial arrangement schematic.

on 11 August 2008. Pig carcasses were purchased from a dead stock company and were euthanized according to industry standards (head bolt; Olfert et al. 1993). Carcasses were buried within several hours of death (c. 1–5 h depending upon site location). Each site consisted of 15 graves containing a carcass and five control graves containing no carcass, to establish a baseline for comparison with the decomposition process of pig carcasses (a total of 20 graves per site).

To more closely represent forensic scenarios, the carcasses were buried at a depth of c. 0.76 m (2.5 ft) in 100% cotton t-shirts and 50% cotton/50% polyester briefs, which are representative of common textiles. The control graves also contained the t-shirts and briefs, in order to determine the natural rate of decomposition of the fibres based upon the soil texture and microbial environment.

GPR data collection

The study was conducted over a 14-month period from 11 August 2008 until October 23, 2009. GPR data were collected during the following months: August, September and October of 2008, and July, August, September and October of 2009. Data collection correlated with the climate of south-central Ontario. This climate experiences temperatures ranging from −35 °C in the winter to 41 °C in the summer and several months of snowfall in autumn, winter and spring. Owing to heavy rain in the area, GPR data was not collected from the Nashville site for the month of July 2009.

A Sensors & Software Inc. Noggin Plus 500 (Mississauga, Canada) ground penetrating radar antenna was used for surveying the graves (Sensors & Software Inc. n.d.). A SmartCart configuration was used to allow for quick and efficient coverage of the sites. A Digital Video Logger was used in the field as a guide for line tracking, to provide real-time display and record data. All data were stored onto a SanDisk Extreme III 1.0 GB CompactFlash, and downloaded to a computer.

A line survey pattern was used at the Nashville site owing to the rough terrain. The use of the line pattern adhered to procedures used by forensic identification officers (Ruffell & Mckinley 2008). A grid pattern was used at the Springwater and Dummer sites for greater detail. The use of the grid pattern adhered to procedures used by researchers and archaeologists (Davis et al. 2000; Neubauer 2001). A total of eight lines were collected at Nashville. Springwater consisted of four grids; three 10 × 10 m and one 2 × 10 m. Dummer consisted of three grids: one 40 × 5 m and two 5 × 15 m. A transect spacing of 0.5 m was used (Davis et al. 2000; Neubauer 2001). The grid, line and spacing parameters used ensured coverage of all sites, and overlap of undisturbed soil.

The software used to view, analyse and qualitatively interpret the GPR data was produced by Sensors & Software Inc.: GFP Edit, EKKO View, and EKKO View Deluxe (Mississauga, Canada). The programs are designed to create, view and edit GFP (GPR Files and Parameters) files.

Results

Pringle et al. (2012) found that target hyperbola(e) for buried pig carcasses were evident in raw 2D data profiles and that 'time slices' need only be produced when the time since burial exceeds 18 months. The burial period for the current study was 14 months. The GPR results are presented as raw data to demonstrate the anomalies observed in real time at the grave sites. Further processing of the data assisted in enhancing the hyperbola and confirming the GPR reflection response for grave sites. A representative line from each site including 10 graves (both experimental and control) is shown to demonstrate the degree of reflection evident at the completion of the study period. Given the large number of grave sites studied, it was not possible to include single, enhanced images for each of the grave sites surveyed at all three locations.

Nashville

Line data were collected in two sections (cross formation) containing 10 graves each: seven experimental and three control. A strong hyperbola was identified for the 10 graves located in each line. Throughout the period of study, the GPR data remained similar in that a hyperbolic-shaped reflection response from all 20 graves was detected on all of the data collection dates. A representative line collected in October 2009 is shown in Figure 2, demonstrating the seven experimental graves and three control graves for one line. A discernible difference in the hyperbolic-shaped reflection response between the control and experimental graves was not observed.

Springwater

Grid data were collected across the two parallel lines containing 10 graves each. Figure 3 shows a representative line from a grid pattern collected in September 2008 (32 days post-burial). Reflection responses were identified as hyperbola with severely reduced reflection amplitudes. Within the parameters of the collected line, four graves should have been detected (three experimental and one control); however, only three experimental graves were identified. The control grave did not produce any hyperbolic-shaped reflection responses. The September 2008 dataset represented the first and only collection date where graves were evident. Reflection responses were not detected for any of the other collection dates at the Springwater grave sites.

Dummer

Grid data were collected over the two lines in the cross formation. The graves could be consistently and clearly identified by strong hyperbola. By the completion of the GPR data collection in October 2009, the grave locations at the Dummer site were still discernible (Fig. 4) although demonstrating reduced reflection amplitudes. Within this figure, 11 hyperbolic shapes representing graves are present. This response was accurate as 11 graves were dug within the section, despite the fact that only 10 were required. A distinct difference in hyperbolic-shaped reflection responses between the control and experimental graves was not identified.

The soil composition, electrical conductivity (EC), moisture content and GPR results are summarized in Table 1. Figure 5 is also included (from Grisso *et al.* 2009) as a reference for the expected ranges in conductivity for different soils.

Discussion

It has been extensively reported that soil properties (including soil texture, moisture and electrical conductivity) will affect the capability of GPR to detect clandestine graves. Results from the current study indicate that GPR provided the most valuable data when used in a silty clay loam soil with medium–low electrical conductivity and moderate–high moisture content (Nashville site). All 20 graves at the Nashville site were detectable by GPR for the entire 14-month duration of the study. The hyperbolae were discernible with consistently strong reflection amplitudes. These findings contradict results presented by Buck (2003), who found that GPR use was not successful in locating an excavated and back-filled trench in silty clay loam soil that was only days old. GPR testing in areas of known soil conditions with clearly defined

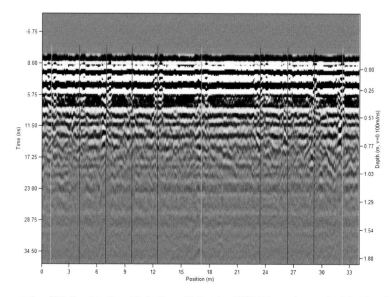

Fig. 2. Representative GPR line data from Nashville – 16 October 2009 (14 months post-burial). Lines bisecting graves represent: ----- control; and ----- experimental.

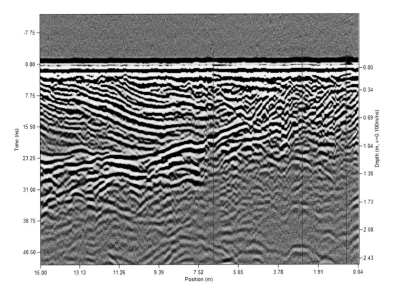

Fig. 3. Representative GPR line data collected from a grid pattern from Springwater – 11 September 2008 (1 month post-burial). Lines bisecting graves represent: ----- experimental graves.

features of known dimensions is important to determine radar applicability based upon soil texture, moisture and conductivity. Long-pulse antennae (300 MHz) are effective for imaging depths of >3.5 m (Miller 1996; Davis *et al.* 2000) whereas 500 MHz antennae are useful for depth investigations >0.5 m (Miller 1996). The study by Buck (2003) involved a trench that was 2.5 m deep. It is possible that the antenna frequency selected did not provide sufficient depth penetration to clearly detect the trench outline in that particular soil environment.

Schultz *et al.* (2006) found that carcasses buried in clayey soils were only generally detectable by GPR for the first 6 months post-burial, and more difficult to discern after that time period. Soils containing a high clay content and with a high electrical conductivity can attenuate EM

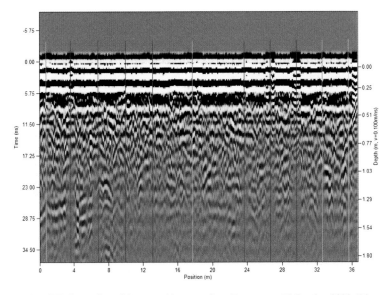

Fig. 4. Representative GPR line collected from a grid pattern from Dummer – 15 October 2009 (14 months post-burial). Lines bisecting graves represent: ----- control; ----- experimental; and ―― extra grave.

Table 1. *Summary of results*

	Conductivity (mS m^{-1})	Soil moisture (%)	Soil composition (%)	Comments on data and GPR response
Nashville	7.5	20–30	Sand: 19.1 Silt: 53.4 Clay: 27.5	Medium–low conductivity (not consistent with Fig. 5), moderate–high moisture, good GPR response over study period*
Springwater	5.9	2–6	Sand: 97.6 Silt: 1.2 Clay: 1.2	Low conductivity (consistent with Fig. 5), low moisture, poor GPR response initially and then no GPR response*
Dummer	39.5	15–18	Sand: 59.9 Silt: 35.2 Clay: 4.9	Medium conductivity (c. 1 order of magnitude higher than in other sites), medium moisture, good GPR response but decreasing with time*

*GPR response refers to the quality and amplitude of hyperbolic-shaped reflection responses.

wave propagation, resulting in a reduction of depth of penetration into the ground and prevention of the detection of burial sites and features contained within them (Schultz *et al.* 2006). Clay can mask the remains by limiting the dielectric permittivity of the body with that of the soil horizon (Schultz *et al.* 2006). The clay content in the silty clay loam soil at the Nashville site was 27.5%. The fact that the Nashville soil was a loam mixture with medium–low conductivity may explain why graves and carcasses were detectable using GPR in the present study, but not by Schultz *et al.* (2006).

In contrast to previous research (Schultz *et al.* 2006; Schultz 2008), the present study found that clandestine graves could not be detected with accuracy by GPR in a fine sandy soil with low electrical conductivity and moisture content (Springwater site). Schultz (2008) found that the degree of skeletonization of buried carcasses appeared to have the greatest effect on whether or not a distinctive hyperbolic shaped response was discernable over the duration of a 21.5-month period. Over time, as a carcass progresses through the stages of decomposition, the dielectric permittivity surrounding the body will equilibrate to the surrounding soil owing to the movement of the soil solution or groundwater (Schultz 2008), making detection by GPR more difficult. It is unclear why, in the present study, graves were not detectable even during the early stage of decomposition (autolysis). The dry, sandy conditions combined with a low electrical conductivity are considered ideal for GPR responses. It is thought that the contrasting textures of sandy soil in the current study compared with those studies conducted by Schultz *et al.* (2006) and Schultz (2008) played a role. The specific properties of the sand within those studies were not stated, and it is possible that those soil environments consisted of more uniform sand particles.

The Springwater site was located at a commercial gravel pit, which represents an extensively disturbed site. The soil environment consisted of fine and very fine sand as well as some gravel identified below the depth of the graves. Nobes (2000) highlights the difficulty in detecting a body or bones in sites that are substantially disturbed because the target response can be readily masked by the background site variation. Within sand, depth of GPR penetration is dependent upon the pore water conductivity, more so than the sand material, and bedding within wet sand deposits can also mask grave detection (www.sensoft.ca). However this was not the case in the current study as the moisture content of the soil only varied between 2 and 6% throughout the entire study period. It is therefore likely that the nature of the disturbed soil caused the greatest attenuation of the EM waves and may explain our contradictory findings.

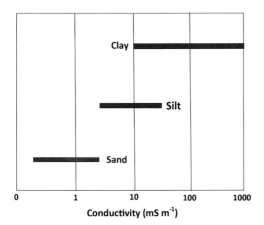

Fig. 5. Expected ranges of soil conductivities for sand, silt and clay (from Grisso *et al.* 2009; http://pubs.ext.vt.edu/442/442-508/442-508_pdf.pdf).

Grave detection in fine sandy loam soil (Dummer site) was successful for the duration of the present study. The hyperbolic-shaped reflection responses from the graves became less defined as the study progressed but were still visible in both experimental and control graves after 14 months' burial. These findings correlate with results by Pringle et al. (2012). who found that GPR could successfully locate buried pig carcasses in a sandy loam soil up to 18 months post-burial. It should be noted that the soil moisture content varied between the two studies, although the background soil conductivity measurements were comparable.

The results of the current study suggest that GPR is most applicable in loam soils even with varying degrees of sand, silt and clay in southern Ontario, Canada. Our findings contradict some of the previously established ideas about the usefulness of GPR in sand v. clay soils. However, it must be highlighted that soil texture alone does not dictate the value of using GPR in a forensic investigation. Soil properties and environmental conditions also need to be considered when determining the likelihood of success in locating a clandestine grave or buried anomaly.

Conclusion

Ground penetrating radar is a useful tool in the location of clandestine graves in areas of known soil conditions, specifically due to its non-invasive nature. Although the use of GPR in forensic scenarios has seen increased interest in recent years, the use of other more traditional non-invasive techniques, such as changes in foliage and soil depression above a grave, should also be considered. We believe that the most effective means of searching for a clandestine grave is a combination of techniques, including GPR. Further controlled research into the applicability of GPR for the detection of clandestine graves based upon soil properties (i.e. texture, moisture and electrical conductivity), rate of carcass decomposition and length of burial is necessary for GPR to remain an effective tool within law enforcement.

References

BUCK, S. C. 2003. Searching for graves using geophysical technology: field tests with ground penetrating radar, magnetometry, and electrical resistivity. *Journal of Forensic Sciences*, **48**, 5–11.

DAVIS, J. L., HEGINBOTTOM, J. A. ET AL. 2000. Ground penetrating radar surveys to locate 1918 Spanish flu victims in permafrost. *Journal of Forensic Sciences*, **45**, 68–76.

DOOLITTLE, J. A., MINZENMAYER, F. E., WALTMAN, S. W., BENHAM, E. C., TUTTLE, J. W. & PEASLEE, S. D. 2007. Ground-penetrating radar soil suitability map of the conterminous United States. *Geoderma*, **141**, 416–421.

GRISSO, R., ALLEY, M., HOLSHOUSER, D. & THOMASON, W. 2009. *Precision farming tools: soil electrical conductivity*. Report produced by Virginia Cooperative Extension, Publication 442-508.

MILLER, P. S. 1996. Disturbances in the soil: finding buried bodies and other evidence using ground penetrating radar. *Journal of Forensic Sciences*, **41**, 648–652.

NEUBAUER, W. 2001. Images of the invisible – prospection methods for the documentation of threatened archaeological sites. *Naturwissenschaften*, **88**, 13–24.

NOBES, D. C. 2000. The search for 'Yvonne': a case example of the delineation of a grave using near-surface geophysical methods. *Journal of Forensic Sciences*, **45**, 715–721.

NOTTER, S. J., STUART, B. H., ROWE, R. & LANGLOIS, N. 2009. The initial changes of fat deposits during the decomposition of human and pig remains. *Journal of Forensic Sciences*, **54**, 195–201.

OLFERT, E. D., CROSS, B. M. & MCWILLIAM, A. A. 1993. *Guide to the Care and Use of Experimental Animals*, **1**, Canadian Council on Animal Care, Ottawa, Ontario, Canada.

OWSLEY, D. W. 1995. Techniques for locating burials, with emphasis on the probe. *Journal of Forensic Sciences*, **40**, 735–740.

PRINGLE, J. K., JERVIS, J. R., HANSEN, J. D., JONES, G. M., CASSIDY, N. J. & CASSELLA, J. P. 2012. Geophysical monitoring of simulated clandestine graves using electrical and ground-penetrating radar methods: 0–3 years after burial. *Journal of Forensic Sciences*, **57**, 1467–1486.

RODRIGUEZ, W. C. & BASS, W. M. 1985. Decomposition of buried bodies and methods that may aid in their location. *Journal of Forensic Sciences*, **30**, 836–852.

RUFFELL, A. 2005. Searching for the IRA 'disappeared': ground-penetrating radar investigation of a churchyard burial site in Northern Ireland. *Journal of Forensic Sciences*, **50**, 1430–1435.

RUFFELL, A. & MCKINLEY, J. 2008. *Geoforensics*. Wiley-Blackwell, Chichester, 77–85.

RUFFELL, A., MCCABE, A., DONNELLY, C. & SLOAN, B. 2009. Location and assessment of an historic (150–160 years old) mass grave using geographic and ground penetrating radar investigation, NW Ireland. *Journal of Forensic Sciences*, **54**, 382–394.

SCHOENLY, K. G., HASKELL, N. H., MILLS, D. K., BIEME-NDI, C., LARSEN, K. & LEE, Y. 2006. Recreating death's acre in the school yard: using pig carcasses as model corpses to teach concepts of forensic entomology and ecological succession. *The American Biology Teacher*, **68**, 402–410.

SCHULTZ, J. J. 2008. Sequential monitoring of burials containing small pig cadavers using ground penetrating radar. *Journal of Forensic Sciences*, **53**, 279–287.

SCHULTZ, J. J., COLLINS, M. E. & FALSETTI, A. B. 2006. Sequential monitoring of burials containing large pig cadavers using ground-penetrating radar. *Journal of Forensic Sciences*, **51**, 607–616.

SENSORS & SOFTWARE INC n.d. Ground penetrating radar (GPR) technology, http://www.sensoft.ca/ (accessed 17 November 2009).

Comparison of magnetic, electrical and ground penetrating radar surveys to detect buried forensic objects in semi-urban and domestic patio environments

J. D. HANSEN & J. K. PRINGLE*

School of Physical Sciences and Geography, Keele University, Keele, Staffordshire ST5 5BG, UK

*Corresponding author (e-mail: j.k.pringle@keele.ac.uk)

Abstract: Near-surface geophysical techniques should be routinely utilized by law enforcement agencies to locate shallowly buried forensic objects, saving human and other resources. However, there has been little published research on optimum geophysical detection method(s) and configurations beyond metal detectors. This paper details multi-technique geophysical surveys to detect simulated unmarked illegal weapons, explosive devices and arms caches that were shallowly buried within a semi-urban environment test site. A concrete patio was then overlaid to represent a common household garden environment before re-surveying. Results showed that the easily utilized magnetic susceptibility probe was optimal for target detection in both semi-urban and patio environments, while basic metal detector surveys had a lower target detection rate in the patio scenario with some targets remaining undetected. High-frequency (900 MHz) ground penetrating radar antennae were optimum for target detection in the semi-urban environment, while 450 and 900 MHz frequencies had similar detection rates in the patio scenario. Resistivity surveys at 0.25 m probe and sampling spacing were good for target detection in the semi-urban environment. 2D profiles were sufficient for target detection, but resistivity datasets required site de-trending to resolve targets in map view. Forensic geophysical techniques are rapidly evolving to assist search investigators to detect hitherto difficult-to-locate buried forensic targets.

Geo-scientific methods are being increasingly utilized and reported upon by forensic search teams for the detection and location of clandestinely buried material in terrestrial environments. Parker *et al.* (2010) provide a comprehensive review of forensic geophysical searches within freshwater bodies. In a law enforcement context, forensic burials are at a maximum of 10 m below ground level (b.g.l.) and usually much shallower (Fenning & Donnelly 2004). Forensic objects needing to be located vary from illegally buried weapons and explosives, landmines and improvised explosive devices (IEDs), drugs and weapons caches to clandestine graves of murder victims and genocide graves (see Pringle *et al.* 2012a). In the USA, neighbourhood criminal gangs often hide used illegal weapons for later recovery (Dionne *et al.* 2011).

Recovery of buried forensic material often results in successful criminal convictions and it is thus critical for it to be located (Harrison & Donnelly 2009). Law enforcement agencies need to have prioritized locations for physical excavation owing to shortages in human and other resources, especially if the search area is large. Specialist trained search dogs have been widely used to identify different buried objects, commonly IEDs (see Curran *et al.* 2010), drugs and human remains, the latter teams sometimes being referred to as cadaver dogs (see Rebmann *et al.* 2000), but are less successful with buried inorganic objects. Metal detector search teams are used during forensic investigations when it is deemed appropriate, especially when there is a high contrast between the target and local background environment (see Nobes 2000).

Geotechnical investigations routinely use near-surface geophysical methods to identify buried locations of, for example, cleared building foundations and underground services (see Reynolds 2011), as well as environmental forensic objects such as illegally buried waste (see Bavusi *et al.* 2006; Ruffell & Kulessa 2009). Magnetic detection methods are commonly used in geotechnical (e.g. Marchetti *et al.* 2002; Reynolds 2004, 2011) and forensic archaeological investigations (see Linford 2004; Hunter & Cox 2005). Acheroy (2007) provides a useful review of field detection of antipersonnel mines using ground penetrating radar (GPR).

However, little control study research has been published in which buried forensic objects are detected using a variety of geophysical methods, other than to confirm metal detection team results (e.g. Davenport 2001; Rezos *et al.* 2010) and for human remains (e.g. Miller 1996; Davenport 2001;

Schultz et al. 2006; Pringle et al. 2008; 2012b; Schultz 2008). Dionne et al. (2011) did conduct a control study with buried weapons and found that electro-magnetic equipment could detect metallic objects buried in a grid distribution in a rural environment, but this study did not have access to a Geonics™ EM38 instrument. The Murphy & Cheetham (2008) control study found that magnetic techniques had difficulty in differentiating between target buried weapons and background materials, even when surface metallic items were cleared from the survey site prior to geophysical data collection. Murphy & Cheetham (2008) also found that GPR methods could locate buried forensic targets but were difficult to locate in certain orientations, so GPR was an obvious technique to trial.

This case study therefore intended to utilize a variety of current commercial, shallow near-surface geophysical equipment to locate hard-to-detect, small-scale buried forensic metallic objects in a semi-urban environment, using survey procedures commonly used in geotechnical and archaeological investigations. The study site was also re-surveyed once a concrete slab patio had been laid to also simulate a common domestic property garden forensic scenario (see Congram 2008; Toms et al. 2008; Billinger 2009). To give the study more of a sense of realism, the survey was performed on heterogeneous soil content, representative of a UK garden, and both target objects and non-target objects (brick, metallic screw and iron plate) were also buried. The locations and orientations of objects were recorded.

The study objectives for both semi-urban and patio environments were to: (1) evaluate and find optimum magnetic detection technique(s) of the target buried forensic material; (2) compare electrical and GPR detection methods; (3) determine optimum GPR detection frequencies; (4) determine optimum respective equipment configuration(s)/survey specifications/optimum processing steps; (5) determine which technique(s) could determine target depth below ground; and (6) determine if different buried metal types could be distinguished. It was also instructive to decide if certain detection techniques could be relatively easily utilized by forensic investigators to acquire, process and interpret forensic geophysical datasets.

Methodology

Test site

The forensic test site was situated on Keele University campus situated near Stoke-on-Trent, in the UK. It was chosen as a representative of a semi-urban UK environment as the site history indicated the

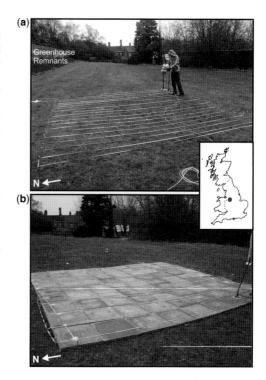

Fig. 1. Photographs of the 5 m by 5 m forensic test site on campus showing (**a**) semi-urban environment and (**b**) simulated domestic concrete patio scenario on the same area with location map (inset). Survey tapes on survey lines are shown. The 0.0 position for all surveys is the SW corner.

presence of greenhouses with remnant cleared foundations still present (Fig. 1). Previous site studies also confirmed this, indicating that the local mixed sand and clay soil was predominantly 'made ground' with Triassic Butterton Sandstone Formation bedrock present at a shallow level, only c. 2.6 m below ground level (or b.g.l.) (see Jervis et al. 2009). The local climate is temperate, which is typical for the UK.

A 5 × 5 m survey area was selected as this was deemed small enough to keep the multi-geophysical techniques data acquisition time feasible, but sufficiently large to allow several targets to be buried and be separately resolvable in the resulting datasets. Permanently marked by plastic tent pegs, survey lines were laid 0.25 m apart (Fig. 1a). Multi-technique geophysical datasets were acquired prior to object burial to give control datasets for comparison purposes (see Table 1). A variety of forensic and mostly metallic objects (see Fig. 2 and Table 2 for details) were then buried c. 15 cm b.g.l. in a non-ordered configuration within the survey area and their locations recorded (Fig. 3).

Table 1. *Summary statistics of geophysical data collected during this 5 × 5 m study area*

Geophysical technique	Survey date (and type)	Equipment setup time (min)	Data acquisition time (min)	Station spacing (m)	Instrument precision	Advantages/disadvantages
Metal detector (Bloodhound Tracker™ IV all-metal)	10 November 2009 (C), 10 December 2009 (S) and 25 February 2010 (P)	1	30	N/A	Unknown	Easy to operate. Picks up all metallic objects. Limited penetration depth.
Magnetic susceptibility (Bartington™ MS.1 with 0.3 m diameter probe)	10 November 2009 (C), 10 December 2009 (S) and 25 February 2010 (P)	1	90	0.25	c. 1 SI	Easy to operate. Limited to c. 8 cm b.g.l.
Fluxgate gradiometer (Geonics™ FM15)	10 November 2009 (C), 10 December 2009 (S) and 25 February 2010 (P)	60	45	0.25	0.1 nT	Can detect subtle targets. Difficult to calibrate and needs careful acquisition.
Magnetic gradiometer (GSMP-40™ potassium-vapour, two sensors 1 m vertical separation)	10 November 2009 (C), 10 December 2009 (S) and 22 March 2010 (P)	60	30	c. 0.05 (collected at 0.05 s)	0.01 nT	Small sample spacing, collects both total field and gradient data. Expensive.
Ground penetrating radar (PulseEKKO™ 1000) using 450 MHz antennae	2 November 2009 (C), 10 December 2009 (S) and 25 February 2010 (P)	30	60	0.05	c. 0.1 m	Resolves fairly small objects and depth to target(s).
Ground penetrating radar (PulseEKKO™ 1000) using 900 MHz antennae	2 November 2009 (C), 10 December 2009 (S) and 25 February 2010 (P)	30	90	0.025	c. 0.05 m	Resolves small objects and depth to target(s). Slow to collect.
Bulk ground resistivity (Geoscan™ RM15-D) using 0.5 m spaced probes	29 October 2009 (C) and 10 December 2009 (S)	10	45	0.5	c. 0.25 m	Relatively quick to collect. Will detect objects up to 1 m b.g.l. Not usable on patios.
Bulk ground resistivity (Geoscan™ RM15-D) using 0.5 m spaced probes	29 October 2009 (C) and 10 December 2009 (S)	10	60	0.25	c. 0.125 m	Will detect objects up to 0.5 m b.g.l. Not usable on patios.

Survey types are: (C) control, (S) semi-urban and (P) patio environments. b.g.l., Below ground level. Survey line spacings were 0.25 m unless otherwise stated.

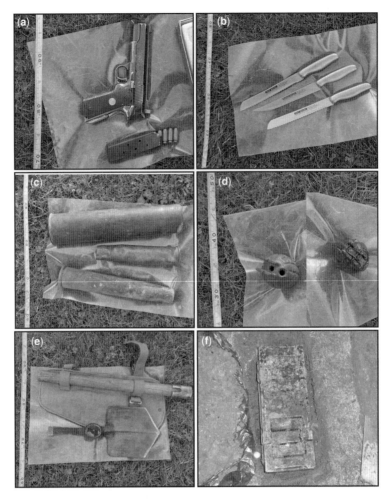

Fig. 2. Selected photographs of forensic buried test objects. (**a**) Colt Government Cup Replica 0.45 calibre automatic handgun with solid brass ammunition; (**b**) three domestic stainless steel kitchen breadknives; (**c**) 1943 75 mm M18 shell and two Second World War smaller diameter spent shells; (**d**) (left) Second World War Allied hand grenade and (right) First World War Allied Mk 1 no. 5 decommissioned hand grenade; (**e**) 1943 Allied wooden-handled entrenchment tool; and (**f**) UK mortar ammunition box (containing the two shell casings shown in c). See Table 2 for details.

Note that the ammunition box (Fig. 2f) had to be dug well below this depth to ensure the top was consistent with other target depths. In addition to these eight target objects, three non-target, non-forensic objects were buried, including a domestic house brick, a steel plate and a metallic bolt for control and comparison purposes (see Fig. 2 and Table 2). This approach therefore significantly differed from the single technique and more ordered target control studies undertaken by Rezos *et al.* (2010) and Dionne *et al.* (2011). The survey area was then geophysically re-surveyed at least 2 weeks after the forensic objects were buried to ensure some settlement of replaced topsoil. Finally a 6 cm thick layer of concrete paving slabs (*c.* 0.5 m by *c.* 0.5 m) was laid over the grid (Fig. 1b) and the area then geophysically re-surveyed for the last time, with the exception of a resistivity survey owing to the inability to insert resistivity probes into the patio slabs.

Metal detector surveys

Standard metal detectors produce an alternating magnetic field that may induce nearby conductive material to produce a secondary field. When the

Table 2. *Description of buried forensic objects used in this study and their known properties (photographs are shown in Fig. 2)*

Number	Forensic buried object	Size (m)	Description
1	Brick	0.17 × 0.11	Clay house-brick, orientated horizontally
2	Bolt and screw	0.08 × 0.05	Unknown metal alloy
3	Steel plate	0.2 × 0.2 × 0.05	Stainless steel, flat, square plate, orientated horizontally
4	Breadknives (Fig. 2b)	0.3 × 0.05	Two domestic stainless steel kitchen breadknives wrapped in thin plastic bag. Orientated north–south
5	Spade (Fig. 2e)	Handle: 0.4 × 0.07; head: 0.32	1943 Allied wooden-handled entrenchment tool with metallic head, orientated NW–SE
6	Knife (Fig. 2b)	0.3	One domestic stainless steel kitchen breadknife, orientated east–west
7	Second World War Grenade (Fig. 2d)	0.08 diameter	Second World War Allied decommissioned metallic hand grenade, orientated vertically
8	First World War Grenade (Fig. 2d)	0.08 diameter	1915 no. 5 Mk 1 Allied decommissioned metallic hand grenade, orientated vertically
9	Handgun (Fig. 2a)	0.18 × 0.14	Colt Government Cup Replica .45 calibre automatic replica handgun with solid brass ammunition. Most likely zinc alloy with stainless steel finish. Wrapped in thin plastic bag and orientated east–west
10	Mortar shell (Fig. 2c)	0.37 × 0.17	Brass spent mortar shell: 1943, 75 mm M18, orientated east–west
11	Ammunition box (Fig. 2f)	0.55 × 0.4 × 0.45	UK mortar ammunition metallic box containing two small Second World War spent mortar shells (Fig. 2c), orientated north–south

Object numbers refer to those shown in Figure 3 and in geophysical datasets.

equipment detects a magnetic field which is in-phase with the transmitted field, it produces an audible (but not usually measured) response (see Dupras *et al.* 2006; Milsom & Eriksen 2011 for theoretical background). The Bloodhound Tracker™ IV all-metal detector was used on the survey site before objects were buried (to act as control), after objects were buried and finally after the concrete patio was laid (Fig. 4a) using a sweep method in parallel transects 0.5 m apart at a constant height of *c.* 5 cm (see Dupras *et al.* 2006; Rezos *et al.* 2010). Any areas where the detector produced an audible signal were then marked on a map of the survey area. These surveys were repeated by three different operators in an attempt to account for any operator technique variations. The survey area was then re-surveyed after forensic objects were buried, and again after the patio was laid (Table 1) with audio target locations again noted each time.

Magnetic susceptibility surveys

Magnetic susceptibility meters generate a low-intensity AC magnetic field and measure the resulting change in positive or negative susceptibilities in SI (dimensionless) units of the sampled medium. This bulk reading is usually due to a combination of highly magnetic minerals (e.g. magnetite), man-made ferro-magnetic material (if present), other materials and background magnetism (see Milsom & Eriksen 2011; Reynolds 2011 for further information). Magnetic susceptibility data were collected using a Bartington™ MS.1 susceptibility instrument with a 0.3 m diameter probe placed on the ground surface at each sampling point (Fig. 4b). Data samples were collected on a 0.25 m grid over the survey area before forensic object burial to act as a control, then resurveyed after burial and finally again after the patio was laid (Table 1). This was a smaller data point sample spacing than typically utilized for clandestine grave surveys (see, e.g. Pringle *et al.* 2008).

Basic data processing was initially undertaken, which involved de-spiking to remove anomalously large isolated data points caused by operator/equipment error. Data were then processed using the Generic Mapping Tools software (Wessel & Smith 1998). To aid visual interpretation of the data, a minimum curvature gridding algorithm was used to interpolate each dataset to a cell size of

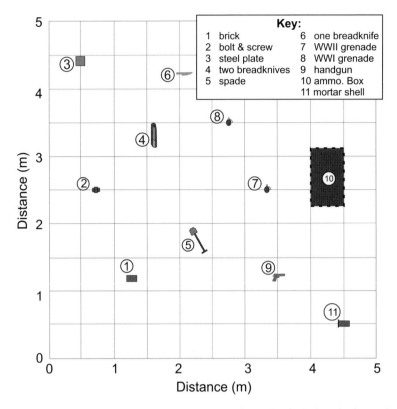

Fig. 3. Site map showing location of buried forensic objects (see key for details) for both semi-urban environment and patio scenarios (Fig. 2 for selected object photographs).

0.0125 × 0.0125 m. In addition, 'de-trending' of the data was conducted to remove long-wavelength site trends to allow smaller, target-sized features to be more easily identified. This was achieved by fitting a cubic surface to the gridded data and then subtracting this surface from the data, as this surface gridding method was found to produce the best results.

Fluxgate gradiometry surveys

Fluxgate gradiometry equipment records only the vertical (Z) component of the Earth's magnetic field that will be affected by proximal ferro-magnetic materials, their orientation, depth b.g.l., and so on (see Milsom & Eriksen 2011 and Reynolds 2011 for more information). Owing to the short data acquisition time (see Table 1), it was deemed not necessary to undertake diurnal correction of the datasets (see Milsom & Eriksen 2011 for further information). Fluxgate gradiometry data were collected using a Geoscan™ FM18 gradiometer held at a constant height (Fig. 4c). For all three surveys (Table 1) the meter was first carefully zeroed over a magnetically 'quiet' area out of the survey area to remove any potential reading differences that may result from positional variation in instrument orientation relative to magnetic north when acquiring data (see Milsom & Eriksen 2011). Survey lines were also orientated to magnetic north to avoid any potential profile line orientation issues (Fig. 1). Basic data processing was again undertaken, which involved de-spiking and de-trending as previously discussed.

Magnetic (potassium-vapour) gradiometry surveys

Magnetic gradiometry data were collected using a GSMP-40 potassium vapour magnetic gradiometer using 1 m vertically separated total field sensors (Fig. 4d and Table 1). As with the fluxgate gradiometry equipment, the potassium vapour gradiometer is another method of measuring the vertical component of the Earth's magnetic field that will be affected by proximal ferro-magnetic materials. The advantages of this equipment are that it collects both upper/lower sensor total magnetic vertical (Z)

Fig. 4. Photographs of geophysical equipment used in this study. (**a**) Bloodhound Tracker™ IV metal detector; (**b**) Bartington™ magnetic susceptibility probe MS.1 with 0.3 m diameter probe; (**c**) Geoscan™ FM-15 fluxgate gradiometer; (**d**) GSMP-40™ potassium-vapour magnetic gradiometer with sensors 1 m vertically separated; (**e**) Geoscan™ RM15-D mobile probe resistivity meter; and (**f**) pulseEKKO™ 1000 Ground Penetrating Radar equipment showing 450 MHz dominant-frequency, bistatic fixed-offset antennae.

field readings as well as gradient measurements between the two sensors and it is the industry standard for geotechnical investigations (see Reynolds 2004, 2011). Owing to the short data acquisition time (see Table 1), it was again deemed not necessary to undertake diurnal correction of the datasets. Data was acquired over the 0.25 m spaced survey lines obtaining readings every 0.2 s, which roughly equated to a sample spacing of c. 0.01 m. The equipment was maintained at a constant height above the ground surface for all surveys (to reduce any data variation owing to variable instrument height) by use of a temporary non-magnetic stick attached to the bottom sensor (Fig. 4d).

Minimal data processing was undertaken, which involved data de-spiking and de-trending as previously discussed.

Fixed-offset resistivity surveys

The inverse of conductivity, electrical resistivity is measured by applying a constant current through a sample (here soil) of known size and measuring the resulting drop in voltage (see Milsom & Eriksen 2011; Reynolds 2011). Bulk-ground resistivity data were collected using a Geoscan™ RM15-D resistance meter mounted on a custom-built frame, which allowed the almost simultaneous acquisition of both 0.25 and 0.5 m spaced, pole–pole probe array measurements using four 0.1 m-long stainless steel electrodes (Fig. 4e). The pole–pole probe array was used as it is rapid, the most popular configuration used and deemed most sensitive to near-surface lateral variations (see Milsom & Eriksen 2011). Remote probes were placed 1 m apart at a distance of 15 m from the survey area to ensure that probe placements did not affect the resulting data (see Milsom & Eriksen 2011). For the control and semi-urban surveys (Table 1), resistivity measurements were made at 0.25 m intervals along survey lines that were spaced 0.25 m apart (Table 1). This sample spacing was smaller than the more typically used 0.5 m spaced resistivity datasets (see, e.g. Pringle & Jervis 2010), but high-resolution datasets were deemed important to acquire for comparison purposes to the magnetic surveys. A post-burial survey was not possible to be acquired over the patio owing to a requirement for probes to be inserted into the ground using the utilized equipment. Minimal data processing was undertaken, which involved data de-spiking and de-trending as previously discussed.

Ground penetrating radar surveys

Ground penetrating radar is a well-documented technique, using an antenna to transmit an electromagnetic pulse into the ground, which reflects at boundaries of contrasting di-electric permittivity and is captured by a receiver antenna, subsequently being converted to a digital image and stored (see Milsom & Eriksen 2011; Reynolds 2011). The signals stored in time formats can be converted to depth if the local site velocity is known. GPR signal penetration depth and resolution are a function of antennae set frequencies; high frequency (450+ MHz) gives relatively high resolution but poor penetration while low frequency gives low resolution but good penetration (see Jol 2009 for background theory and operational detail). GPR datasets were collected using pulseEKKO^{TM} 1000 equipment using both 450 MHz (Fig. 4f) and 900 MHz dominant-frequency bi-static, fixed-offset (0.34 and 0.17 m respectively) antennae along 0.25 m spaced lines and having trace sample intervals of 0.05 and 0.025 m respectively (Table 1). The survey area was surveyed three times: the first to provide a control dataset, the second over the buried forensic objects and the third over the buried forensic objects in the patio scenario.

The resulting GPR datasets were sequentially processed using Reflex-Win^{TM} Version 3.0 (Sandmeier) software using the following steps: (1) 'dewow' (low-cut filter) to remove nonlinear effects associated with the antennae; (2) movement to constant start-time; (3) 1D bandpass filter (Butterworth) to remove high frequency noise; (4) 2D filter to make anomalous features more prominent; (5) Stolt migration to collapse hyperbolae to point sources (only used for time-slices); and finally (6) horizontal time-slice generation of each dataset to produce plan-view, relative-amplitude images of the test site.

Results

Metal detector

For the post-burial semi-urban environment survey, all eight target objects and one non-target object were detected. The two undetected objects were: the (1) brick (as might be expected); and (2) the metallic bolt (Fig. 3 and Table 2). For the post-burial patio survey, the brick and metallic bolt non-target objects remained undetected and, of the target objects, the (5) entrenching tool and both the (7) Second and (8) First World War hand grenades were also not detected. Therefore 100% (semi-urban) and 63% (patio) total target detection success rates were calculated for the respective metal detector surveys. For both surveys, six additional anomalies were noted.

Magnetic susceptibility

Magnetic susceptibility datasets (441 data points for each survey) for the control, post-burial semi-urban and patio environment scenarios were highly variable between surveys, having respective median and 2σ values of 55.0 SI and 214.8 2σ (control), 93.0 SI and 412.2 2σ (semi-urban) and 42.0 SI and 110.8 2σ (patio), respectively. The 2σ (two standard deviations) given here and throughout represents a 95% confidence limit and gives the variance of each respective dataset. The control and semi-urban survey results indicated significant heterogeneous ground conditions, as would be expected as the test site was a semi-urban environment.

Magnetic susceptibility data for the post-burial, semi-urban environment also showed significant site variations, with the same magnitude of high- and low-susceptibility readings as obtained in the control dataset. In addition to the control isolated high anomalies again being present, several other isolated high anomalies were present that could be correlated with two non-target object locations – (2) the bolt and (3) the steel plate – and four target object locations – (4) the two breadknives, (5) the entrenching tool, (6) the single breadknife and (7) Second World War hand grenade. Low isolated anomalies, with respect to background values,

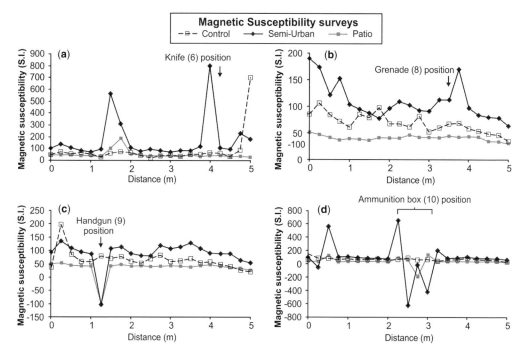

Fig. 5. Magnetic susceptibility selected 2D profiles for control, semi-urban and patio surveys with respective target positions marked. (**a**) Profile 9 ($X = 2$ m) over target (6), single knife; (**b**) profile 12 ($X = 2.75$ m) over target (8), First World War hand grenade; (**c**) profile 15 ($X = 3.5$ m) over target (9), handgun; and (**d**) profile 18 ($X = 4.25$ m) over target (10), ammunition box (all marked). See key for survey type and Table 1 for details.

could also be correlated with the remaining three target object locations – (9) the handgun, (10) the ammunition box and (11) the spent mortar shell (Figs 5 & 6). Magnetic susceptibility data for the post-burial patio environment had significantly fewer site variations, ranging from −242 to 496 SI units. In addition to the control isolated high anomalies again being present, several other isolated high anomalies were present that could be again correlated with two non-target object locations – (2) the bolt, (3) the steel plate – and now three target object locations – (4) the two breadknives, (5) the entrenching tool and (7) the Second World War hand grenade (Figs 5 & 6). Low isolated anomalies, with respect to background values, could also be correlated with (9) the handgun, (10) the ammunition box and (11) the spent mortar shell locations (Figs 5 & 6). Selected 2D profiles are shown in Figure 6. Target detection rates with magnetic susceptibility are therefore 100% (semi-urban) and 88% (patio) respectively.

Fluxgate gradiometry

Fluxgate gradiometry datasets (441 data points in each survey) for the control, post-burial semi-urban and patio environment scenarios were very variable and geophysically 'noisy', having respective survey median and 2σ values of −56.6 nT and 145 2σ (control), −3.1 nT and 157 2σ (semi-urban) and −45.8 nT and 144 2σ (patio) surveys, respectively. This would be expected in such heterogeneous ground conditions, with a significant proportion of the datasets (32, 31 and 30%, respectively) not recording data at sampling positions. However these non-sample areas were consistent, which suggested that the instrument was not faulty nor calibrated incorrectly. With such a high proportion of the survey area not recording values, the resulting gridded and contoured map view plots of the control, post-burial semi-urban and patio environment scenarios were not that useful, having significant large areas of high and low magnetic gradiometry areas with respect to background values. However, 2D data profiles acquired over the forensic objects did allow estimation of target detection to be undertaken, and some selected 2D survey profiles are shown in Figure 7.

Within the post-burial semi-urban environment, high magnetic anomalies, with respect to background values, could be correlated with one non-target object location – (3) the steel plate – and

Within the post-burial domestic patio environment, high magnetic anomalies, with respect to background values, could again be correlated with (3) the steel plate, and the same four target object locations – (4) two breadknives, (6) the single breadknife, (8) the First World War hand grenade and (10) the ammunition box (Fig. 7).

Fluxgate gradiometry survey results therefore gave a 50% (semi-urban) and 50% (patio) total target detection success rate respectively.

Magnetic (potassium-vapour) gradiometry

Magnetic (potassium-vapour) gradiometry data for the three surveys (total data points of 5437 (control), 3729 (semi-urban) and 4050 (patio), respectively) were also geophysically 'noisy'. Respective survey medians and 2σ of lower sensor total field data were 49 172.7 nT and 450 2σ (control), 49 182.4 nT and 1112 2σ (semi-urban) and 49 184.5 nT and 1106 2σ (patio). Survey medians and 2σ of gradiometry data were 81.7 nT and 860 2σ (control), 88.5 nT and 742 2σ (semi-urban) and 94.8 nT and 708 2σ (patio), indicating a generally good survey repeatability. Magnetic gradiometry map view plots of the control, post-burial semi-urban and patio environment scenarios are shown in Figure 8, and de-trended datasets are displayed in Figure 9 for comparison. It was found considerably easier to use the 2D profiles for estimation of target detection (selected examples shown in Fig. 10) owing to the high variability of gradiometry measurements within the survey area, which made subtle anomalies difficult to identify in plan-view plots (Fig. 8), even after de-trending (Fig. 9).

Within the post-burial semi-urban environment magnetic dataset, high magnetic anomalies, with respect to background values, could be correlated with, of the non-target object locations, (3) the steel plate, and of the target object locations, (6) the single breadknife, (7) the Second World War hand grenade, (8) the First World War hand grenade, (9) the handgun and (10) the ammunition box positions (Figs 8–10). Within the patio scenario magnetic dataset, high magnetic anomalies, with respect to background values, could be correlated with, of the non-target object locations, (2) the bolt and (3) the steel plate, and of the target object locations, (4) the two breadknives, (6) the single breadknife, (7) the Second World War hand grenade, (8) the First World War hand grenade, (9) the handgun and (10) the ammunition box locations (Figs 8–10). Selected 2D survey profiles are shown in Figure 10. Potassium vapour gradiometry survey results therefore gave a 63% (semi-urban) and 75% (patio) total target detection success rate, respectively.

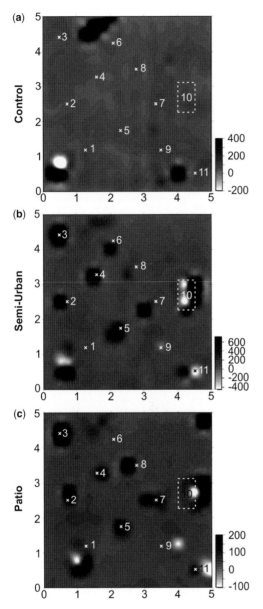

Fig. 6. Magnetic susceptibility processed, gridded and contoured map view data plots of (**a**) pre-burial control with interpreted isolated anomalies, with respect to background values, marked (see text); (**b**) post-burial semi-urban environment; and (**c**) post-burial patio garden environment. Scales for (a) and (b) are the same. SI (dimensionless) units are used (see text). See Table 2 for target descriptions.

four target object locations – (4) two breadknives, (5) the entrenchment tool, (8) the First World War grenade and (10) the ammunition box (Fig. 7).

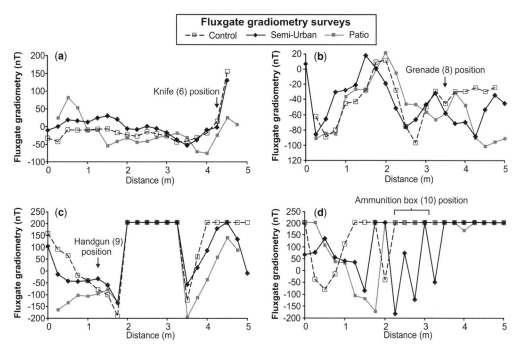

Fig. 7. Fluxgate gradiometry selected 2D surveys profiles for control, semi-urban and patio surveys with respective target positions marked. (**a**) Profile 9 ($X = 2$ m) over target (6), single knife; (**b**) profile 12 ($X = 2.75$ m) over target (8), First World War hand grenade; (**c**) profile 15 ($X = 3.5$ m) over target (9), handgun; and (**d**) profile 18 ($X = 4.25$ m) over target (10), ammunition box (all marked). See key for survey type and Table 1 for details.

Resistivity

Fixed-offset (0.5 m) resistivity data for the control dataset (441 data points) had resistance maximum/minimum values of 111.7 Ω/47.3 Ω with median of 75.0 Ω and 25.4 2σ value, therefore confirming that the site was relatively electrically heterogeneous. The post-burial (semi-urban) 0.25 and 0.50 m fixed-offset repeat surveys had resistance maximum/minimum values of 194.5 Ω/76.0 Ω (25 cm) and 129.5 Ω/51.5 Ω (50 cm), with median values of 121.6 Ω (25 cm)/78.8 Ω (50 cm) and 37.2 2σ (25 cm)/27.2 2σ (50 cm), respectively. Data repeatability for the 0.5 m fixed-offset surveys was therefore generally good, which can also presumably be said for 0.25 m surveys, despite the lack of a control dataset.

Within the post-burial semi-urban environment, high-resistance anomalies in the 0.25 m fixed offset survey, with respect to background values, could be correlated with target object locations of (5) the entrenching tool, (6) the single knife, (7) the Second World War hand grenade, (9) the handgun, (10) the ammunition box and (11) the spent shell (Figs 11 & 12). Low resistance anomalies, with respect to background value, could be correlated with non-target object locations, (1) the brick and (3) the steel plate.

Within the semi-urban environment resistivity (0.5 m fixed-offset) survey, only high-resistance anomalies, with respect to background values, could be correlated with (10) the ammunition box and (11) the spent shell locations (Figs 11 & 12). Selected 2D profiles are shown in Figure 12. This therefore gave a 63% (25 cm) and 25% (50 cm) total target detection success rate.

Ground penetrating radar

Both the 450 and 900 MHz dominant-frequency GPR control datasets showed that a number of non-target objects were located within the survey area; this therefore provides confirmation that the study site is representative of a semi-urban, heterogeneous site. Within the post-burial semi-urban environment dataset, half-parabolae isolated anomalies in the 450 MHz frequency dataset could be correlated with (3) the steel plate, (7) the Second World War hand grenade, (9) the handgun, (10) the ammunition box and (11) the spent mortar shell locations (Figs 13 & 14). Within the 900 MHz frequency dataset, half-parabolae isolated anomalies

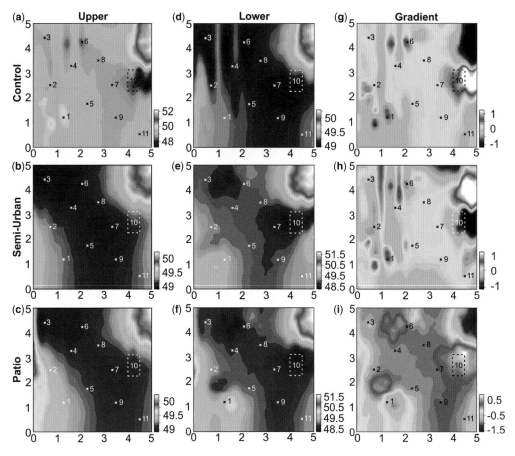

Fig. 8. Magnetic (potassium-vapour) gradiometry processed, gridded and contoured map-view plots using upper sensor, lower sensor and gradient for pre-burial, post-burial semi-urban and patio environments (**a–i**, respectively). Units in nT. See Table 2 for target descriptions.

could be correlated with (3) the steel plate, (4) the two breadknives, (6) the single breadknife, (7) the Second World War hand grenade, (9) the handgun, (10) the ammunition box and (11) the spent mortar shell locations (Figs 13 & 15). Selected 2D profiles are shown in Figures 14 and 15. This therefore gave a 50% (450 MHz) and 75% (900 MHz) total target detection success rate.

Within the post-burial patio environment dataset, half-parabolae isolated anomalies in the 450 MHz frequency dataset could be correlated with (3) the steel plate, (6) the single breadknife, (8) the First World War hand grenade, (9) the handgun, (10) the ammunition box and (11) the spent mortar shell locations (Figs 13 & 14). Within the 900 MHz frequency dataset, half-parabolae isolated anomalies could be correlated with (3) the steel plate, (4) the breadknives, (5) the entrenching tool, (6) the single breadknife, (9) the handgun, (10) the ammunition box and (11) the spent mortar shell locations (Figs 13 & 15). Selected 2D profiles are again shown in Figures 14 and 15. This therefore gave a 63% (450 MHz) and 75% (900 MHz) total target detection success rate.

Discussion

This section has been deliberately organized to answer and discuss the study objectives.

Evaluate and find optimum magnetic detection technique(s) of the target buried material

The metal detector survey results for post-burial, semi-urban surveys of the forensic targets were very successful, with a target detection success rate of 100%. However, the addition of the patio material

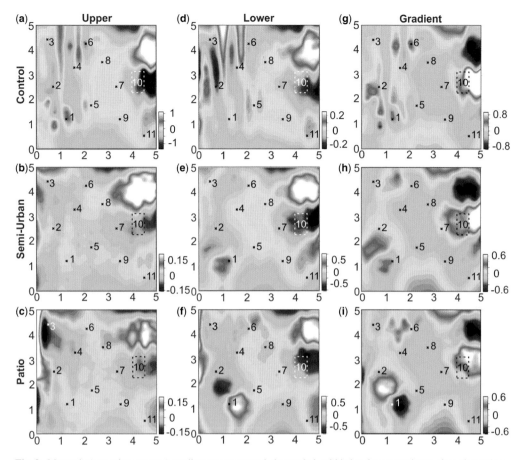

Fig. 9. Magnetic (potassium-vapour) gradiometry processed, de-trended, gridded and contoured map view plots using upper sensor, lower sensor and gradient for pre-burial, post-burial semi-urban and pre-burial patio environments (a–i, respectively). Units in nT. See Table 2 for target descriptions.

over the survey area significantly reduced the success of target detection to 63%. The success rate reduction over the patio was presumably due to the difficulty of the electro-magnetic waves penetrating the concrete paving slabs. These results would be a cause for concern if metal detectors were the sole magnetic detection method in a forensic search within a semi-urban or patio environment as simulated by this study. These results also provide a contrasting metal detector study to Rezos et al. (2010) within a rural environment, which gained a 100% target detection success rate (Fig. 16).

The magnetic susceptibility survey results after burial of forensic targets proved very good, with target detection success rates of 100% (semi-urban) and 88% (patio), respectively (Fig. 16). In fact all the forensic buried target objects were found in the semi-urban environment scenario; it was just the two control buried objects, (1) the brick and (2) the bolt and screw, that were not detected.

Both magnetic gradiometry methods compared poorly against the metal detector and magnetic susceptibility equipment. The fluxgate gradiometry survey results after burial of forensic targets were generally poor, with target detection success rates of 50% for both semi-urban and patio surveys (Fig. 16). The grouped breadknives, the entrenching tool, the ammunition box and one hand grenade were successfully located, although a key target, the handgun, was not detected. This technique may also be problematic to utilize in urban environments owing to the high percentage of the survey area (averaging 31% over the three surveys) having out-of-range data recorded, as other authors have discussed (Reynolds 2011).

The magnetic (potassium-vapour) gradiometry survey results after burial of forensic targets were

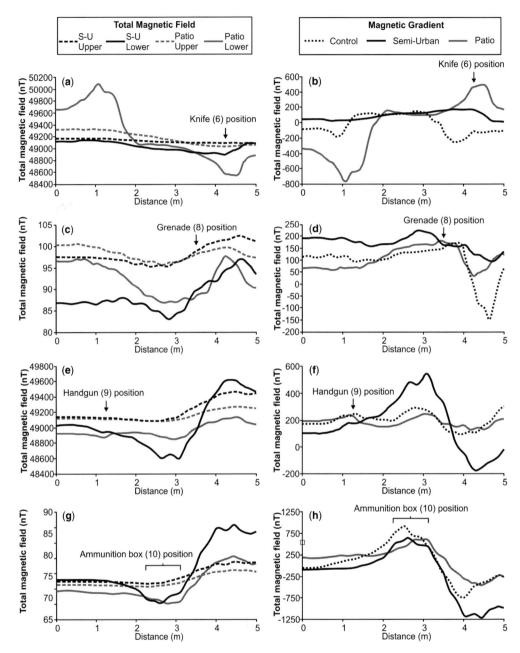

Fig. 10. Magnetic (potassium-vapour) gradiometry with total magnetic (left) and gradient (right) selected 2D survey profiles for control, semi-urban and patio surveys with respective target positions marked. (**a/b**) Profile 9 ($X = 2$ m) over target (6), single knife; (**c/d**) profile 12 ($X = 2.75$ m) over target (8), First World War hand grenade; (**e/f**) profile 15 ($X = 3.5$ m) over target (9), handgun; and (**g/h**) profile 18 ($X = 4.25$ m) over target (10), ammunition box (all marked). See key for sensors, survey type and Table 1 for details.

relatively good, with considerably better target detection success rates than the fluxgate gradiometry equipment, of 63% (semi-urban) and 75% (patio) respectively (Fig. 16). Interestingly, the target detection success rates increased over the patio v. the semi-urban environment – perhaps owing to

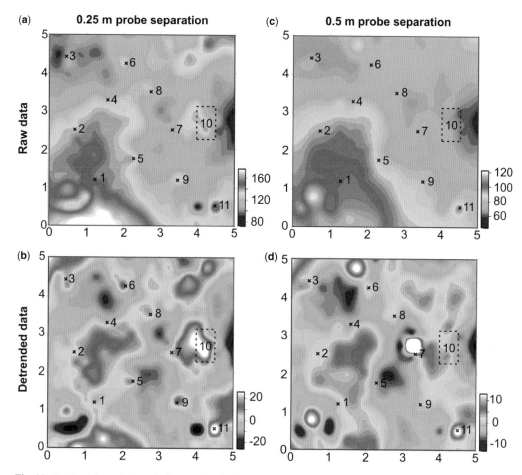

Fig. 11. Post-burial, semi-urban, bulk ground-resistivity contour plots using raw and de-trended datasets with 0.25 m (**a** and **b**, respectively) and 0.5 m (**c** and **d** respectively) probe spacings. Units in ohms. Note the relatively high anomalies corresponding to the knife (6), handgun (9) and mortar shell (11). See Table 2 for target descriptions.

less geophysical 'noise' as the patio had a damping effect on low-intensity, background anomalies. A small sampling increment spacing suggests that data had good resolution but target detection success rates were not higher than the magnetic susceptibility surveys, which had a much wider sampling point separation. Data repeatability was reasonable, with similar 2σ values for both post-burial surveys. With the instrument utilized, however, it was often difficult to obtain a 'lock' between sensors to gain usable data, which may prove problematic in forensic surveys where limited survey time may be a significant issue. One suggestion may be for equipment to be cart-mounted to improve data quality (see Reynolds 2004).

Considering that the magnetic methods measure-related properties, it would not have been surprising if the techniques had yielded similar results. However, the success of the techniques is quite variable, which can be attributed to the differences in the way each piece of equipment acquires data, for example, at different heights above ground level from the target objects.

Compare magnetic methods with electrical and GPR detection methods

The variability in the control resistivity dataset confirmed the heterogeneous ground conditions of the survey site. The post-burial dataset target location success rates for the 0.25 and 0.5 m fixed-offset probe spacings were very different: 63 and 25%, respectively (Fig. 16). The 0.25 m spaced probe survey data are therefore less favourable to the magnetic survey techniques, although both the handgun and single knife were detected. However, this technique could not be utilized over the patio owing to the inability to inset the steel

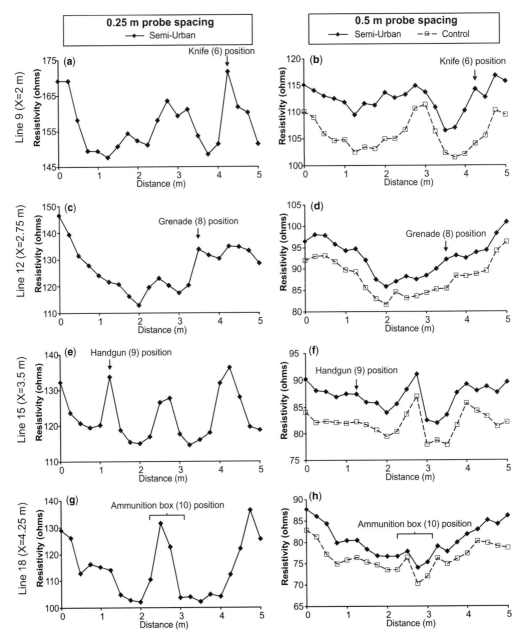

Fig. 12. Bulk-ground resistivity selected 2D survey profiles for control and semi-urban 0.25 and 0.5 m probe separation surveys with respective target positions marked. Units in Ohms (Ω). Note generally high-resistivity anomalies associated with targets with 0.25 m probe separation surveys generally showing better results. See key for survey type and Table 1 for details.

probes inserted into the ground. Equipment from other manufacturers does have the ability to record data from hard ground by having a flat probe end, which may be worth exploring in future research.

The GPR survey results were mixed, with only 50 and 63% of targets found using 450 MHz dominant-frequency antennae over the urban and patio environments, respectively. This contrasted with 75% of targets found using 900 MHz

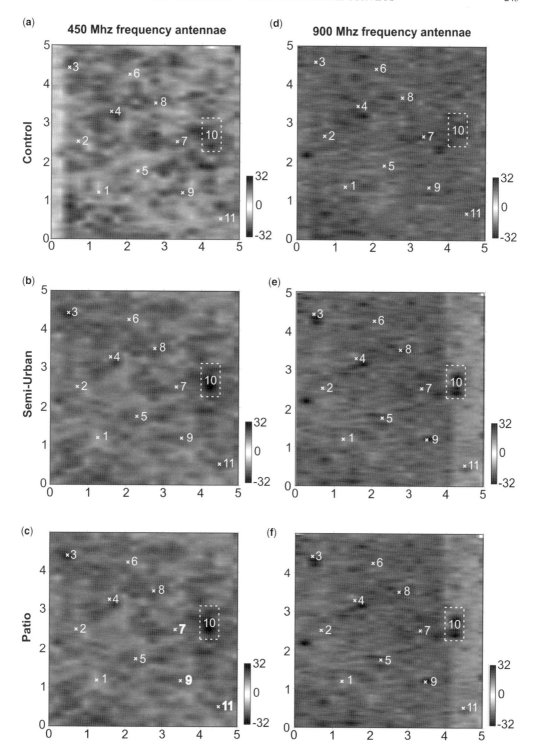

Fig. 13. GPR time slices over the test site using 450 MHz (**a–c**) and 900 MHz (**d–f**) dominant-frequency antennae with units in relative amplitudes. Some relatively high- and relatively low-amplitude anomalies correspond to target positions. See Table 2 for target descriptions.

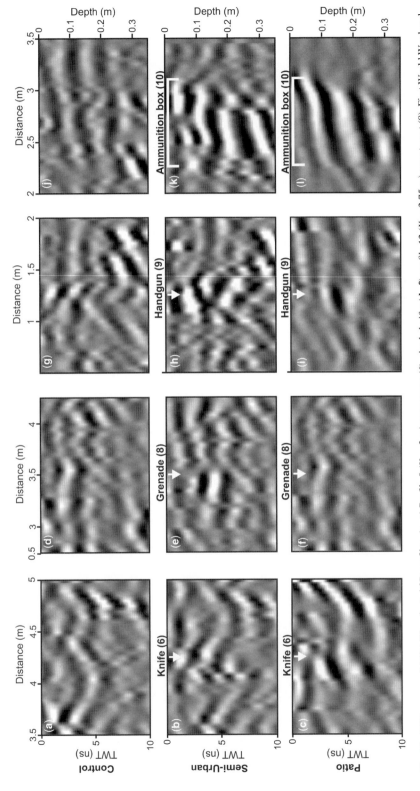

Fig. 14. GPR processed 450 MHz selected 2D profiles. (a–c) Profile 9 ($X = 2$ m) over target (6), single knife; (d–f) profile 12 ($X = 2.75$ m) over target (8), First World War hand grenade; (g–i) profile 15 ($X = 3.5$ m) over target (9), handgun; and (j–l) profile 18 ($X = 4.25$ m) over target (10), ammunition box for control, semi-urban and patio environment scenarios, respectively (all marked). See Table 1 for details.

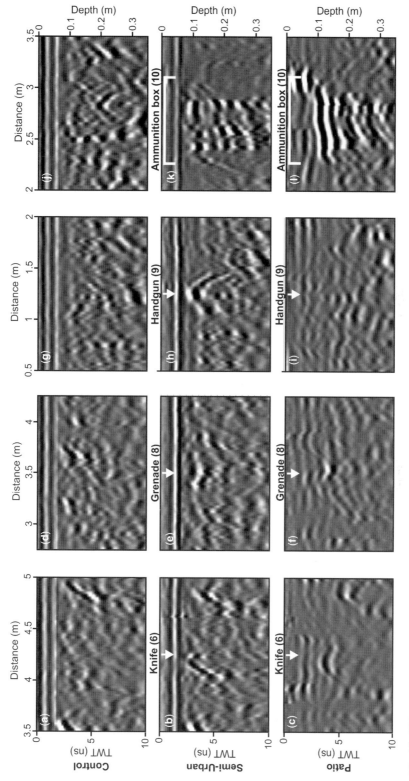

Fig. 15. GPR processed 900 MHz selected 2D profiles. (**a**–**c**) Profile 9 ($X = 2$ m) over target (6), single knife; (**d**–**f**) profile 12 ($X = 2.75$ m) over target (8), First World War hand grenade; (**g**–**i**) profile 15 ($X = 3.5$ m) over target (9), handgun; and (**j**–**l**) profile 18 ($X = 4.25$ m) over target (10), ammunition box for control, semi-urban and patio environment scenarios respectively (all marked). See Table 1 for details.

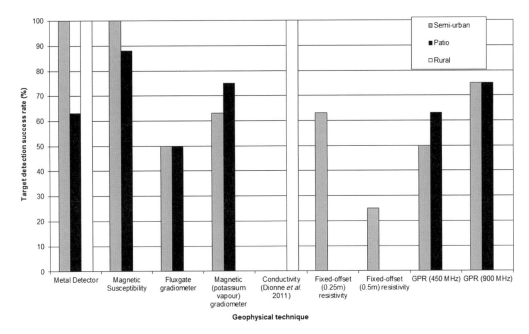

Fig. 16. Summary graph showing percentage total of target detection success rates for the different geophysical techniques trialled in semi-urban, patio and rural environments (see key). Note rural environment results are from Rezos *et al.* (2010) and Dionne *et al.* (2011) for metal detector and conductivity surveys respectively.

dominant-frequency antennae over both the semi-urban and patio environments.

Determine optimum GPR detection frequencies

From the detail shown in this study, it was suggested that 900 MHz dominant-frequency antenna was the optimal set frequency. Murphy & Cheetham (2008) also found that higher frequency (800 v. 400 MHz) GPR antennae were optimal in buried handgun detection in rural environments.

Determine optimum respective equipment configuration(s)/survey specifications/ optimum processing steps

Magnetic susceptibility datasets showed that 0.25 m spaced gridded sampling points were sufficient to resolve even the smallest objects with little data processing required, and this arrangement was deemed optimal in this study – simply creating 2D graphical summaries of survey lines was sufficient to gain a high target detection success rate. Fluxgate gradiometry datasets were geophysically 'noisy' and required significant time removing erroneous data points and de-trending data to gain usable data to interpret from. Magnetic (potassium-vapour) gradiometry equipment proved useful at 1 m sensor separations orientated vertically in order to obtain gradient data. There were, however, significant amounts of data generated that needed to be processed and de-trended before being usable. However, even after de-trending of the datasets, fluxgate gradiometry and magnetic (potassium-vapour) gradiometry results were difficult to interpret in plan-view plots owing to the subtle anomalies caused by the target objects. In fact, it could be argued that many of the target locations would not have been identifiable at all in these scenarios, had the control data not been collected for comparison. Equipment operators also needed to be careful that a constant height was maintained between the sensors and the ground surface to improve data quality, which may be problematic in forensic search scenarios on uneven ground.

The electrical resistivity 0.25 m fixed-offset probe spacing data were vastly superior to the 0.5 m offset probe spaced datasets even when using the same sampling spacings, making the closer probe spacing the more obvious one to utilize for such small and high-resolution surveys. However, the amount of ground covered in larger forensic search surveys using this configuration and 0.25 m grid sample spacings may make this technique more problematic.

As mentioned, 900 MHz dominant-frequency GPR antennae proved optimal, with a 0.025 m trace sampling interval on 0.25 m spaced survey lines. Basic 2D profile data processing of gain filters and background removal would prove sufficient for target detection, although it would be deemed worthwhile to generate horizontal 'time-slices' if targets were more subtle in comparison to heterogeneous ground, and if processing time is allowed.

Determine which technique(s) could determine target depth below ground level

Only GPR data could definitively determine the depth of buried forensic target below ground level. Total field magnetic data, such as from the potassium-vapour gradiometer and the bulk electrical resistivity data, could both be forward modelled to gain simple estimations of target depths if sufficient time and specialist resources were available (see Juerges *et al.* 2010; Reynolds 2011 for examples).

Determine if different metal types could be distinguished

Distinguishing between different buried metallic object types was difficult using the equipment utilized; Rezos *et al.* (2010), for example, used a higher specification metal detector which did allow some metal differentiation to be determined. The resistivity survey results did differentiate between conductive (the metal plate) and non-conductive (the brick) buried forensic targets, which may be useful information for forensic search investigators. 2D magnetic forward modelling of total field magnetic data would allow the relative magnetic susceptibility contrast between the target object and the background material to be assessed (see, e.g. Scott & Hunter 2004), but these would not be definitive values.

Finally it was determined that the metal detector, magnetic susceptibility meter, resistivity meter (if in semi-urban environments) and a commercial GPR unit would be relatively easy for forensic search investigators to acquire, process and interpret for buried forensic targets. Metal detector equipment is relatively cheap but also arguably the simplest to use and to generate data from that forensic search teams could use to interpret buried target locations. When considering both the semi-urban and patio scenarios, however, the magnetic susceptibility equipment provided the best target detection rates, with relatively few additional non-target anomalies. The equipment was also relatively cheap and easy to process into a visual data-plot.

The magnetic susceptibility dataset from the patio scenario showed very low variability at points other than at target and non-target object locations, and so would be optimal in this environment considering the low number of false positives. GPR data could be viewed in real time and suspected burial positions marked during the field work. Resistivity data would need to be downloaded and line profiles generated in any data graphical packages, of which there are many. The fluxgate gradiometer and magnetic (potassium-vapour) gradiometer are only recommended to be utilized by experienced operators owing to the difficulty of calibration, operation and data processing.

It should, however, be noted that the success rates from these surveys are alone not enough to determine optimum techniques and equipment configurations for detection of buried metallic objects. One must also consider that a technique that is capable of detecting all target objects may also be overly sensitive to background anomalies. For example, the metal detector, although capable of detecting all eight target objects, also detected an additional six background anomalies. This means that only 57% of the anomalies can be attributed to buried targets.

Conclusions

From the results of this study, usable geophysical techniques gaining the highest buried forensic object target success rates in semi-urban environments were (in descending order): magnetic susceptibility, metal detection, 900 MHz GPR and electrical resistivity (0.25 m fixed-offset probes), magnetic (potassium vapour) gradiometry, 450 MHz GPR, fluxgate gradiometry and electrical resistivity (0.5 m fixed-offset probes) (Fig. 16). Usable geophysical techniques gaining the highest buried forensic object target success rates in patio environments were (in descending order): magnetic susceptibility, magnetic (potassium vapour) gradiometry, 900 MHz GPR, metal detection, 450 MHz GPR and fluxgate gradiometry (Fig. 16). Note that resistivity surveys were not utilized in the patio environment. It was worth noting that the magnetic susceptibility had a considerably higher success rate than the other magnetic equipment utilized, that is, compared with the metal detector and the gradiometers, despite them measuring similar properties and the potassium-vapour gradiometer having a closer sample point spacing.

Concerns were raised in this study over the use of metal detectors and GPR detection equipment solely for detection of buried forensic targets, as important objects such as knives and hand grenades were not detected by even the higher frequency

GPR configuration, particularly beneath the patio. It is therefore recommended that the magnetic susceptibility equipment, with is ease of utilization and high target success rates, should be used as a complementary tool for forensic search investigators in the search for buried objects such as those used in this study. The bulk electrical resistivity technique also showed potential owing to its relatively quick collection time and reasonably high detection rate. Unlike GPR data processing, resistivity data processing is relatively straightforward (given available software and operator experience) and can produce either 2D profiles or a single mapview image that can then be interpreted.

Keele University are thanked for land donation for the test site. L. Ore, S. Reid, L. Patrick and E. Postlethwaite are thanked for field assistance. M. Hannah is thanked for replica handgun donation. Geophysical equipment was funded by a 2003 SRIF2 equipment bid. James Hansen is funded by a Keele University Acorn grant.

References

ACHEROY, M. 2007. Mine action: status of sensor technology for close-in and remote detection of anti-personnel mines. *Near Surface Geophysics*, **5**, 43–56.

BAVUSI, M., RIZZO, E. & LAPENNA, V. 2006. Electromagnetic methods to characterize the Savoia di Lucania waste dump in southern Italy. *Environmental Geology*, **51**, 301–308.

BILLINGER, M. S. 2009. Utilizing ground penetrating radar for the location of a potential human burial under concrete. *Canadian Society of Forensic Science Journal*, **42**, 200–209.

CONGRAM, D. R. 2008. A clandestine burial in Costa Rica: prospection and excavation. *Journal of Forensic Sciences*, **53**, 793–796.

CURRAN, A. M., PRADA, P. A. & FURTON, K. G. 2010. Canine human scent identifications with post-blast debris collected from improvised explosive devices. *Forensic Science International*, **199**, 103–108.

DAVENPORT, G. C. 2001. Remote sensing applications in forensic investigations. *Historical Archaeology*, **35**, 87–100.

DIONNE, C. A., SCHULTZ, J. J., MURDOCK, R. A. II & SMITH, S. A. 2011. Detecting buried metallic weapons in a controlled setting using a conductivity meter. *Forensic Science International*, **208**, 18–24.

DUPRAS, T. L., SCHULTZ, J. J., WHEELER, S. M. & WILLIAMS, L. J. 2006. *Forensic Recovery of Human Remains: Archaeological Approaches*. Taylor & Francis, Abingdon.

FENNING, P. J. & DONNELLY, L. J. 2004. Geophysical techniques for forensic investigations. *In*: PYE, K. & CROFT, D. J. (eds) *Forensic Geoscience: Principles, Techniques and Applications*. Geological Society, London, Special Publications, **232**, 11–20.

HARRISON, M. & DONNELLY, L. J. 2009. Locating concealed homicide victims: developing the role of geoforensics. *In*: RITZ, K., DAWSON, L. & MILLER, D. (eds) *Criminal and Environmental Soil Forensics*. Springer, Berlin, 197–219.

HUNTER, J. & COX, M. 2005. *Forensic Archaeology: Advances in Theory and Practice*. Routledge, Abingdon.

JERVIS, J. R., PRINGLE, J. K. & TUCKWELL, G. W. 2009. Time-lapse resistivity surveys over simulated clandestine graves. *Forensic Science International*, **192**, 7–13.

JOL, H. M. 2009. *Ground Penetrating Radar: Theory and Applications*. Elsevier, Amsterdam.

JUERGES, A., PRINGLE, J. K., JERVIS, J. R. & MASTERS, P. 2010. Comparisons of magnetic surveys over simulated clandestine graves in contrasting burial environments. *Near Surface Geophysics*, **8**, 529–539.

LINFORD, N. 2004. Magnetic ghosts: mineral magnetic measurements on Roman and Anglo-Saxon graves. *Archaeological Prospection*, **11**, 167–180.

MARCHETTI, M., CAFARELLA, L., DI MAURO, D. & ZIRIZZOTI, A. 2002. Ground magnetometric surveys and integrated geophysical methods for solid buried waste detection: a case study. *Annals of Geophysics*, **45**, 563–573.

MILLER, P. S. 1996. Disturbance in the soil: finding buried bodies and other evidence using ground penetrating radar. *Journal of Forensic Sciences*, **41**, 648–652.

MILSOM, J. & ERIKSEN, A. 2011. *Field Geophysics*, 4th edn. John Wiley & Sons, Chichester.

MURPHY, J. & CHEETHAM, P. 2008. *A comparative study into the effectiveness of geophysical techniques for the location of buried handguns*. Abstract presented at the Geoscientific Equipment & Techniques at Crime Scenes, Forensic Geoscience Group Conference, 17 December, Geological Society, London.

NOBES, D. C. 2000. The search for 'Yvonne': a case example of the delineation of a grave using near-surface geophysical methods. *Journal of Forensic Sciences*, **45**, 715–721.

PARKER, R., RUFFELL, A., HUGHES, D. & PRINGLE, J. 2010. Geophysics and the search of freshwater bodies: a review. *Science & Justice*, **50**, 141–149.

PRINGLE, J. K. & JERVIS, J. R. 2010. Electrical resistivity survey to search for a recent clandestine burial of a homicide victim, UK. *Forensic Science International*, **202**, e1–e7.

PRINGLE, J. K., JERVIS, J., CASSELLA, J. P. & CASSIDY, N. J. 2008. Time-lapse geophysical investigations over a simulated urban clandestine grave. *Journal of Forensic Sciences*, **53**, 1405–1417.

PRINGLE, J. K., RUFFELL, A. ET AL. 2012a. The use of earth science methods in terrestrial forensic searches. *Earth Science Reviews*, **114**, 108–123.

PRINGLE, J. K., JERVIS, J. R., HANSEN, J. D., CASSIDY, N. J., JONES, G. M. & CASSELLA, J. P. 2012b. Geophysical monitoring of simulated clandestine graves using electrical and Ground Penetrating Radar methods: 0–3 years. *Journal of Forensic Sciences*, **57**, 1467–1486.

REBMANN, A., DAVID, E. & SORG, M. H. 2000. *Cadaver Dog Handbook: Forensic Training and Tactics for the Recovery of Human Remains*. CRC Press, Boca Raton, FL.

REYNOLDS, J. M. 2004. Environmental geophysics investigations in urban areas. *First Break*, **22**, 63–69.

REYNOLDS, J. M. 2011. *Applied and Environmental Geophysics*, 2nd edn. John Wiley & Sons, Chichester.

REZOS, M. M., SCHULTZ, J. J., MURDOCK, R. A. II & SMITH, S. A. 2010. Controlled research utilizing a basic all-metal detector in the search for buried firearms and miscellaneous weapons. *Forensic Science International*, **195**, 121–127.

RUFFELL, A. & KULESSA, B. 2009. Application of geophysical techniques in identifying illegally buried toxic waste. *Environmental Forensics*, **10**, 196–207.

SCHULTZ, J. J. 2008. Sequential monitoring of burials containing small pig cadavers using ground-penetrating radar. *Journal of Forensic Sciences*, **53**, 279–287.

SCHULTZ, J. J., COLLINS, M. E. & FALSETTI, A. B. 2006. Sequential monitoring of burials containing large pig cadavers using ground-penetrating radar. *Journal of Forensic Sciences*, **51**, 607–616.

SCOTT, J. & HUNTER, J. R. 2004. Environmental influences on resistivity mapping for the location of clandestine graves. *In*: PYE, K. & CROFT, D. J. (eds) *Forensic Geoscience: Principles, Techniques and Applications*. Geological Society, London, Special Publications, **232**, 33–38.

TOMS, C., ROGERS, C. B. & SATHYAVAGISWARAN, L. 2008. Investigations of homicides interred in concrete – the Los Angeles experience. *Journal of Forensic Sciences*, **53**, 203–207.

WESSEL, P. & SMITH, W. H. F. 1998. New improved version of Generic Mapping Tools. *EOS Transactions*, **55**, 293–305.

The effectiveness of geophysical techniques in detecting a range of buried metallic weapons at various depths and orientations

THOMAS RICHARDSON & PAUL N. CHEETHAM*

School of Applied Sciences, Bournemouth University, Poole, Dorset BH10 5BB, UK

*Corresponding author (e-mail: pcheetham@bmth.ac.uk)

Abstract: The use of weapons in violent crime in the UK is well attested by crime statistics. It has been demonstrated by previous experimental studies that geophysical techniques have the potential for use in the search and detection of buried metallic weapons. Previous research has focused on only one weapon type, or the depths at which weapons can be detected, while this study also considers weapon orientation. This study employs magnetometry (fluxgate gradiometry), Slingram electromagnetic (both phases in both horizontal magnetic dipole (HMD) and vertical magnetic dipole (VMD) orientations) and ground-penetrating radar (500 and 800 MHz) methods to investigate the detectability of a wide variety of weapon types when buried in the ground in different orientations and depths. This study demonstrates that there were limitations for each instrument employed resulting from the orientation of the weapon in the ground, with none of the techniques in this work being able to detect all of the weapons in all orientations. A second limitation was the inability of all these instruments to detect smaller weapons. This study shows that a combination of techniques is required to increase the chances of successful detection. Blind searching trials in a range of environments are now required to quantify the level of successful detection that can be achieved in practice. The influence of metal composition of the weapons also needs to be further investigated.

Weapons are used in around 20% of incidents of violent crime in the UK (Chaplin *et al.* 2011). The weapon used in a crime is not only a vital piece of evidence as the murder weapon, and so cause of death, but it may also possess additional physical evidence, such as fingerprints and DNA, that can help link a suspect to a victim and a crime. Development of more efficient and accurate search techniques to locate buried metallic weapons would help to reduce the time required for searches while increasing detection rates. Other techniques and more general approaches to best practice in geoforensic search are well documented (e.g. Harrison & Donnelly 2009) and will not be considered further here, as this study focuses solely on the imaging of the weapons by geophysical techniques. Historically, the bulk of work that has been published on forensic geophysics has focused on the detection of clandestine graves (e.g. Pringle *et al.* 2008), with other classes of forensic evidence that may be found buried not receiving the same attention (Cheetham 2005). A number of recent studies have shown promising results when using geophysics as a method for detecting buried metallic weapons (e.g. Mann 2003; Murphy 2008; Murphy & Cheetham 2008; Schultz 2009; Rezos *et al.* 2010, 2011; Dionne *et al.* 2011).

However, in respect of weapon orientation, these studies were limited by looking at the orientation of one type of weapon (Mann 2003; Murphy 2008) or the depth of the weapons at one orientation (Schultz 2009; Dionne *et al.* 2011). Any technique must be thoroughly tested against the greatest range of variables so it can be used with confidence, and effectively, within real forensic search operations. This study aims to build upon the previous work by using a larger weapon set than Mann (2003) and Murphy (2008) – whilst investigating a greater number of geophysical instruments and weapon orientations than Schultz (2009) and Dionne *et al.* (2011) – through the inclusion of the EM38B response in both phases, and in both vertical (VMD) and horizontal magnetic dipole (HMD) orientations. Conventional metal detectors were not included in this study owing to the wide product range variations in terms of the coil configurations, the frequencies employed, and the proprietary electronics and software analysis systems. To provide a valid comparison with the results from this study would require an extensive further study as a subject in its own right. Recent work by Rezos *et al.* (2010) has started to address the issue of the effective basic metal detectors through controlled experiments.

Weapons studied and burial parameters

The weapons chosen for this study are representative of weapons used in violent crime. The weapons used are listed below and shown in Figure 1:

- 680 g hickory-shafted hardened steel ballpein hammer;
- 567 g hickory-shafted hardened steel hatchet;
- three stainless steel serrated kitchen knives (76, 152 and 228 mm);
- magnetic chrome vanadium No. 2 100 mm cross-head screwdriver;
- non-magnetic chrome vanadium No. 2 100 mm cross-head screwdriver;
- replica Glock 17 pistol (Bruni GAP);
- replica Smith and Wesson 0.38 calibre styled revolver (Bruni Olympic 5).

The frames, slides, chambers and barrels of the replica weapons are made of Zamak, an electrically conductive but non-magnetic zinc alloy with 4% aluminium content. However, they also have tungsten carbide blocks in the barrels and some small steel parts, such as the springs, and, in the case of the replica Glock 17, a steel magazine. While differing significantly metallurgically from the real weapons, such replica weapons are used in crimes and, therefore, their geophysical responses are of forensic interest.

The weapons were buried at depths of 0.1, 0.3 and 0.5 m below ground level (bgl), with a 2 m gap between each hole to ensure that the potential geophysical anomalies produced by the targets would not overlap (Fig. 2). The 0.1 m bgl equates to just under the turfline, while the 0.5 m depth is around the maximum that a small hole can be comfortably dug using a hand trowel. The weapons were buried in holes relative to their size. A 0.3×0.3 m control excavation containing no weapon was also surveyed at each of the depths in which the weapons were placed (0.1, 0.3 and 0.5 m) to ensure that any geophysical anomalies being produced were not from the soil disturbance caused by digging the holes. The weapons were then successively reburied and surveyed in six different orientations at a depth of 0.3 m (Table 1). For the orientation study, they were buried at 0.3 m bgl as this was shown to give the greatest number of successful results from each instrument at the three depths trialled, and therefore was considered appropriate for the weapon orientation study.

Geophysical instrumentation

The geophysical instruments in this study were chosen to allow for comparison of results with previous studies cited, as well as being instruments that would be widely available to law enforcement

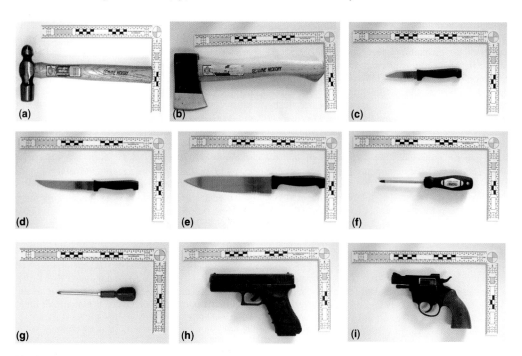

Fig. 1. The weapons used in this study: (**a**) 680 g ballpein hammer; (**b**) 567 g hickory shaft hatchet; (**c**) 76 mm kitchen knife; (**d**) 152 mm kitchen knife; (**e**) 228 mm kitchen knife; (**f**) 100 mm magnetic cross-head screwdriver; (**g**) 100 mm non-magnetic cross-head screwdriver; (**h**) replica Glock 17 pistol; and (**i**) replica Smith and Wesson 0.38 calibre revolver.

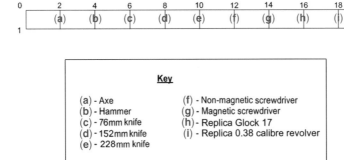

Fig. 2. The positions of the weapons within the 20 × 1 m study area. Each weapon (**a–i**) was spaced at 2 m intervals along the survey traverse and placed centrally within the 1 m-wide survey strip.

personnel. This research employs one magnetic and two electromagnetic techniques. The instruments employed and shown in use in Figure 3 are:

- Geoscan Research FM256 fluxgate gradiometer (magnetic) (Fig. 3a);
- Geonics EM38B (Slingram electromagnetic) (Fig. 3b, c);
- MALA RAMAC X3M ground-penetrating radar with 500 and 800 MHz shielded antennas (electromagnetic) (Fig. 3d).

For each instrument, the reading intervals and survey methodology selected were based on each instrument's spatial resolution and the practicality of utilizing such reading intervals in actual searches. Traverse and reading intervals along traverses are summarized in Table 2.

Magnetic technique

The Geoscan Research FM256 consists or two fluxgate magnetometers, one placed 0.5 m vertically above the other (Fig. 3a). The instrument passively detects changes to the Earth's ambient magnetic field strength. These changes are caused by either permanent or induced magnetic fields in the ground interacting both constructively and destructively with the Earth's field (Clark 1996; Cheetham 2005). The short 0.5 m gradiometer arrangement allows for background magnetic noise, including problematic urban ferrous clutter, to be filtered out more effectively than increased sensor separation gradiometers, making it particularly suited for forensic applications. The intensity of an induced magnetic field produced by a material depends

Table 1. *Weapon orientation details*

Orientation	Description
Perpendicular and flat	The weapon placed with the longest axis perpendicular to the survey traverse and on the flat of the blade, head or side
Parallel and flat	The weapon placed with the longest axis parallel to the survey traverse and on the flat of the blade, head or side
Perpendicular and on edge	The weapon placed with the longest axis perpendicular to the survey traverse and on the edge of the blade or the head. The guns were placed with the length of the barrel down and the hand grip pointing upwards
Parallel and on edge	The weapon placed with the longest axis parallel to the survey traverse and on the edge of the blade or the head. The guns were placed with the length of the barrel down and the hand grip pointing upwards
Perpendicular and on point	The weapon placed with the longest axis perpendicular to the survey traverse and on the point of the blade. The hammer was placed with the pein down and face up. The axe was buried with the cutting edge facing up and the back of the head down. The guns were buried with the muzzle facing downwards
Parallel and on point	The weapon placed with the longest axis parallel to the survey traverse and on the point of the blade. The hammer was placed with the pein down and face up. The axe was buried with the edge facing up and the back of the head down. The guns were buried with the muzzle facing downwards

Fig. 3. The geophysical survey instruments used in this survey on the test site: (**a**) Geoscan Research FM256 fluxgate gradiometer; (**b**) & (**c**) Geonics EM38B in vertical and horizontal dipole orientations, respectively; (**d**) MALA RAMAC X3M ground-penetrating radar with 500 and 800 MHz shielded antennas, shown with the 800 MHz antenna in use.

upon its magnetic susceptibility. The high magnetic susceptibility of iron (Breiner 1999) means that magnetometry is a useful tool in the search for ferrous weapons, such as those used in this study. However, non-magnetic metals such as copper, gold or silver are not detectable by such instruments, so their use for locating caches of stolen materials is limited. In practice, many metallic objects are made from alloys and so there has to be careful consideration before deciding whether or not to employ magnetometry. Even iron in the form of wrought, cast, steel and stainless steel has differing magnetic properties, with high-grade stainless steel being practically non-magnetic (Breiner 1999). Ordinary steel can also possess permanent magnetic properties either as a result of the manufacturing process or induced during use (thermo or shock remanent magnetism) or even

Table 2. *Traverse and reading intervals employed*

Instrument	Traverse interval (m)	Reading interval (m)
Magnetometry (FM256)	0.125	0.125
Slingram EM (EM38B)	0.25	0.25
GPR (MALA X3M)	0.10	0.01

purposefully magnetized, as in magnetically tipped screwdrivers.

For the FM256, a reading interval of 0.125 m along traverses that were spaced 0.125 m apart was employed. The samples were taken using the automatic time-based trigger to provide a realistic rate of the survey rather than stopping and manually taking readings at each survey point. A parallel survey of the traverses was employed to give better results due to there being no rotation of the sensors, so minimizing heading errors (Gaffney & Gater 2003; Aspinall et al. 2008) and also so that any systematic distance offset in the time-based triggering should be constant between survey lines. The results were processed using DW Consulting ArcheoSurveyor 2 software (DW Consulting 2012).

Slingram electromagnetic technique

The EM38B Slingram system consists of transmitter and receiver coils fixed at 1 m apart. By passing an alternating current through the transmitter coil a time-varying magnetic field is created, which passes into the ground. If this electromagnetic field encounters a conductive material in the ground it will create currents within that material that, in turn, will generate a secondary electromagnetic field which will be detected by the receiver coil (McNeill 1980; Simms 1995; US Army Corps of Engineers 1995; Beamish 2011). The EM38B is capable of taking both quadrature (conductivity) and in-phase (magnetic susceptibility) readings, as well as being used in VMD and HMD coil orientations (Fig. 3b, c). This study investigates both the quadrature and in-phase readings as each is sensitive to different materials, together with both the available dipole orientations as these provide different target depth (McNeill 1980; Geonics 1999) and orientation sensitivities. Unlike the FM256, the EM38B detects objects as a result of their conducting properties, and so it will detect all conducting metals and alloys in the quadrature response. Metals that also have magnetic properties will serve to enhance their detectability by producing strong effects in the in-phase response.

A traverse and sample reading interval of 0.25 m was employed for the EM38B. The surveys were conducted along zigzag traverses using a manual trigger to take readings. The results were processed using DW Consulting ArcheoSurveyor 2 software (DW Consulting 2012).

Electromagnetic (GPR) technique

GPR works by transmitting electromagnetic pulses into the ground from an antenna and measuring the amount of time between their transmission, reflection off an electromagnetic discontinuity in the ground, and being received by the antenna at the ground surface (Clark 1996; Gaffney & Gater 2003; Conyers 2004). The ability of GPR to detect small buried metallic objects has been shown by studies looking into the use of GPR to detect landmines (Bruschini et al. 1998; Stanley et al. 2002; Zoubir et al. 2002), which would suggest it is an applicable technique to the search for buried metallic weapons.

The GPR survey was conducted at a slow walking pace along 0.1 m traverses, with readings being taken every 0.01 m employing auto stack mode (Fig. 3d). The surveys employed a wheel odometer, and were performed in parallel traverses to reduce positional errors and to simplify the processing of the data (Conyers 2004). The results were processed using MALA RAMAC GroundVision 2 and MALA Easy3D software.

General survey methodology, geology and soil analysis

In order to ensure that any anomalies observed in the surveys were being produced by the weapons, and not due to any other objects already present in the ground, a 20 × 20 m area was initially surveyed with each of the instruments at the same sample intervals as the post-burial surveys. From these survey results, geophysically 'quiet' areas were selected for the weapon burial. Full-area surveys were undertaken using all the instruments, with the additional weapon orientation surveys achieved by surveying a 1 m corridor over the weapons. An empty control hole was also created within the survey area to verify that it was not merely the soil disturbance that was being detected by any of the geophysical techniques.

The test area is within a region of superficial Quaternary period river terrace sands and gravels that overlie the Branksome Sand Formation (Eocene) (BGS 2013). It is located on the Bournemouth University campus at OSGR SZ 0756 9351. The river terrace deposits, being described from a number of boreholes in the immediate vicinity of the test area, have a maximum proved thickness of 5.1 m. They consist dominantly of angular, brown flint gravel in a matrix of clayey sand, within which interbeds of coarse-grained locally cross-bedded sand up to 0.4 m thick occur (Bristow et al. 1991). The soil profile in the test area (Fig. 3), which was on former agricultural pasture, was turf overlying 0.30–0.35 m of sandy loam topsoil (67:32:1 sand: silt:clay by wt). This was overlying a uniform subsoil of loamy sand (78:21:1 sand:silt:clay by wt) extending to an unknown depth. Both contained 8% by wt of small stones up to 20 mm in size. The organic matter determined by loss on ignition was 8% by wt for the topsoil and 4% for the immediate

Table 3. *Response classification criteria*

Strong	A clear response on the geophysical visualization that is a good indication of the target sought
Moderate	Visible but within the range that would not clearly stand out from the background and other irrelevant anomalies as the target sought
Weak	While weakly visible, such anomalies would be difficult to interpret as the potential forensic targets that were sought
None	Undetectable in the visualizations
N/A	Not applicable owing to the target not being included in that trial

subsoil. Determination of the apparent resistivity of the soil by employing a Wenner array at separations of 0.25–1.0 m (so to a nominal depth of 0.5 m) gave an average of 365 Ω m, which equates to a conductivity of 2.74 mS m^{-1}.

Results and discussion

In this section the results for each geophysical survey instrument are presented and then discussed in the context of the general operational characteristics of the instruments used in this study to provide a broader review of their effectiveness for successfully detecting buried weapons. The results were classified from no response to a strong response according to the criteria given in Table 3, and are presented in Tables 4–10. This classification is inevitably subjective in the respect that it is dependent on the level of background geophysical soil noise, site contamination, and instrument and survey noise, the effect of which will vary from instrument to instrument, and from site to site.

Magnetic (FM256) surveys

The FM256 was able to detect all of the weapons at all depths and orientations apart from the knives and non-magnetic screwdriver (Table 4). As would normally be expected, the deeper buried weapons produced weaker magnetic intensity anomalies (Fig. 4) because the strength of the anomaly produced by any object will decrease when the distance from the magnetometer is increased (Breiner 1999). For dipole targets that these weapons broadly represent, this fall-off should be of the order of the inverse cube of the distance from the sensor. In practice, the fall-off rate with depth bgl is also dependent on the shape of the target, as the negative responses of the more complexly constructed and shaped guns demonstrate (Fig. 4).

In this study, the strongest anomalies were produced when the weapons were placed on their point (Fig. 4). When considering the induced magnetic field, the anomaly created will be greater when there is a greater distance between the induced north and south poles within an object.

Table 4. *The FM256 results*

	Axe	Hammer	228 mm knife	152 mm knife	76 mm knife	Magnetic screwdriver	Non-magnetic screwdriver	Replica Glock 17 pistol	Replica 0.38 calibre revolver
10 cm	Strong	Strong	None	None	None	Strong	N/A	Strong	Strong
30 cm	Strong	Strong	None	None	None	Strong	N/A	Moderate	Moderate
50 cm	Moderate	Moderate	None	None	None	Strong	N/A	Weak	Weak
Orientation at 30 cm									
Perpendicular and flat	Strong	Weak	None	None	None	Strong	None	Moderate	Moderate
Parallel and flat	Strong	Strong	Weak	None	None	Strong	None	Strong	Weak
Perpendicular and on edge	Weak	Weak	Weak	None	None	Strong	None	Weak	Weak
Parallel and on edge	Strong	Moderate	Strong	None	None	Strong	None	Moderate	Weak
Perpendicular and on point	Strong	Strong	Strong	None	None	Strong	None	Strong	Moderate
Parallel and on point	Strong	Strong	Strong	None	None	Strong	None	Strong	Moderate

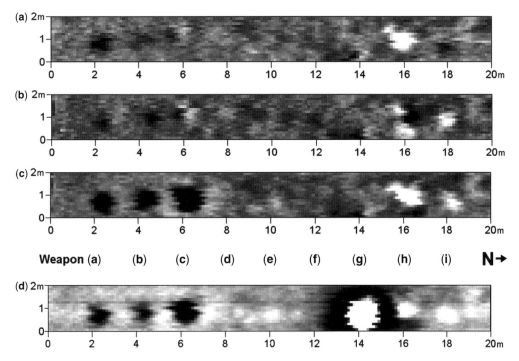

Fig. 4. FM256. Comparison greyscale plots of weapon responses, with all of the weapons at 0.3 m depth bgl: (a) weapon parallel to the survey direction and flat; (b) weapon perpendicular to the survey direction and on edge; and (c) weapon perpendicular to the survey direction and on point. All plotted on a clipped range of −5 to 5 nT, with black positive and white negative. (d) Weapon orientation as in (c) but including the magnetic screwdriver and clipped −5 to 10 nT.

This is because when the magnetic poles are very close together, as in laminar objects that are aligned perpendicular to the inducing field direction, the magnetic flux leaving the object is detected at the same time as the returning flux, which cancel each other out (Breiner 1999; Clark 1996), together with an effect known as demagnetization (Aspinall et al. 2008). Because the Earth's field direction is nearer to the vertical in the UK (dipping around 68° to the horizontal), stronger anomalies are produced when the weapons are positioned vertically on their points as their axis is broadly aligned with the field (Aspinall et al. 2008). This effect was also demonstrated in a more detailed study of knife responses using a similar FM36 instrument undertaken by Mann (2003), and observed in Murphy (2008) for the FM256 results of horizontal v. vertical weapon orientation using the same two replica handguns used in this study.

The distance of the lower gradiometer sensor from the ground surface is an important survey parameter. Murphy (2008) conducted surveys on this same site with the same two handguns buried at the two depths 0.3 and 0.5 m, which were depths repeated in this latest study. However, in his study, he observed weaker anomalies than in this present one. This could be due to the fact that Murphy's methodology employed a wider traverse interval of 0.25 m (but with an identical 0.125 m sample interval along the traverse) and so the targets may have been passed over slightly more off to one side, or that the magnetometer was being held further from the ground surface. The latter is a fundamental issue with this instrument, as the distance of the lower sensor off the ground when using an FM256 is largely determined by the physical characteristics of the individual surveyor, an effect that is amplified by the rapid fall-off of anomaly intensity with increasing distance between the target and sensor. In fact, the distance of the lower sensor from the ground should be considered in the light of a number of considerations. For small traverse intervals, then, a lower sensor height will produce both a greater resolution and a higher anomaly intensity. However, this has to be balanced against an increase in the magnetic response caused by the often disturbed immediate topsoil. The latter, if too close to the sensor, can produce topsoil noise in the data that can mask out deeper anomalies. To avoid this, the lower sensor must be held far enough off the ground to reduce these topsoil effects but not so far above the surface that it will fail to detect deeper targets.

In this study the main limitation of the FM256 was the inability to detect the smaller knives or the non-magnetic screwdriver at any depths or orientations. This is particularly disappointing when the high incidence of knife crimes is taken into consideration. In England and Wales in 2010–2011 there were 32 714 incidents involving a knife or sharp instrument (Chaplin *et al.* 2011), all with the potential for the weapon being disposed of or concealed by burial. The lack of observable anomalies being produced by these weapons was likely to be due to their size. Generally, an object with less mass will produce a smaller anomaly (Clark 1996). The strong anomaly produced by the magnetic screwdriver in each survey was due to the fact it has a permanent magnetism (Fig. 4d). This permanent magnetism was far more intense than any induced magnetism would be in an object of this size, and indeed was far greater than the magnetism of any of the other weapons in this study, thus making such magnetized weapons easily detectable by magnetometry. This shows the effect that orientation has on the gradiometer's ability to detect the weapon. The consistently negative responses of the more complexly constructed and shaped replica handguns are also of note.

Slingram electromagnetic (EM38B) surveys

Overall, the EM38B VMD quadrature responses provided the best results for this instrument. However, the results were complex owing to the different orientations of the instrument and the differing depth sensitivities that result (Fig. 5). These differences were best shown by the quadrature responses in each orientation. The HMD results showed that the strongest anomalies were produced when the weapons were buried at shallower depths, whilst the VMD responses generally show stronger responses being produced at 0.3 m. For the EM38B, the quadrature sensitivity profiles show that the HMD orientation was most sensitive to shallow anomalies but this drops off sharply as depth increases, whilst the VMD orientation was most sensitive at about 0.3 m depth and drops off both above and below this depth (Geonics 1999) (Figs 2 & 3). The responses given by the weapons reflect these sensitivity profiles very well and this highlights the necessity of using both orientations to get the best from this instrument (Tables 5–8).

When the in-phase responses were assessed there was little difference between the two dipole orientations, with both showing strong anomalies at shallow depths, getting weaker as the depth increases. The only difference was that the anomalies were stronger in the VMD orientation. Again, these responses reflect the published data on in-phase sensitivity with depth (Geonics 1999) (Figs 2 & 3). Furthermore, the depth sensitivities are similar for each dipole orientation, with the vertical dipole having a stronger positive response compared with the negative response of the horizontal dipole (Geonics 1999) (Figs 2 & 3). The difference observed in depth sensitivity was due to the magnetic field produced by the coils. With the coils in the VMD orientation, the magnetic field flux density is greater than when arranged horizontally (Abdu *et al.* 2007). This greater flux density gives greater sensitivity at around 0.2–0.3 m but, unlike the

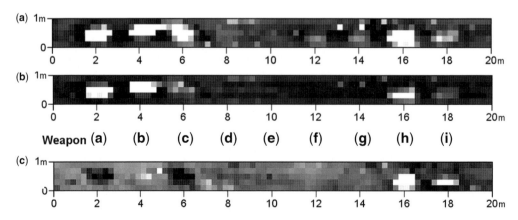

Fig. 5. EM38B. Quadrature responses showing the differences between weapon orientations and an in-phase response plot with black high/positive and white low/negative. (**a**) The VMD quadrature response with the weapon orientation parallel to the survey direction and flat. Clipped at 1 SD. (**b**) The VMD quadrature response with the weapon orientation perpendicular to the survey direction and on edge. Clipped at 2 SD. (**c**) The VMD in-phase response with the weapon orientation parallel to the survey direction and on point. Clipped at 2 SD.

Table 5. *The EM38B HMD quadrature results*

Depth	Axe	Hammer	228 mm knife	152 mm knife	76 mm knife	Magnetic screwdriver	Non-magnetic screwdriver	Replica Glock 17 pistol	Replica 0.38 calibre revolver
10 cm	Strong	Strong	Moderate	None	None	Weak	N/A	Strong	Moderate
30 cm	Strong	Strong	Moderate	None	None	None	N/A	Strong	Strong
50 cm	Moderate	Moderate	Weak	Weak	Weak	None	N/A	Moderate	Weak
Orientation at 30 cm									
Perpendicular and flat	Strong	Strong	Moderate	None	None	None	None	Strong	Strong
Parallel and flat	Strong	Strong	Weak	None	None	None	None	Strong	Moderate
Perpendicular and on edge	Strong	Strong	Strong	None	None	None	None	Strong	Strong
Parallel and on edge	Strong	Strong	Moderate	None	None	None	None	Strong	Moderate
Perpendicular and on point	Strong	Strong	Moderate	None	None	None	None	Strong	Weak
Parallel and on point	Strong	Strong	Weak	None	None	None	None	Strong	Weak

HMD coil orientation, at a depth of about 0.6 m the response sensitivity is zero and then goes negative.

When looking at the results from the orientations of the weapons, the axe, hammer, 228 mm knife and both replica handguns show little difference in the anomalies resulting from each orientation. This was also shown on a larger scale with oil drums by Westphalen & Rice (1992). This shows an advantage of the EM38B as it was able to effectively detect these weapons regardless of orientation. However, when looking at the 152 and 76 mm knives and the two screwdrivers there was a clear difference. At most orientations, these smaller weapons do not produce a visible anomaly, and only when parallel and flat did they produce small anomalies. The knives also produce small anomalies in the parallel and on-edge orientation. These results would concur with the expected responses according to the way in which the electromagnetic field is induced horizontally through the ground (McNeill 1980). Although the EM38B was successful in detecting every weapon in at least one orientation, it was shown to be limited in its detection of small knives and screwdrivers.

The results for depth sensitivity can be compared with those of Dionne et al. (2011). In their

Table 6. *EM38B HMD in-phase results*

Depth	Axe	Hammer	228 mm knife	152 mm knife	76 mm knife	Magnetic screwdriver	Non-magnetic screwdriver	Replica Glock 17 pistol	Replica 0.38 calibre revolver
10 cm	Strong	Moderate	Strong	Weak	Weak	Weak	N/A	Strong	Strong
30 cm	Strong	Moderate	Strong	None	None	None	N/A	Strong	Strong
50 cm	Moderate	None	None	None	None	None	N/A	Moderate	Moderate
Orientation at 30 cm									
Perpendicular and flat	Weak	Weak	Strong	None	None	None	None	Strong	Strong
Parallel and flat	Strong	Moderate	Weak	Weak	Weak	None	None	Strong	Strong
Perpendicular and on edge	Weak	Weak	Weak	Weak	Weak	None	None	Moderate	Moderate
Parallel and on edge	Strong	Weak	Weak	None	None	None	None	Moderate	Moderate
Perpendicular and on point	Strong	Weak	Weak	None	None	None	None	Strong	Strong
Parallel and on point	Strong	Weak	Weak	None	None	None	None	Strong	Strong

Table 7. *The EM38B VMD quadrature results*

Depth	Axe	Hammer	228 mm knife	152 mm knife	76 mm knife	Magnetic screwdriver	Non-magnetic screwdriver	Replica Glock 17 pistol	Replica 0.38 calibre revolver
10 cm	Moderate	Moderate	Moderate	None	None	None	N/A	Moderate	Moderate
30 cm	Strong	Strong	Weak	None	None	Weak	N/A	Strong	Strong
50 cm	Strong	Strong	None	None	None	None	N/A	Strong	Moderate
Orientation at 30 cm									
Perpendicular and flat	Strong	Strong	Weak	None	None	Weak	Weak	Strong	Strong
Parallel and flat	Strong	Strong	Strong	Weak	Weak	Weak	Weak	Strong	Moderate
Perpendicular and on edge	Strong	Strong	Strong	Weak	Weak	Weak	Weak	Strong	Strong
Parallel and on edge	Strong	Strong	Strong	Weak	Weak	Weak	Weak	Strong	Moderate
Perpendicular and on point	Strong	Strong	Strong	None	None	None	None	Strong	Moderate
Parallel and on point	Strong	Strong	Weak	None	None	None	None	Moderate	Weak

VMD quadrature surveys (the only orientation of the EM38 used by Dionne *et al.* 2011), comparable weapons show similar responses. Both studies located the hammer and firearms to a depth of at least 0.5 m. The 228 mm knife in this study can be compared with the buck knife from the work by Dionne *et al.* (2011), both of which were located down to a depth of 0.3 m. However, there is a difference in the clarity of the responses. The results shown in the work of Dionne *et al.* (2011) appeared to have much more striping and background noise to it than in this study. This is likely to be due to differences in data processing and temporal changes as the surveys were conducted over a number of days.

GPR surveys

GPR provided the best detection across the whole range of weapons (Tables 9 & 10). When looking at the depth sensitivity of the two antennas used in this study it was expected that the 500 MHz antenna would be able to produce better responses at a depth of 0.5 m than the 800 MHz antenna. This was due to the lower frequency antenna being able to penetrate to greater depths, albeit at a loss in target resolution. Thus, the 800 MHz has better responses for the smaller weapons owing to the shorter wavelength increasing the resolution (Gaffney & Gater 2003; Conyers 2004). The results confirmed this, with the 500 MHz antenna

Table 8. *The EM38B VMD in-phase results*

Depth	Axe	Hammer	228 mm knife	152 mm knife	76 mm knife	Magnetic screwdriver	Non-magnetic screwdriver	Replica Glock 17 pistol	Replica 0.38 calibre revolver
10 cm	Strong	None	Strong	None	None	None	N/A	Strong	Strong
30 cm	Moderate	None	Moderate	None	None	None	N/A	Strong	Strong
50 cm	Weak	None	Weak	None	None	None	N/A	Strong	Strong
Orientation at 30 cm									
Perpendicular and flat	Moderate	Strong	Moderate	None	None	None	Weak	Strong	Strong
Parallel and flat	Moderate	None	Moderate	Weak	Weak	None	None	Strong	Strong
Perpendicular and on edge	Moderate	Weak	Weak	None	None	None	None	Strong	Strong
Parallel and on edge	Weak	None	Strong	Weak	Weak	Weak	Weak	Moderate	Moderate
Perpendicular and on point	Moderate	Strong	Weak	None	None	None	None	Strong	Strong
Parallel and on point	Moderate	Weak	Moderate	Weak	Weak	None	None	Strong	Moderate

Table 9. *The 500 MHz GPR results*

Depth	Axe	Hammer	228 mm knife	152 mm knife	76 mm knife	Magnetic screwdriver	Non-magnetic screwdriver	Replica Glock 17 pistol	Replica 0.38 calibre revolver
10 cm	Weak	Weak	Strong	Strong	Strong	Strong	N/A	Strong	Strong
30 cm	Weak	Weak	Strong	Strong	Moderate	Strong	N/A	Strong	Strong
50 cm	Weak	Weak	Strong	Strong	Weak	Weak	N/A	Strong	Strong
Orientation at 30 cm									
Perpendicular and flat	Moderate	Moderate	Strong	Strong	Moderate	Strong	Weak	Strong	Strong
Parallel and flat	Weak	Weak	Weak	None	None	None	None	Moderate	Moderate
Perpendicular and on edge	Moderate	Moderate	Strong	Strong	None	Strong	None	Moderate	Moderate
Parallel and on edge	None	None	None	None	None	None	None	Weak	None
Perpendicular and on point	None	None	None	None	None	None	None	Moderate	Moderate
Parallel and on point	None	None	None	None	None	None	None	Weak	Weak

producing much clearer anomalies for all of the weapons buried at a depth of 0.50 m, while the 800 MHz antenna produced clearer responses for the weapons buried at depths of 0.1 and 0.3 m. When looking at the weapon orientations, the results of the GPR surveys were not exactly as expected. It was expected that the best weapon orientation for detection would be parallel and flat, as this would provide the greatest surface area for reflection of the electromagnetic pulses (Conyers 2004). However, the results show that the weapons produce better anomalies in the perpendicular and flat orientation. This is possibly due to the weapons lying in-between the survey traverses and therefore not being imaged by the GPR, or because the antennas in the MALA system are orientated perpendicular to the survey traverse. This highlights a limitation of the GPR, as if it was not detecting the weapons at such narrow traverse intervals it would therefore not be practical for use in forensic searches owing to the unreliability at 0.1 m traverses, and the added survey time it would take for smaller traverse spacings and/or orthogonal surveys.

Table 10. *The 800 MHz GPR results*

Depth	Axe	Hammer	228 mm knife	152 mm knife	76 mm knife	Magnetic screwdriver	Non-magnetic screwdriver	Replica Glock 17 pistol	Replica 0.38 calibre revolver
10 cm	Weak	Weak	Moderate	Moderate	Weak	Weak	N/A	Moderate	Moderate
30 cm	Weak	Weak	Moderate	Moderate	Weak	Weak	N/A	Moderate	Moderate
50 cm	Weak	Weak	Weak	Weak	Weak	Weak	N/A	Weak	Weak
Orientation at 30 cm									
Perpendicular and flat	Moderate	Weak	Moderate	Moderate	Weak	Weak	Weak	Moderate	Moderate
Parallel and flat	Weak	Weak	Weak	None	None	None	None	Moderate	Moderate
Perpendicular and on edge	Weak	Weak	Moderate	Weak	Weak	Weak	Weak	Moderate	Moderate
Parallel and on edge	None	Weak	None	None	None	None	None	Weak	None
Perpendicular and on point	Weak	Weak	None	None	None	None	None	Moderate	Strong
Parallel and on point	None	None	None	None	None	None	None	Weak	Weak

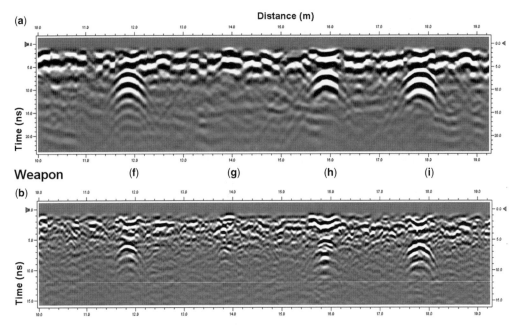

Fig. 6. (a) 500 MHz and (b) 800 MHz antenna two-dimensional (2D) GPR profiles, with the screwdrivers and guns positioned perpendicular to the line of the direction of survey and on edge. Note that the 800 MHz profile image shows the screwdriver (G), which is not visible in the 500 MHz profile, and shows the disturbance caused by the digging more clearly.

A greater survey limitation was revealed when other orientations were analysed. When the weapons were orientated parallel and on edge, and on their point, it was only the replica handguns that were detected by the 500 MHz antenna, and the axe and replica handguns by the 800 MHz antenna. This was an expected result due to the small surface available to reflect the electromagnetic pulses in these orientations. This would clearly be a problem in a search scenario as it rules out the possibility of detecting most of the weapons in half of the orientations. However, an encouraging observation that can be taken from it was the ability to detect both of the replica handguns, which produced clear anomalies at all orientations with both antennae (Fig. 6).

The major advantage that GPR has over the other two instruments used in this study was the ability of GPR to detect the soil disturbance caused by the digging of the holes for the weapons. This allows differentiation between an object that has been buried in the ground recently and less recently buried targets that may be irrelevant to an investigation. This would allow the prioritization of anomalies, thus potentially saving vital time in a forensic search scenario. The results demonstrate that the 800 MHz antenna shows the soil disturbance most clearly but it can still be identified in the 500 MHz antenna responses (Fig. 6). However, such disturbances may not be as clearly defined in other survey environments or if longer periods of time between burial and survey have resulted in the compaction and consolidation of the fills.

The results of this study were very similar to those of Murphy (2008), who used the same replica handguns and GPR antenna. Murphy showed that both guns could be detected at 0.5 m with both antenna, and that stronger anomalies were produced with the weapons laid flat. This shows the reproducibility of the results between different users. However, the results produced by Schultz (2009) differ slightly from those of this study. Schultz found stronger anomalies being produced with the 500 MHz antenna at a greater depth (60 cm) than in this study. This is probably due to different soil types at each survey site, as this can greatly affect the attenuation of the electromagnetic pulses employed by GPR (Conyers 2004) or it may be due to the differing metallic compositions of the real and replica firearms.

Conclusions

The main point of concern raised by this study is that the instruments employed were unable to detect smaller weapons such as the screwdrivers and knives. The inability of any instrument to detect them at all depths and orientations would suggest that these geophysical techniques and/or instruments are not appropriate for the forensic search for these types of weapons. However, there were more promising results when looking at the larger weapons, particularly the replica handguns. The EM38B VMD quadrature response produces clear anomalies for the axe, hammer and both handguns at every depth and orientation investigated in this study. The 500 MHz GPR produced good results when looking at the two replica handguns, and has the added advantages of being able to detect the soil disturbance and help in the differentiation of shallow non-forensic anomalies. Although these results are promising, there needs be further investigation looking at more burial depths, orientations at different depths, a wider weapon set, and differing geology and ground conditions, together with extensive blind searching trials before it is possible to confidently suggest that any technique is appropriate for use in any specific forensic search.

The authors would like to thank the reviewers for their helpful comments, which have been acted upon wherever possible to improve this paper. This research is a combination of work undertaken as the dissertation elements of an undergraduate BSc Archaeological and Forensic Sciences degree and an MSc Recovery and Identification of Human Remains degree within the School of Applied Science Bournemouth University, UK.

References

ABDU, H., ROBINSON, D. A. & JONES, S. B. 2007. Comparing bulk soil electrical conductivity determination using the DUALEM-1S and EM38-DD electromagnetic induction instruments. *Soil Science Society of America Journal*, **71**, 189–196.

ASPINALL, A., GAFFNEY, C. & SCHMIDT, A. 2008. *Magnetometry for Archaeologists*. Altamira Press, Plymouth.

BEAMISH, D. 2011. Low induction number, ground conductivity meters: a correction procedure in the absence of magnetic effects. *Journal of Applied Geophysics*, **75**, 244–253.

BREINER, S. 1999. *Applications Manual for Portable Magnetometers*. Geometrics, San Jose, CA.

BRITISH GEOLOGICAL SERVICE 2013. *1:50 000 Scale Deposits Descriptions*. Geology of Britain Viewer. http://mapapps.bgs.ac.uk/geologyofbritain/home.html [accessed 22 May 2013].

BRISTOW, C. R., FRESHNEY, E. C. & PENN, I. E. 1991. *Geology of the Country Around Bournemouth*. British Geological Society, Memoir for 1:50 000 Geological Sheet 329 (England and Wales). HMSO, London.

BRUSCHINI, C., GROS, B., GUERNE, F., PIECE, P. & CARMONA, O. 1998. Ground penetrating radar and imaging metal detector for antipersonnel mine detection. *Journal of Applied Geophysics*, **40**, 59–71.

CHAPLIN, R., FLATLEY, J. & SMITH, K. (eds). 2011. *Crime in England and Wales 2010/11. Findings from the British Crime Survey and Police Recorded Crime*. Home Office Statistical Bulletin (HOSB) **10/11**. Home Office, London.

CHEETHAM, P. 2005. Forensic geophysical survey. In: HUNTER, J. & COX, M. (eds) *Forensic Archaeology: Advances in Theory and Practice*. Routledge, Abingdon, 62–95.

CLARK, A. 1996. *Seeing Beneath the Soil Prospecting Methods in Archaeology*, 2nd edn. Routledge, New York.

CONYERS, L. B. 2004. *Ground-Penetrating Radar for Archaeology*. Altamira Press, Walnut Creek, CA.

DW CONSULTING 2012. *ArcheoSurveyor 2*. Available from http://www.dwconsulting.nl/.

DIONNE, C. A., SCHULTZ, J. J., MURDOCK, R. A., II & SMITH, S. A. 2011. Detecting buried metallic weapons in a controlled setting using a conductivity meter. *Forensic Science International*, **208**, 18–24.

GAFFNEY, C. & GATER, J. 2003. *Revealing the Buried Past Geophysics for Archaeologists*. Tempus, Stroud.

GEONICS 1999. *Application of 'Dipole–Dipole' Electromagnetic Systems for Geological Depth Sounding*. Technical Note TN-31. Geonics, Mississauga, Ontario.

HARRISON, M. & DONNELLY, L. J. 2009. Locating concealed homicide victims: developing the role of geoforensics. In: RITZ, K., DAWSON, L. & MILLER, D. (eds) *Criminal and Environmental Soil Forensics*. Springer, London, 197–219.

MANN, R. 2003. *An investigation of geophysics for locating buried bladed weapons*. MSc dissertation, Bournemouth University.

McNEILL, J. D. 1980. *Electromagnetic Terrain Conductivity Measurement at Low Induction Mumbers*. Technical Note 6. Geonics, Mississauga, Ontario.

MURPHY, J. 2008. *A comparative study into the effectiveness of archaeo-geophysical techniques for the location of buried firearms*. MSc dissertation, Bournemouth University.

MURPHY, J. & CHEETHAM, P. 2008. A comparative study into the effectiveness of geophysical techniques for the location of buried handguns. In: *Geoscientific Equipment & Techniques at Crime Scenes: The Geological Society Forensic Geosciences Group FGG 2008 Conference*, 17 December 2008. Conference Abstracts. Geological Society, London.

PRINGLE, J. K., JERVIS, J., CASSELLA, J. P. & CASSIDY, N. J. 2008. Time-lapse geophysical investigations over a simulated urban clandestine grave. *Journal of Forensic Sciences*, **53**, 1405–1417.

REZOS, M. M., SCHULTZ, J. J., MURDOCK, R. A. & SMITH, S. A. 2010. Controlled research utilizing a basic all-metal detector in the search for buried firearms and miscellaneous weapons. *Forensic Science International*, **195**, 121–127.

REZOS, M. M., SCHULTZ, J. J., MURDOCK, R. A. & SMITH, S. A. 2011. Utilizing a magnetic locator to search for buried firearms and miscellaneous weapons at a controlled research site. *Journal of Forensic Sciences*, **56**, 1289–1295.

SCHULTZ, J. J. 2009. *Detecting Buried Firearms Using Multiple Geophysical Technologies (2007-DN-BX-K304)*. Document 227935. National Institute of Justice, Washington, DC.

SIMMS, J. E. 1995. *Locating Grave Sites Using Geophysics at Walton Cemetery, Picatinny Arsenal, New Jersey.* US Army Corps of Engineers, Vicksburg, MS.

STANLEY, R. J., GADER, P. D. & HO, K. C. 2002. Feature and decision level sensor fusion of electromagnetic induction and ground penetrating radar sensors for landmine detection with hand-held units. *Information Fusion*, **3**, 215–223.

US ARMY CORPS OF ENGINEERS 1995. *Geophysical Exploration for Engineering and Environmental Investigations.* EM110-1-1802. US Army Corps of Engineers, Washington, DC.

WESTPHALEN, O. & RICE, J. 1992. Drum detection: EM vs Mag: some revealing tests. *In: Proceedings of the Sixth National Outdoor Action Conference on Aquifer Restoration, Ground Water Monitoring, and Geophysical Methods*: May 11–13, 1992, Riviera Hotel, Las Vegas, Nevada. National Ground Water Association, Westerville, OH, 665–688.

ZOUBIR, A. M., CHANT, I. J., BROWN, C. L., BARKAT, B. & ABEYNAYAKE, C. 2002. Signal processing techniques for landmine detection using impulse ground penetrating radar. *IEEE Sensors Journal*, **2**, 41–51.

Index

Figures are shown in *italic* font and Tables in **bold**

adipocere 210
aerial imagery 174, *182–184, 187,* 188–192
aerobic conditions 210
age of sample source 17, 19
age, metals in tree rings *129*
agricultural and land use maps 22–23
airborne emission, industrial trespass 81–85
Alabama, airborne emission study 81, 84
ammunition box, search for *232,* **233**, *234,* 237–247
anaerobic conditions 210
Anglo-Saxon gold 195–207
animal, provenance 14, 22–23
ANOVA test (L*a*b* indices) 99–102
anthropogenic
  deposits **175**, *189,* 191
  heavy metal release 27, 32, 40
  provenance analysis 15, *16,* 23
anthropogenic particles 66
  automated mineralogy 47, 48, 52, 57, 62
archaeological survey 196–207
  geoforensic search strategy 198–206
  post-search monitoring 206–207
arsenic poisoning 121
automated mineral analysis 3, 156, **158**, 160
automated soil mineralogy 47–63
  experiment design and sampling 48–52
  lithotype data 58–59
  modal mineralogy 52, **53**–58

balloon bombs, tracing source 11, 14
barite 62
basalt 166–167, *168*
  forensic provenance 115–117
Ba-Sb particles 57, 62
bauxite *116*, 117
Bayesian statistical approach 110, 115
beach burial 191–192
Bethlehem Steel Plant 40
biological materials, provenance analysis 13–15, 20–24
biomarkers, decomposition 5, 209, 217
biomonitoring, lichen 133–134, 144
biotite 92
bolt, target **233**, *234, 238, 240, 241, 243, 245*
botanical taxa/macerals 13–14, 20–23, **93, 94**
brake shoe 62

brick, buries **233**, *234, 238, 240, 241, 243, 245*
British Geological Survey
  G-base survey 109, 110, 111
  map development 178, *179,* 181, *183, 188,* 189
Buffalo–Niagara Metropolitan Area 27–28, *32*
bulk ground resistivity statistics **231**
burial search 173–194
burial sites, case studies 164–172
buried objects, detection methods 229–250
buried weapons, geophysical techniques 253–265

cadaver decomposition island 209–217
  perimeter measurements *215*
  trials 211, 223–224
calcite **123**, 170
  automated analysis **52**, 53, **55**–**60**, *158*
  trace evidence 76, 77, **78**, *159*
  whitener 3, 85
  X-ray diffraction 30, *33*
calcium, cadaver decomposition 217
carbonate component, soil samples 32, 40, 44
carbonate rocks 102–103
carbonate sand, urban forensics *152*
carborundum 15, 23
case studies, drift geology 167–172
cathodoluminescence 17
ceramics, neutron activation analysis 123, **124**, *125*
chemical composition
  beach and dune sand 98–99, 102–107
  bulk analysis 47, 62
chemistry of cadaver decomposition 210, 214, 216, 217
chemistry, neutron activation analysis 122–125
chewing gum as transfer material 156
clay minerals 48
  X-ray diffraction 34, *39,* 44
clay soil, ground penetrating radar 222–223, 226, 228
climate recording 210–213, 224
climate suitability model 20, *21*
clothing, soil recovery 53
cluster analysis, beach and dune sand *102, 104, 106, 107*
coastal sediment characteristics 97–107
  sampling 98

INDEX

colour analysis 99–102
   beach and dune sand 98–101, 106, 107
   river beach 89, 90–91
colour-coded RAG (red-amber-green) maps 173, 178–188, 191–192
conceptual geological model
   archaeological survey 196, 201, *202*
   criminal burials 173, 176, 191, 192
concrete slabs 5, 230, 232, 241
concrete, urban forensics 148
contaminant trespass 81–85
control samples 112, 117, 160
copper smelter, pollution 135–144
copper wire, provenance 121
criminal burials 174–175
criminal damage to graves 75–79
criminal forensic database, use 109–117
criminal investigations 1–2, 122

databases 18, 19, 23
   and map integration 112–113
   Portugal, soils 95, 97–98
   use in geoforensics 109–117
decomposition stages 211, 213
depth sensitivity 260–262, 265
diamagnetism 66
diamond finger-printing 130
diatoms 14
diggability 4–5, 200–201, *202*
   beach deposits 191–192
   burial site 175–178, *181*, 183, *184, 186–188, 190*
dioxin 84
DNA analysis 13–14, 22, 23
dog, search 164–172, *191*, 192, 221, 229
Douro River, beach study 88
drift geology 200
   geoforensics review 163–172
   see also superficial deposits
dust analysis 10, 24, 82
dust from smelter 142–144
dust, urban forensics *152*

ecosystem maps 21–22
EDS/EDX energy dispersive spectroscopy
   see under scanning electron microscopy
Ehrenberg, Hans Christian 1, 110, 117
electrical resistivity 235–236, 248–250
   comparison 243–248
electromagnetic equipment 230
electromagnetic minimal metal detector 203
electromagnetic technique 257, 260–262

element analysis 31–39, 44, 102–107
   in lichen 136, 139, 140, *142*
element mobility 42
engineering properties
   diggability 175–178, *184–187,* 190–192
   rating **181**
environment protection 165
environmental contamination 127–130
environmental forensics 1, 4
   database use 109–117
   lichen monitoring 133–144
   neutron activation analysis 121, 125
excavators and site search 174
exotic construction material 153, 154
explosives, criminal burial 164–165

fatty acids, decomposition experiment 209–217
   analysis 213–215
   methods 211–212
   results 215–217
   site 210–211
ferromagnetism 66
fibres *152*
   decomposition 223, 224
field management 114, *115*
field portable X-ray fluorescence 31, 32
firearm discharge, Sb-Ba 62
fireworks, Ba-Sb 62
fixed offset resistivity survey 235–236
fluorescent minerals 127–128
fluxgate gradient magnetometer 203
fluxgate gradiometer 234, 237–238, *239*, 248–249, 255–257
   statistics **231**
fly ash 15, *16*
   magnetite 83, *84*
   trace elements 85
foliage around graves 221
footmat particulates 157–160
footwear, automated mineral analysis 49, *50*, 54–62
forensic geology 163–172
   database use 109–117
   Staffordshire Hoard 195–207
forensic geophysics 253–265
forensic geoscience groups/conferences 1, 5
forensic geoscience investigation 2
   buried objects 229–250
   buried weapons 253–256
   current research 3–5
   environmental 133–144
   historical context 1–3

neutron activation analysis 121–130
   provenance 9–24
   urban areas 147–160
fungi 14
fuzzy membership, soil classification 114

gamma ray 122
gas chromatography 215–216
G-Base surveys 109, 110, 111
geochemistry, soils 109–117
geoforensic publications 163–164
geoforensics, map interpretation 173–194
geographical information systems, datasets 19, 23
geological materials
   dust source 82–83
   provenance analysis 10–13, 19–20, 24
   trace evidence 147, 155–160
geological RAG (red-amber-green) maps 178–182
   case examples 183–188
geological trace evidence in criminal damage 75–79
   headstones 77–78
   samples 75–76
   soil 76
geologist, military 179–180
geology in forensic investigation 163–172
   automated analysis 47, 50–52, 59, 62
   Buffalo–Niagara area 30–32, 34, *38*
   coastal sediments, Portugal 97–98, *99*, 104–106
   databases **110**, 111, 112, 113
   Douro River site 87–88
   magnetic susceptibility 66, *67*, 68, 69
   Serengeti 125–127, *128*
   site recognition 173–194
   Staffordshire Hoard 198–200
   urban forensics 147–160
geomorphology 164, 200
   maps and forensics 173–194
geophysical detection, method comparison 229–250
   methodology 230–236
   results 236–250
   techniques **231**
geophysical instrumentation, comparison 254–258
geophysical survey, archaeology 196, 198, *199*
   control measures 204
   instruments 203
   search strategy 204–206
geophysical techniques 5, 163, 192
geophysical techniques, in weapon search
   burial parameters 253–254
   instrumentation 254–258
   results 258–265
geoscience evidence in murder investigation 155–160
geostatistics, soil properties 114–117
glass in nuclear forensics 130
glass shard, urban debris *152*
glazes, chemistry **123**
gold, Anglo-Saxon hoard 5, 182, 195–207
grain size, magnetic susceptibility 69, 71–72
grass, neutron activation analysis 125–127, *128*
grave stones, damage to 3, 75–79
grave
   concealment of 47, 48, 49, 58
   detection 169–171, 221–228
   locating 221
   mass 5, 222
grenade, search for *232*, **233**, *234*, 236–247, 249
Gross, Hans 1
ground penetrating radar 5
   comparison 230, 236, 239–240, 243–250
   statistics **231**
   survey 262–264
ground penetrating radar in grave detection 169–171
   materials and methods 223–224
   physics and limitations 221–223
   soil texture 224–228
ground search 4–5
groundwater pollution 165
growth threshold 20
gun, buried *232*, **233**, *234*, 237–247
   search techniques *254, 255,* 258–264
gunshot residue 121, 151
gypsum, urban forensics 156–160

hair, arsenic poisoning 121
hammer, search techniques *254, 255,* 258–264
hatchet, search techniques *255,* 258–264
hazard, mapping 180–181
hazardous waste 82
headstones, damage to 75, 77–79
health problems 82, 84, 85
heavy elements, detection 31
heavy metal dispersal *134*
heavy mineral analysis 89, 91–92, 95
heliflux magnetic locator 203
historical burial grounds 5, 222
honey, tracing source 13
hydrogen sulphide 84
hydrogeology and buried waste 165

INDEX

ICP *see* inductively coupled plasma
improvised explosive devices 229
INAA *see* instrumental neutron activation analysis
inductively coupled plasma
  emission spectroscopy (ICP–ES) 126
  optical emission spectrometry (ICP–OES) 135
inductively coupled plasma mass spectrometry (ICP–MS) 69
  coastal sediments 98, 101, 102, 106
  grass 126
  lichen 135
  maple syrup 124
industrial trespass
  paper mill 83–85
  steel mill 81–83
inorganic components of soil 47–48
instrumental neutron activation analysis 4, 121–122
  application to forensics 122–130
  detection limits **122**
Irish famine, burial grounds 222
isotope analysis 135, 139, 142–144
isotope variation neutron activation analysis 121, 127

jeans, soil particles 49–50
Jewish tombs, damage to 3, 75–79

kaolinite 40, 54, 57
Karabash copper smelter 135–137
  element analysis 137–142
  isotope data 142–144
knife search geophysical techniques
  evaluation *254, 255*, 258–264
  in domestic/urban setting *232*, **233**, *234*, 236–247, 249
Kraft process 83–84

landslide *185*, 186, 188, *189*, 191
laterite 116
lazurite *11*
lead in lichens 139, 140–144
leaf litter samples 51, 52, **54**, 57
lichen, environmental forensics 4, 133–144
  particle analysis 137, 141
  sampling/analysis 135
  statistics and modelling 139–140, 142
  transplant methodology 134, 135, 137
limestone, headstone damage 77, 78, 79
liquid and waste, urban forensics 154
lithotype 48, 52
  data **58–60**, *61*
  grouping **57**

loam soil, ground penetrating radar 223, 225, 228
Locard, Edmond 1
Locard's Exchange Principle 147, 150, 151, 153
low field magnetic susceptibility 65–66, 98

magnetic anomaly *199*, 200
magnetic gradiometer 234–235, 238, *240–242*, 248–249
  statistics 231
magnetic grain concentration 66
magnetic methods, comparison 243–248
magnetic screwdriver, search for 260
magnetic susceptibility 3–4
  beach and dune sand 99, **100**, 101, 106–107
  comparison 233–237, 241–243, 248–250
  evaluation 68–69
  statistics **231**
magnetic susceptibility, soil forensics 65–79
  analysis 68–69
  low field 65–66
  values **70–71**
magnetic techniques, buried weapons 255–260
magnetism 66
magnetite 40, *92*
  from electric arc furnace 82–83
manganese 85
maple syrup, neutron activation analysis 123–125, *126*, *127*
maps and geoforensics 173–194
mass burial sites 5, 222
  animals 111
mass susceptibility 66
Mercian Kingdom, gold 195
metal content in lichens 135
metal detector 229, 232–233, 236, 240–243, 249
  MD4 203–204
  statistics **231**
metal types, distinction 249
metallic weapons, geophysical techniques 253–265
methane source 164, 169, 170
microfossils 12–13, 17
military
  diggability map 176
  RAG (red-amber-green) map 179–180
mineralogy, automated soil analysis 47–63
mineralogy, soils 27–32
modal mineralogical data **52–54**
modal mineralogy, soil 52–58, 59, 60
Montserrat, volcanic hazard map *179*
mortar shell, search target **233**, *234, 238–241, 243, 245*

Murray, Ray 2
myristic acid 210, 214, *215,* 216

New Jersey, zinc mining 127–128
ninhydrin reactive nitrogen 217
Niton field portable X-ray fluorescence **32,** **35–37**, *41*
nitrate fertilizer, buried 165
nitrate, cadaver decomposition 217
non-destructive techniques 9, 24, 27, 65, 76
    instrumental neutron activation analysis 121, 127
non-invasive methods 221
Northern Ireland, soil geochemistry 113, 116
Northern Ireland, gang murder 154–160
nuclear forensics 130
nuclear tests, biomonitoring 133

Observed CDI Perimeter 212, *213,* **215**, 216
    [CDI, cadaver decomposition island]
oleic acid 210, 214, *215,* 216
Ontario, carcass case study 223–228
opal phytoliths 14
orientation of weapons and detection 253–265

Palmerton smelter 127–130
palmitic acid 210, 214, *215,* 216
palynological characterization, river beach 87–95
palynomorphs 13
paper mill, airborne emission 81, 83–85
paramagnetism 66
particle analysis, lichens 137, 141
particle size analysis
    beach and dune sand 98, 99, 101–102, 106, 107
    river beach 89, **90**, *91*–92, 95
particulate from clothing 49–50, *152*
patio, comparing detection methods 229–250
peat, burial site 167–170
peat, diggability **177**
petrographical techniques 10
pH soil 212, 213, *214,* 215, 216
pig carcass, fatty acid dispersal 209–217
    ground penetrating radar 222–228
plagioclase zoning 12
plant survey **93**
plasterboard, particulates 157–160
plastic bag and magnetic signal 68, 69, **71**, 72
polarizing light microscopy 60
    and provenance 9–24
police search team 195, 201, 203, 205, 207
police use of RAG (red-amber-green) maps 180–181
pollen 13, 15, 20, *21*

pollen analysis river beach 87, 89–90, 92–95
pollutants, biomonitoring 133
polymer foam *152*
Popp, Georg 1
portlandite 78, 79
Portuguese sites
    geology 66, *67*
    river beach palynology 87–95
    soil database 97–98, 107
postmortem interval 210
potassium vapour magnetic gradiometer 234–235, 238, 241, 248–249
protocols 87, 144
    automated mineralogy 61, 62
    magnetic susceptibility 69, 72
provenance
    lithotyping 62
    X-ray diffraction 27
provenance, polarizing light microscopy 9–24
    anthropogenic materials 15, *16*
    biological materials 13–15
    geological materials 10–13
    other techniques 16–18

QEMSCAN 147, 156, **158**
    particle analysis 48, *51*, 52, 58, 59, *62*
quantitative clay mineral analysis 44
quartz grains 12, *16*

RAG (red-amber-green) map 4–5, 201
Raman microspectroscopy 18
rare earth elements 98–99, 102–106
    ceramics *125*
reflected light microscopy 48
resistivity **231**, 235–236, 239, 243–250
river beach, palynology 87–95
road and building materials, urban forensics 150–153
rock fragments 10, 19, 62
    lithotyping 48
rock, burial sites 174–175, **181**

sample preparation for forensics 44
sand and burial site **175**, 176, 191–192
sandy soil, decomposition fluids 210
sandy soil, ground penetrating radar 222, 227, 228
Saxon gold 5, 182, 195–207
Sb-Ba particles 62
scanning electron microscopy 16
    energy dispersive X-ray spectroscopy 17, 31, 48, 76, 79
screwdriver, search techniques *254, 255,* 258–264

INDEX

search, geoforensic 173–175, 195, 198–206
sediment provenance, polarizing light microscopy 9–24
seismic shock from industrial process 82
SEM *see* scanning electron microscopy
Serengeti, grass sample 125–126, *129*
   geology map *128*
silver 195, 197, 207
silver theft 12
simulated forensic samples
   automated mineralogy 49
   sample collection/preparation 28, 29
   X-ray diffraction/fluorescence 34, 38–39, 40
Slingram electromagnetic technique 257, 260–262
smelter, copper 134
smokestack emission 82
software 18, 68, 89, 99, 233
   QEMSCAN 48
soil
   analysis 76, *77*, 78
   automated mineral analysis, 47–63
   cadaver decomposition 209, 213
   classification 114
   database, Portugal 97–98, 107
   disturbance 223, 254, 257, 264–265
   Douro River 87
   electrical conductivity 225, 226, **227**
   geochemical mapping 110–117
   magnetic susceptibility 65–79
   maps 19, *20*
   mineralogy 3–4
      industrial contaminant 81–85
   moisture content 225, **227**, 228
   probe 204, 206
   properties 114–117, 212, 213
   provenance, polarizing light microscopy 9–24
   sampling programmes 107
   texture, use of ground penetrating radar 221–228
   types 175, **181**
   urban geoforensics 148, 150
soil screening, X-ray diffraction/fluorescence project design
   data 32–39
   materials and methods 27–32
   methods evaluation 39–44
SoilFit project, urban soils 111, 112
spade, search for **233**, *234,* 236, *238, 240, 241, 243, 245*
spatial analysis and mapping 18–23
spatial coverage, databases 111–112, 117
Spearman's rank correlation 32, *37*
Staffordshire Hoard 197

starch grain *15*
statistical analysis
   beach and dune sand 99–104, *106–107*
   river beach sediments **89–90**, *91–92*
   soil properties 114–115, 117
stearic acid 210, 214, *215,* 216
steel mill, airborne emission 81–83
steel plant and tree rings 127–130
steel plate, search 236, *238,* 240, *241,* 245
sulphuric acid 84, 85
superficial deposits 169, 175, **177**, *190, 191,* 200
   weapon burial study 257–258

tape samples, X-ray diffraction patterns 38, *39, 40*
tax evasion and quarrying 166–167, *168*
tektites 130
Tellus project (soil geochemistry) 110–115, 117
temperature recording 210–213, 223, 224
tephra *18*, 19
time, magnetic susceptibility 69
total field magnetics 249
trace elements 3, 126
   coastal sediments 98–99, 102–106
   determination 27, 32, 44
trace evidence
   automated analysis 27, 47, *50,* 62, 160
   criminal damage 75–79
transfer material in urban forensics 148–154, 156
transferred soil materials 27, 42, 44, 47, 62
transmitted light microscopy 48
Treasure Act (1990) 195, 197
tree rings 4
   neutron activation analysis 127–130
trespass, emissions 81–85
trinitite 130

U-Pb dating 17
urban forensic geology 147–148
   analysed materials 148–154
   case study, turf-war murder 154–160
urban landscape
   comparing detection methods 229–250
   soil screening 27–28, 32, 34, *38,* 40, 41, 44
   soil survey 111, 117
   trace evidence analysis 160
   transferable material *149–152*

validation, powder X-ray diffraction 30, 48
volatiles 84–85
volcanic rocks *10, 19*

Ward hierarchical clustering dendrogram X-ray diffraction/fluorescence 42, *43*
waste disposal 1
waste dump, illegal 115–117
waste dumps, metal mining 137
water bodies, forensic geophysics 229
water, urban forensics *155*
weapon detection 229
   geophysical techniques 232–248, 253–265
   small 264–265
wind direction in trespass emission 82, 85
wind-blown particles, urban forensics 153

X-ray diffraction (XRD) 3, 18, 48
   compare, automated mineralogy 60
   gypsum 156, *157*
   headstone damage 78, 79
   soil geostatistics 115, 116, 117
X-ray diffraction, soil screening 27–44
   bulk soil analysis **33**
   data 32
   sample analysis 29–32
   sample collection and preparation 28–29
X-ray fluorescence (XRF) 3
   soil geostatistics 114
X-ray fluorescence, soil screening 27–44
   bulk soil analysis **33**, 38
   data 32–33
   sample analysis 29–32
   sample collection and preparation 28–29

zinc mining 127–128
zircon 12, *17*, 54, 104